U.S. STABLING GUIDE

Eighth Edition

BALZOTTI PUBLICATIONS
5 Barker Street, Pembroke, Massachusetts 02359

1-800-829-0715
1-781-829-0710
www.jimbalzotti.com

U.S. STABLING GUIDE

THE COUNTRY'S COMPREHENSIVE GUIDE FOR HORSE TRANSPORTATION

5 Barker Street, Pembroke, MA 02359
800-829-0715

Publisher: James D. Balzotti

Editor: Lisa Doubleday

Graphic Artist: Cherry Dahlgreen, Graphics Plus
Duxbury, Massachusetts

Cartoonist: Fred Leary, Leary Cartoon Associates
Pembroke, Massachusetts

PUBLISHER'S NOTICE
Every effort is made to compile and print this directory as accurately as possible. However, errors or omissions may occur. In the event of an error or omission, Balzotti Publications assumes no obligation to correct the same in this directory or by any special notice of any kind to the individuals and businesses listed herein.

TABLE OF CONTENTS

TABLE OF CONTENTS

ADVERTISERS DIRECTORY

INTRODUCTION

There are over five million horses in the United States. Horse ownership often changes hands and therefore the horses must be transported to their new home and owners. In addition, there are over 100 different breeds of horses now registered in the country. Each of the breeds has its own national organization with each sponsoring numerous local, regional, state, and national horse shows, not to mention the rodeos, team-penning, and other competitions that also require travel. There are many horse trailers on the road whose drivers need reliable information as to where they can stop for the night, a week, or longer. This book contains listings of over 900 stables, ranches, farms, equestrian centers, and fairgrounds that will accept overnight boarders. It also contains many bed & breakfasts with stable facilities where you can stay with your horse. The U.S. Stabling Guide is in the business of both providing overnight stabling locations for the traveling horse owner and promoting nationally the businesses of our listed stables.

I have transported horses cross-country many times and have experienced first-hand the difficulties of finding overnight accommodations for both my horses and myself after a long day of driving. This book is designed to help you plan your trip. With my own experiences in mind, my goal in putting together this book was to have enough conveniently located facilities along the major highways to prevent the need to be on the road for a dangerously long period of time because the next stable that would take overnight boarders was too far away. I never want to hear any more horror stories of having to stay overnight on the side of the road with the horses in a portable round pen because there was not a stable nearby. I feel that this book accomplishes that by allowing you, the traveler, to make choices as to the best routes to follow, based on the kind of facilities that best suit your needs.

In the course of compiling the listings contained in this guide, I have spoken to many wonderful horse people across the country and found that the primary reason for accepting overnight boarders is to provide a much needed service for their fellow horse owners. It seems that most of us have traveled with our horses at one time or another and know the necessity of a layover location, whether it be planned or in an emergency. For that reason, I hope that all of you travelers will respect the courtesy and service being extended to you by the stable owners and/or managers and will act in kind by calling ahead and making reservations with them and then, just as importantly, arriving at the expected time. Please do not abuse this service by not showing up and/or not informing the stable of your change in plans. I am sure that all of you can understand the inconvenience to the stable owner who prepares a stall or pen and waits patiently for someone who does not arrive. It would not be long before that stable owner no longer offers a facility to those of us who need it.

I enjoy traveling with my horses and hope that this book makes your travels enjoyable too.

Jim Balzotti
Publisher

HOW TO USE THE GUIDE

1. States are listed alphabetically with cities and towns within the state also listed alphabetically. Canadian provinces follow Wyoming and are also listed alphabetically.

2. Accompanying state maps provide quick visual locations of stables and major highways but should not be used as your map while you are traveling. Use accurate road maps such as the ones from AAA for that purpose.

3. Each page has a written reminder that all stables require current health papers. Some stables require proof of additional vaccinations, and that is included in their listing. Our advice to all of the listing stables is that they not allow the unloading of horses until those papers are produced. Please discuss all of the traveling medical requirements with your veterinarian well before you begin your trip.

4. Stable listings provide the owner and/or manager's name, telephone number(s), address, a description of facilities, current overnight rates per horse, and, in many cases, concise directions to the stable. Unless otherwise stated, do not drive directly to a stable without prior notification of your arrival. Preparations usually have to be made before overnight boarders arrive.

5. "Accommodations" lists the nearest motels to the stable for your own overnight stay. This has been included so you can make your reservations at a motel or bed & breakfast and be assured of a good night's rest before beginning another day on the road.

6. For your interest, we have included additional information about breeding programs, unique services or products available, local activities or sights to see while you are in a new area, etc.

7. We are not a reservation service and request that you deal directly with the stables in making your overnight arrangements. However, we would be happy to answer any general questions about traveling with your horse.

8. We have made every effort we can at this time to assure that the stables in our book will provide a clean and safe environment for your horse. However, it is impossible to inspect each stable. If you feel that any particular stable fails to provide a clean and safe environment, notify us. After receiving two or more similar complaints, we will arrange for an inspection of that stable.

TRAILERING TIPS

by Jim Balzotti

Since we first published this book, we have spoken to hundreds of horse owners who are either planning to travel for the first time with their horse, or owners who are going to be traveling a longer distance than they have before. A lot of the questions are the same: How far should I travel in one day? (It depends.) And some new ones: Can I leave my horse in the trailer while going through a car wash? (Definitely not a good idea.) I have written some helpful hints here but if there is any question that is not answered, please feel free to call and ask us!

The keys to traveling with a horse trailer are practice and preparation. If you are hauling horses for the first time, do not wait until you have a horse on board to "try it out." There is much to learn and become familiar with before risking an accident with your horse on board. Driving with a trailer is very different. Visibility is drastically affected as are turning ratios, braking, and driving in reverse. I suggest you hook up the trailer, empty, and drive to a large shopping center or open area in the early morning when you will encounter few cars. Practice turning and backing up. Slow, gradual turns on the wheel will work better than severe, sharp turns. Get to know the turning radius of your truck and trailer, and find out what they will and will not do. Inevitably in your travels, you will pull into a gas station, parking lot, or rest area and be forced to negotiate an unexpected and difficult turn. That is not the time to learn. If you have the time, take your empty trailer on a drive on the road surfaces you may encounter on your planned trip, such as an interstate highway (practice passing and changing lanes), narrow roads, unpaved and rough road surfaces, etc. Before the trip, practice with the horse on board. Take your horse for some local trips, gradually taking longer ones. It never hurts to reward a horse with both treats and praise for being so brave. Again, do not have your horse's first trip on a trailer be an across-the-country journey.

Now for the preparation. Make sure the trailer is in proper working condition. All lights should also be in working order. Carry spare light bulbs and fuses. The braking system should be checked by a mechanic (and periodically checked when on the road), and the wheel bearings should always be packed with grease to prevent their locking up. A solid trailer hitch should be welded onto the towing vehicle directly, never attached to the bumper. The ball hitch should be checked to ensure that it is the proper size and your trailer should also be equipped with safety chains should the hitch fail. On some rigs, anti-sway bars will add a greater sense of stability. Seek the advice of a reputable trailer sales and service dealer to see what equipment you may need for your trailer. Always travel with a spare tire for both your truck and trailer and a complete tool box. Flashlights, fire extinguisher, first aid kit, tire air gauge, and a knife to free an entangled horse are also essential and should be easily accessible.

Inside the trailer, check for any sharp objects that could injure or panic your horse. Make sure you have trailering straps with quick release snaps. Extra lead ropes, halters, blankets, hay nets, water tubs, and feed buckets should be taken along on the trip. I usually take two large 6-gallon Rubbermaid water containers filled with the water my horse is used to drinking. Also take a supply of your horse's grain and hay and some bran and molasses to make the traveling transition easier for your horse.

Because I often travel across the country with my horses, I feed them Purina grain because Purina is the only national feed company with dealers located in every state. I personally do not think it is a good idea to change a horse's feed if it can be avoided.

When you know of your departure date, call your veterinarian two weeks in advance to have him/her prepare a negative coggins and health certificate. Plan your route in advance, taking into consideration the time of year and the weather conditions you could encounter in different parts of the country. For example, when I traveled from Massachusetts to Arizona in January, I tried to head south as soon as I could and hit the southern east-to-west interstate highways so as not to be caught on any icy, mountainous roads.

Right before leaving, I visit my local AAA office and pick up road maps of the states I plan to be driving through and ask the travel consultants there if any major roads in those states are under construction and what alternative routes would be available.

When I am finally ready to leave, I will usually pull my horse's shoes off and trim the hoofs. I put plenty of shavings on the trailer floor. It is always advisable to wrap your horse's legs (but not too tightly) and possibly put bell boots on to prevent injuries. You may also want to wrap the horse's tail to prevent rubbing.

Now it is time to go. If traveling with only one horse, always load the horse on the left side of the trailer for balance. If loading two horses, load the heavier of the two on the left side. Make sure your trailer is parked on level ground in an area that has plenty of light, especially if you are loading for the first time. If possible, load a seasoned, quiet horse first. The second horse will want to join the first horse and then be easier to load. Horses are social animals and like to be with other horses.

If you are having some difficulty loading a horse, a little bit of grain and some verbal encouragement will go a long way. However, if you are in a hurry and have a stubborn horse, it may be necessary to use more forceful measures. What works for me is to attach a lead shank around the halter on a horse's nose to prevent backward motion and have a helper tap on the horse's hind hock with a long buggy whip to encourage forward motion. In the most severe cases of a stubborn or panicky horse, tranquilizers may be necessary, but, in my opinion, only as a last resort. It is far better to take the time and teach every horse to trailer properly. This takes time and patience but will pay off in having horses that walk quickly and easily onto the trailer.

The most asked question I hear is how far do I drive each day when trailering horses. As far as the horse is concerned, most horses can travel all day (meaning 12 hours) easily if they are given breaks. When on the road, I try to stop every two hours for a break of about 10-15 minutes. I find this reduces fatigue for both myself and my horses. This allows the horses to relax from their locked-legs position. I offer water at that time and monitor their water consumption. I always recheck my truck, trailer, and hitch before getting back on the road.

As I am an early riser, I try to get on the road as soon as possible and plan to stop in the early evening at the latest. In the winter months, when darkness comes earlier, I make certain that I have arrived at my overnight stable location in plenty of time to unload my horses and let them stretch out in a round pen or pasture if one is available and/or take them out for a short ride so they can get some exercise.

Since not all of the overnight stables have outside lighting, I think it is important to arrive in the daylight to give the stable a quick inspection for cleanliness and safety and then get my horses squared away in the light. You should also remember that most stable owners live on the property and are probably not thrilled to see a trailer with a load of horses pull in late at night.

Finally, I recommend that you allow yourself plenty of time and always have a layover reservation. Remember to either keep your reservation or be courteous enough to call the stable to cancel or to let them know you will be late or early. Whenever I pull into a host stable, I always feel very grateful that these folks have been good enough to put my horses up for a night.

I find being a member of AAA and carrying a portable CB radio provides a much needed sense of security should I have any mechanical problems on the road. I would even go so far as to recommend a cellular telephone to call AAA or the local police in an emergency if you are going to do a lot of traveling. Should you have an unexpected mechanical breakdown and need to off-load your horses, call the local police, sheriff, or veterinarian. I have always found them to be very helpful.

FEEDING THE TRAVELING HORSE

By Tara Devine, P.C.H.A.

In order to make your horse's journey comfortable, you need to look at certain factors that will affect the horse during the trip. First, how far are you traveling with the horse? Is the trip going to involve 8 hours of traveling or 8 days? A long trip will require more planning and caution when feeding your horse than a short trip will. Second, what are the weather conditions going to be like? Will it be 90 degrees and humid or 30 degrees and raining? Intense heat and humidity can really add to the stress of travel, just as extreme cold temperatures will require the horse to create more body heat. Third, how do you expect the horse to handle the stress of traveling? Is the horse a seasoned campaigner who is used to long hauls or is the horse a nervous Nelly who goes off feed when upset? Your answers to these questions are clues that will help you care for your horse so he/she will be comfortable and in good condition when you arrive at your destination.

Providing enough water is a crucial part of keeping the horse healthy during the ride. Water should be offered every two to three hours. If possible, try to carry some of your horse's current water supply with you in spill-proof containers (new, clean gas jugs work well). Many horses may refuse to drink different water because it smells or tastes funny to them. If this happens, you can add small amounts of flavoring to the new water (try molasses or powdered drink mixes). The horse may also be more comfortable if you bring his/her old familiar water bucket along. Be sure to monitor the amount of water that your horse drinks. Compare the quantity to what he/she normally drinks at home. Generally speaking, a 1000-lb. idle horse will drink about 8-10 gallons of water a day. Realize that if the weather is very hot and humid the horse will have a higher water requirement in order to replace the water that is lost during sweating. If you are concerned that the horse is not drinking enough water, you can help by giving a sloppy bran mash or by adding water to the regular ration of grain. Water is crucial to a horse's digestion and ability to regulate body temperature. It is probably the most critical factor in keeping your horse in good health during a trip.

One of the biggest components of feeding a horse on the road (or at home) is offering hay. In general, you should provide hay in a hay net at all times during the trip. By munching on hay constantly the horse will keep essential bacteria in the gut alive and healthy. These bacteria are necessary for the proper digestion of food. Providing free choice hay will help keep the horse occupied and will also provide the horse with a substantial portion of the calories and nutrients he/she needs for the day. It is best if you bring your own supply of hay with you so that the horse's digestive system is not subjected to different hay. If you cannot bring hay along, then be sure to buy bales along the way that are similar to what you were feeding. For example, try not to go from a local grass hay to a straight alfalfa hay. It is too drastic a change. If you are concerned that the new hay will be very different, save enough of your old hay to blend with the new. This will help your horse gradually become accustomed to the new hay.

Whether or not you feed grain during your trip is really dependent on the questions presented in the first paragraph. During a short trip when the horse would only be missing a meal or two, you would probably be better off not to feed any grain. The hay that you are offering will provide them with the calories and nutrients he/she needs for the day. But if it is a longer trip, it may be necessary to feed the horse some of his/her normal ration. How much depends on what the normal ration is based on. If you normally ride your horse a couple of hours a day, his/her grain ration contains enough calories to support that amount of work. But once the horse is on the trailer, he/she will not be burning as many calories so you must cut back on the feed. Also take a look at the horse's personality and how he/she is reacting to the trip. Nervous horses present a problem because they tend to need the extra calories that the grain provides. On the other hand, they are more prone to colic and other digestive upsets. If your horse seems to be extremely agitated, you would be better off not to feed grain and just offer hay. If you do feed grain, offer it in small feedings at least three times a day. Smaller portions offered more frequently are the best way to go to avoid stomach upsets. Be sure to bring your grain with you or use a commercial feed like Purina that is available throughout the country. It is critical not to switch your horse's grain during the trip (even more so than the hay) to avoid serious digestive problems. If you know that you will have to switch feeds during the trip, bring enough of the old feed along so you can gradually blend in the new feed.

By planning ahead and observing your horse's reaction to the stress of traveling, you will be able to feed your horse so he/she will arrive healthy and happy. Happy trails!

[Publisher's note: Tara Devine is a Purina Certified Horse Advisor. She also has a B.A. degree in Management, has been a horse nutritionist for ten years, conducts horse nutrition seminars, and owns a feed and tack store. Tara has been a horse owner all of her life.]

TRAVELING MANAGEMENT TIPS

Most people who show and compete with their horses find themselves on the road frequently. Here are some management tips that may be of help.

- Before leaving home, let your horses have free exercise for a day or two.
- Stop occasionally while traveling to let horses relax from jarring and motion.
- Be sure trailer has adequate ventilation.
- Offer water during rest stops. If feasible, carry some water from home since water odor and palatability can change from place to place.
- If possible, get to your show in time to allow horses to relax . . . up to a day before working out.
- Since horses are often stabled on concrete, take along rubber stall mats and plenty of bedding for comfort.
- As much as possible, keep the same feeding and exercise schedule on the road as home.
- After a trip, rub and wash horses down to relax them.
- When bringing horses home, let them rest and enjoy free exercise for a couple of days.

COMPLIMENTS OF PURINA MILLS, INC.

HORSE TRANSPORTATION COMPANIES

The following is a list of individuals or companies that offer local, regional, and/or national transportation for one or more horses. Please refer to the page number following their names and addresses where further information about their services can be found.

STATE TRANSPORTATION REQUIREMENTS

Below is basic information on the health requirements of the 50 states for transient equine. For detailed information or questions, please call that state's veterinarian's office. "E.I.A." is Equine Infectious Anemia and "Within" is number of months within which the E.I.A. test must have been done. "Hlth. Cert." is whether or not a health certificate is required and "Special" is any other tests, information, etc. needed.

STATE	E.I.A.	WITHIN	HLTH.CERT.	SPECIAL
ALABAMA	yes	12 mos.	yes	None
ALASKA	yes	6 mos.	yes	H.C. to state no ectoparasites
ARIZONA	no	N/A	yes	H.C. to include brands &/or tattoos
ARKANSAS	yes	12 mos.	yes	Temp. reading & date/lab on H.C.
CALIFORNIA	yes	6 mos.	yes	Note on H.C. if horse is transient
COLORADO	yes	12 mos.	yes	Permit if E.I.A. pending
CONN.	yes	12 mos.	yes	Temp. req. on H.C.
DELAWARE	yes	12 mos.	yes	Temp. reading req.
FLORIDA	yes	12 mos.	yes	LV Rhinopneumonitis vacc. not eligible for 21 days after vacc. Temp. reading on H.C.
GEORGIA	yes	12 mos.	yes	Equidae (6 mos.) not w/dam - test
HAWAII	yes	3 mos.	yes	Call 808-487-5351
IDAHO	no	N/A	yes	None
ILLINOIS	yes	12 mos.	yes	Check w/ race track if that is destination
INDIANA	yes	12 mos.	yes	Foals w/dam exempt from test
IOWA	yes	12 mos.	yes	E.I.A. date/lab on H.C.
KANSAS	no	N/A	yes	None
KENTUCKY	yes	6/12 mos.	yes	Horses for sale tested w/in 6 mos., others w/in 12
LOUISIANA	yes	12 mos.	yes	E.I.A. test info on H.C.
MAINE	yes	6 mos.	yes	None
MARYLAND	yes	12 mos.	yes	Temp. reading on H.C.
MASS.	yes	6 mos.	yes	Name of lab req.
MICHIGAN	yes	6 mos.	yes	Preapproved H.C. for exhibitions
MINNESOTA	yes	12 mos.	yes	Permit if E.I.A. pending. Suckling foals exempt

STATE TRANSPORTATION REQUIREMENTS
(continued)

STATE	E.I.A.	WITHIN	HLTH.CERT.	SPECIAL
MISSISSIPPI	yes	12 mos.	yes	Orig. test cert. req.
MISSOURI	yes	6 mos.	yes	Call 314-751-3377 for info. on VEE vacc.
MONTANA	no	6 mos.	yes	Prior approval of H.C. req.
NEBRASKA	yes	12 mos.	yes	Name of lab on H.C.
NEVADA	no	6 mos.	yes	Neg. E.I.A. & permit req. if going to rodeo finals
N. HAMPSHIRE	yes	6 mos.	yes	None
NEW JERSEY	yes	12 mos.	yes	E.I.A. test date/lab on H.C.
NEW MEXICO	yes	12 mos.	yes	E.I.A. test date/lab on H.C.
NEW YORK	yes	12 mos.	yes	E.V.A. test w/in 30 days on TB stallions req.
N. CAROLINA	yes	12 mos.	yes	None
N. DAKOTA	yes	12 mos.	yes	Suckling foals exempt; AZ horses exempt from E.I.A. test
OHIO	yes	6 mos.	yes	Temp. reading on H.C.
OKLAHOMA	yes	6 mos.	yes	E.I.A. test date/lab info. on H.C.
OREGON	yes	6 mos.	yes	Permit: 503-378-4710
PENN.	yes	12 mos.	yes	Suckling foal exempt
R. I.	no	N/A	yes	Prior approval of H.C. 401-277-2781
S. CAROLINA	yes	12 mos.	yes	None
S. DAKOTA	yes	12 mos.	yes	H.C. within 10 days
TENNESSEE	yes	6 mos.	yes	None
TEXAS	yes	12 mos.	yes	Permit req. on slaughter equine. 512-479-6697
UTAH	yes	12 mos.	yes	Suckling foals exempt from test
VERMONT	yes	12 mos.	yes	Permit 802-828-2450
VIRGINIA	yes	12 mos.	yes	None
WASH.	yes	6 mos.	yes	None
W. VIRGINIA	yes	12 mos.	yes	Suckling foals exempt
WISCONSIN	yes	12 mos.	yes	Suckling foals exempt All E.I.A. test info on H.C
WYOMING	yes	12 mos.	yes	All E.I.A. test info. on H.C.

CAMPFIRE COOKING

RECIPES FOR THE TRAVELING HORSEPERSON!
(Send us your favorite recipe for our next edition)

We thought it would be fun (and really different!) to share some food recipes with our readers that we have found to be "tried and true travelers" - that is: easy to make either at home or on the road and/or stay pretty fresh! Let us know if they work well for you and we'd love to receive some of your recipes from around the country!

CREAM CAN DINNER
(For 20 hungry cowhands or guests)

The original cream can dinner was prepared in an old cream can, banded with a steel strap and tossed into the fire for several hours. It was a favorite at round ups and brandings in the West. Old cream cans are hard to find and are known to explode in the cooking process resulting in some exciting dinners! It is just as good and results are more consistent if done in a large roaster oven. The aroma while cooking is wonderful. When the lid comes off, stand back out of the way of the stampede to dinner!

15 lbs. variety of link sausage-garlic, Italian, brats, etc. - cut into 3" pieces

10 lbs. medium New Potatoes - washed well

20 large carrots - washed and cut into 3" pieces

3 large heads of cabbage - washed and cut into wedges

10 ears of corn - shucked and cut in half

5 large white onions, peeled and cut into quarters

2 cans or bottles of dark beer

2 cans or bottles of regular beer (Alcohol free beer can be used, but is not necessary as all alcohol will evaporate in the cooking process)

Pepper to taste (sausage will usually be enought salt)

Roaster should have rack in bottom to keep bottom layer from overcooking in juice. Layer ingredients in roaster as follows; bottom layer- potatoes, next carrots, then cabbage, then onions, then corn and lastly place sausage over the top. Sprinkle with pepper. Pour beer evenly over all ingredients. Roaster needs to have tight fitting lid. Cook 2 to 2 1/2 hours at 350 or until vegetables are fork tender. (Cooking time varies depending on ovens and if ingredients are cold or at room temperature.) Serve in deep plates that will accommodate the juice from the bottom. Serve with corn bread or home made rolls.

-from Carnahan Ranch, Fort Laramie WY

DAD'S CHILE
(Plan outdoor activities)

2 lbs. hamburger browned
(3) 15 oz. cans red kidney beans
(2) 15 oz. cans dark kidney beans
1-1/2 tsp. chile powder to taste
(2) 14 oz.cans italian tomatoes
(2) 28 oz. canswhole tomatoes
2 medium onions chopped
salt and pepper to taste

Brown hamberger and onion in a big skillet. Rinse canned beans Add these to the pot along with the canned tomatoes and seasoning to taste. Gently simmer for 40 minutes stirring once or twice to heat and mix everything up. Serve with corn bread Makes 8 servings.

-from Firmin Bishop's Ranch-

TO HELL WITH TEXAS CORNBREAD

2-1/2 cups cornmeal
2 cups sugar
2 tbsp. baking powder
1 pint milk
2-1/2 cups flour
4 eggs
2 tbsp.. salt

Mix all ingredients for 15 seconds in a mixer, add 1/2 cup liquid shortening. Pour into 13"x9" sheet pan and bake for 45 minutes in a 325 degree oven.

-from Wilderness Trails Ranch in Colorado

QUICK CHICKEN
(When it says "quick", we like it!)

4 or 6 chicken breasts, skinned
1 tsp. paprika
salt and pepper
3 tbsp. butter
1 can cream of chicken soup
1/2 cup mayonnaise
2 tsp. lemon juice
1/2 tsp. curry powder
1 cup grated Cheddar cheese

Wash chicken and pat dry with paper towels. Sprinkle with paprika, salt and pepper. Melt butter in skillet. Slowly brown chicken on both sides, about 10 minutes. Combine soup with mayonnaise, lemon juice and curry powder. Pour over chicken and stir. Cover and gently simmer for 20 minutes. Add cheese and stir to blend. Cover and cook 5 minutes longer. Makes 4-6 servings.

MONKEY BREAD
(Quick and easy and sooo good)

1/2 cup white sugar
1/2 cup brown sugar
3 small packages buttermilk biscuits
2 tsp cinnamon
1 stick butter or margarine

Cut biscuits in quarters. Roll in mixture of sugar and cinnamon. Put into Bundt pan. Boil butter and brown sugar and pour over top. Bake at 350° for 1/2 hour.

ACE OF HEARTS RANCH FLAMING FAJITAS

2 lbs chicken or beef strips
2 onions
2 bell peppers
8 oz. sour cream
1/2 head lettuce
1 tbsp. cumin
8 oz. cheddar cheese
3 tomatoes
aluminum foil
16 flour tortillas
10 oz. salsa
8 oz. guacamole
seasoning to taste

Place chicken or beef on heavy foil. Sprinkle with seasonings and top with onions, tomatoes and bell peppers. Wrap tightly and place on coals. Cook for 20 minutes each side. Remove from fire and place on heated tortillas. Top with shredded lettuce, grated cheddar cheese, sour cream, guacamole and salsa. Meats may be marinated ahead of time for extra flavor.

-from Ace of Hearts Ranch, Florida

PEPPERONI PIE
(Quick, easy, travels well and tastes like pizza)

3/4 cup diced pepperoni
1 cup flour
2 eggs
3/4 cup cubed Muenster cheese
1 cup milk

Grease 9" pie plate. Put pepperoni and cheese on bottom. Whip together eggs, flour and milk and pour over the pepperoni and cheese. Bake in 400 oven for 25 minutes or until lightly browned.

COWBOY BAKED BEANS

(A stick to your ribs bean dish)

1/2 lb. ground beef
1/2 cup chopped onion
1/2 cup brown sugar
1/4 cup ketchup
2 tbsp. molasses
1/2 tsp chili powder
1/2 tsp salt 1 can (16 oz.) kidney beans, drained
1 can (16 oz.) bean of your choice, drained

5 strips bacon, diced
2 cans (16 oz.) pork & beans
1/4 cup sugar
1/4 cup barbecue sauce
2 tbsp. prepared mustard

In a large skillet brown beef, onion, bacon and drain. Add beans. Combine remaining ingredients; stir into bean mixture, Pour into greased 2 1/2 qt. casserole and bake uncovered at 350 for 1 hour.

CHEX MUDDY BUDDIES

(We call this Puppy Chow, great to munch on the road)

9 cups Chex cereal (any variety)
1/2 cup smooth peanut butter
1 tsp vanilla
1 cup peanuts (optional)

6 oz. (1 cup) chocolate chips
1/4 cup margarine
1 1/2 cups powdered sugar

Measure cereal (and nuts) into bowl and set aside. Melt chocolate chips, peanut butter and margarine until smooth and stir in vanilla. Pour chocolate mix over cereal and stir. Put all in large plastic bag and add powdered sugar. Seal bag and shake until well coated. Spread on wax paper to cool. Store in airtight container in refrigerator. Makes 9 cups.

THAT CHERRY STUFF

(Couldn't be easier or taste better)

1 can cherry pie filling
1 can sweetened condensed milk
1 cup chopped nuts

1 large container cool whip
1 can crushed pineapple, drained

Mix everything together. Refrigerate and ENJOY!

A COWBOY'S GUIDE TO LIFE

IF YOU FIND YOURSELF IN A HOLE, THE FIRST THING TO DO IS STOP DIGGIN'.

DON'T SQUAT WITH YOUR SPURS ON.

DON'T INTERFERE WITH SOMETHING THAT AIN'T BOTHERIN' YOU NONE.

IF IT DON'T SEEM LIKE IT'S WORTH THE EFFORT, IT PROBABLY AIN'T.

NEVER ASK A BARBER IF YOU NEED A HAIRCUT.

IF YOU GET TO THINKIN' YOU'RE A PERSON OF INFLUENCE, TRY ORDERIN' SOMEBODY ELSE'S DOG AROUND.

THE EASIEST WAY TO EAT CROW IS WHILE IT'S STILL WARM.
THE COLDER IT GETS, THE HARDER IT IS TO SWALLER.

IT DON'T TAKE A GENIUS TO SPOT A GOAT IN A HERD OF SHEEP.

THE BIGGEST TROUBLEMAKER YOU'LL PROBABLY EVER HAVE TO DEAL WITH WATCHES YOU SHAVE HIS FACE IN THE MIRROR EVERY MORNING.

TIMING HAS A LOT TO DO WITH THE OUTCOME OF A RAIN DANCE.

DON'T WORRY ABOUT BITIN' OFF MORE'N YOU CAN CHEW;
YOUR MOUTH IS PROBABLY A WHOLE LOT BIGGER'N YOU THINK.

ALWAYS DRINK UPSTREAM FROM THE HERD.

GENERALLY, YOU AIN'T LEARNIN' NOTHING WHEN YOUR MOUTH'S A-JAWIN'.

TELLIN' A MAN TO GET LOST AND MAKIN' HIM DO IT ARE TWO ENTIRELY DIFFERENT PROPOSITIONS.

IF YOU'RE RIDIN' AHEAD OF THE HERD, TAKE A LOOK BACK EVERY NOW AND THEN TO MAKE SURE IT'S STILL THERE WITH YA.

GOOD JUDGMENT COMES FROM EXPERIENCE,
AND A LOTTA THAT COMES FROM BAD JUDGMENT.

WHEN YOU GIVE A PERSONAL LESSON IN MEANNESS TO A CRITTER OR TO A PERSON, DON'T BE SURPRISED WHEN THEY LEARN THEIR LESSON.

LETTIN' THE CAT OUTTA THE BAG IS A WHOLE LOT EASIER THAN PUTTIN' IT BACK IN.

THE QUICKEST WAY TO DOUBLE YOUR MONEY IS TO FOLD IT OVER AND PUT IT BACK INTO YOUR POCKET.

NEVER MISS A GOOD CHANCE TO SHUT UP.

U.S. STABLING GUIDE

STATE LISTINGS

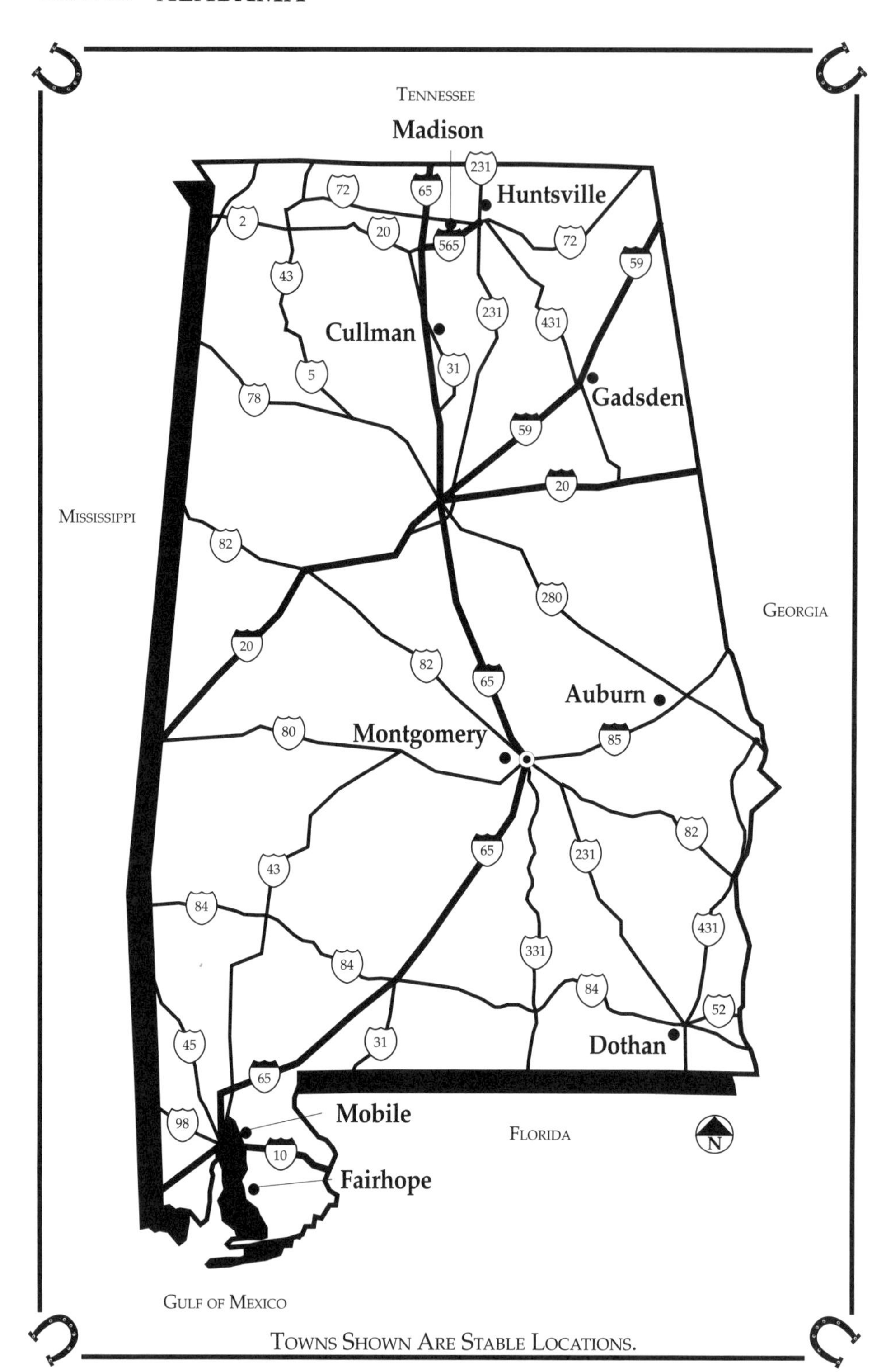

TOWNS SHOWN ARE STABLE LOCATIONS.

ALL OF OUR STABLES REQUIRE CURRENT NEG. COGGINS, CURRENT HEALTH PAPERS & OWNERSHIP PAPERS.

AUBURN

Diamond D Stables **Phone: 334-821-6664**

Richard Garcia or Peggy Durell

610 Lee Road 395 [36830] Directions: From I-85: Take Exit 51, Auburn University. Go south on Hwy 29 to mile marker #177. Follow signs to "Diamond D." **Facilities:** 5 indoor stalls, pasture/turnout; 10 minutes from large animal clinic. **Rates:** $25 per day, $100 per week. **Accommodations:** Auburn University Hampton Inn 2 miles from stable.

CULLMAN

Caudle's Farm **Phone: 205-796-2400**

2235 County Road, RD #1527 [35055] Directions: From Hwy 65: Exit 157 to Hwy 69. Turn left onto County Road. **Facilities:** 10 indoor stalls, feed/hay, riding ring, trailer parking. **Rates:** $20 per day. **Accommodations:** Motels in Cullman, 6 miles from stable.

DOTHAN

Houston County Farm Center **Phone: 334-792-5730**

Jerry Miller

1701 E. Cottonwood Road [36301] Directions: On S.E. side of Ross Clark Circle and Hwy 53. Call for directions. **Facilities:** 100 indoor 10' x 10' stalls, riding areas on 50 acres. Feed store nearby. Open 24 hours a day. Four registered quarter horse sales per year, monthly grade horse sales, and rodeos held at Center. **Rates:** $10 per night. **Accommodations:** Motels within 1 mile of stable.

FAIRHOPE

Lakewood Stables **Phone: 334-928-6711**

Bill Adams **after 6 P.M.: 334-928-3612**

654 Fairhope Ave. [36532] Directions: I-10 to 98 South to Fairhope. Call for directions. **Facilities:** 44 indoor stalls, 30 acres fenced pasture/turnout, feed/hay, trailer parking. Reservations required. Call by 6:00 P.M. **Rates:** $15 per day; $75 per week with advance notice. **Accommodations:** Marriot Grand Hotel, 2 miles from stable; Baron's Motel, 3 miles; Holiday Inn Express, approx. 3 miles.

GADSDEN

Lazy K Stables **Phone: 205-546-1087**

Danny Kidd

128 Lakehardin Road [35901] Directions: From I-59 take Hwy 211 to Noccalula Falls Park. Call for further directions. **Facilities:** 10 indoor box stalls, round pen, 5,000 acres of riding trails. Call for reservations. **Rates:** $20 per day. **Accommodations:** Holiday Inn 8 miles from stable.

ALL OF OUR STABLES REQUIRE CURRENT NEG. COGGINS, CURRENT HEALTH PAPERS & OWNERSHIP PAPERS.

HUNTSVILLE
Flint Ridge Farm, Inc. **Phone: 205-776-3635**
Robert & Diana Rose
3616 Maysville Road [35811] Directions: Approx. 4 miles from Hwy 72 off of I-65. Call for directions. **Facilities:** 2 indoor stalls, feed/hay available, trailer parking, trail riding, hunter/jumper, dressage instruction. No smoking, no alcohol, reservations required. **Rates:** $25 per day, weekly available with advance notice. **Accommodations:** Comfort Inn 15 minutes from stable.

MADISON
Rainbow Riding Academy, Inc. **Phone: 256-830-2911**
Patricia Whitfield **or: 256-837-7758**
212 Capshaw Road [35757] Directions: Approx. 15 miles from I-65. Take I-65 to I-565 East to Rideout Road North to Hwy 72 West. Right on Jeff Road, left at first traffic light onto Capshaw Road. **Facilities:** 6-10 indoor 10′ x 16′ stalls, 4 outside paddocks, 40 acres pasture/turnout, wash stall (hot and cold water), outside working pen, walker, lighted outside arena, hay/feed available, trailer parking. 10 minutes to several vets. Western lessons, parties, hay rides; carousel with ponies for rent. **Rates:** $20 per day. **Accommodations:** Close to numerous motels/hotels, restaurants, and shopping centers. 10 minutes to space and rocket center.

MOBILE
Raintree Farms **Phone: 334-633-9029**
Kay Massey **Barn: 334-633-9028**
7991 Yorkhaven [36695]
Directions: Call for directions. **Facilities:** 15 indoor stalls. 4 turnout paddocks, riding ring, electric camper hookup, feed/hay. Call for reservations. **Rates:** $15 per day. No weekly boarding. **Accommodations:** Motels within 6 miles.

MONTGOMERY
Alabama Agricultural Center **Phone: 334-242-5597**
Bill Johnson
1555 Federal Drive [36107] Directions: Call for directions. **Facilities:** 600 - 12' x 12' wood stalls, trailer parking. Stalls are located at the Agricultural Center, which is the home of the Southeastern Livestock Exposition.
Rates: $10 per day. **Accommodations:** Call for information.

MONTGOMERY
Double D Stables **Phone: 334-284-4982**
Benjamin & Connie D'Amico
124 Mizell Drive [36116] Directions: Call for directions. **Facilities:** 4 - 12' x 12' stalls, 12' x 24' also available, turnout runs with stalls, 1 acre pasture per horse, trailer parking, secured premises, feed/hay, vet within 10 minutes of facility. **Rates:** 12' x 12': $15 per day, $100 per week; 12′ x 24′: $35 per day.
Accommodations: Econo Lodge within 6 miles.

ALL OF OUR STABLES REQUIRE CURRENT NEG. COGGINS, CURRENT HEALTH PAPERS & OWNERSHIP PAPERS.

Harness Hill Stables **Phone: 334-272-9059**
Keith Valley
#2 Harness Hill Drive [36116] Directions: Taylor Road South exit off of I-85. Call for directions. **Facilities:** 16 indoor 12' x 12' box stalls, show barn, outdoor covered arena, covered round pen, 14' x 14' stallion stall, sky lights in every stall, ventilation fans, and natural gas heat in barn. Stable is on 150 acres and has Western tack & saddle shop on premises. Training of horses for Western pleasure and dressage. Call for reservations. **Rates:** $15 per day; ask for weekly rate. **Accommodations:** Mariott 4 miles from stable.

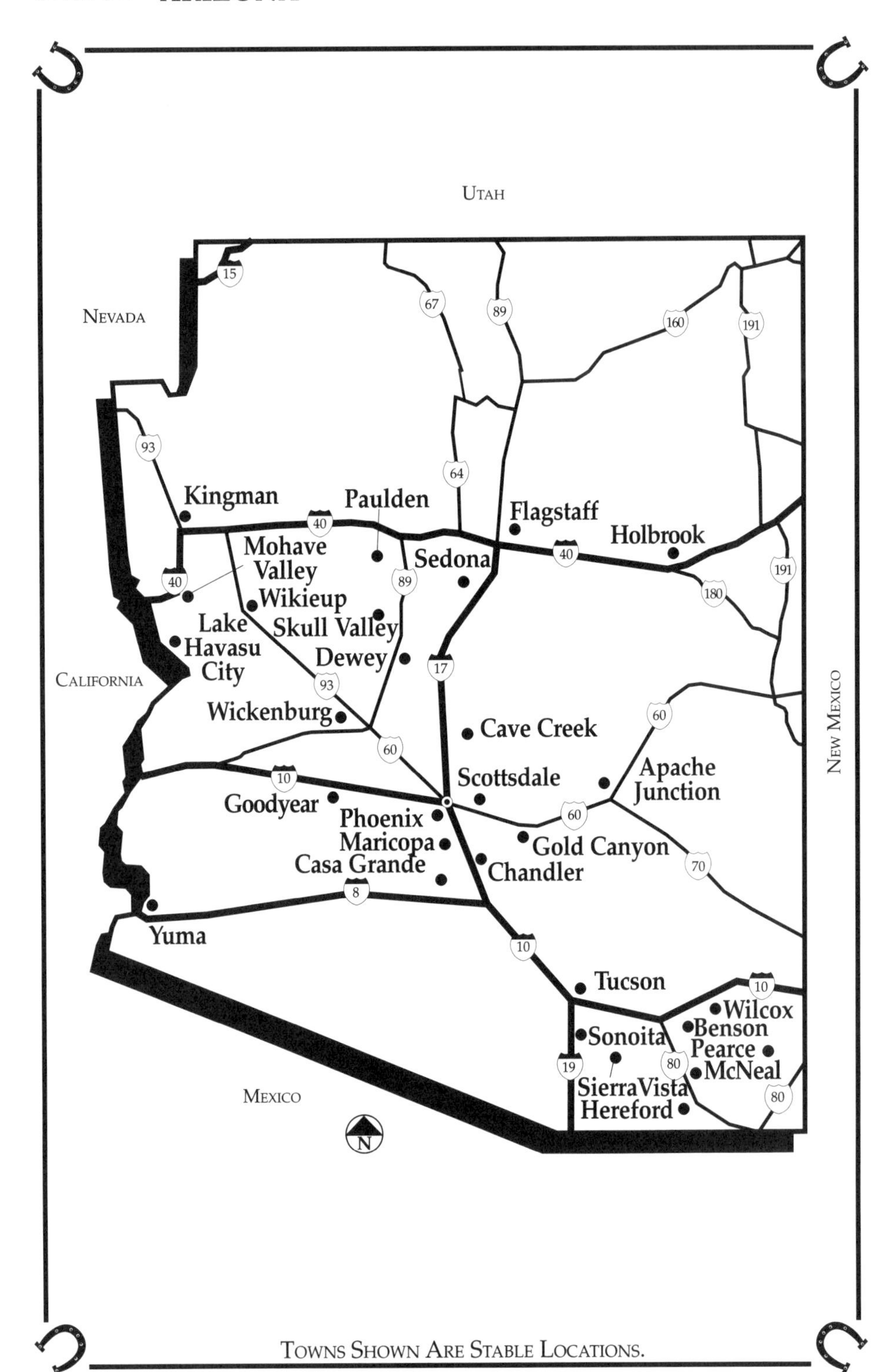

Towns Shown Are Stable Locations.

ALL OF OUR STABLES REQUIRE CURRENT NEG. COGGINS, CURRENT HEALTH PAPERS & OWNERSHIP PAPERS.

APACHE JUNCTION

OK Corral Stables **Phone: 480-982-4040**
Ron & Jayne Feldman **Web: www.okcorrals.com**
P.O. Box 528 [85217] Directions: Highway 60 to N. Tomahawk Exit. North on Tomahawk 5 miles. **Facilities:** 25 - 12' x 24' corrals, 70" round pen, miles of riding trails nearby, easy access to Superstition Mountain Wilderness. Feed/ hay, horses bought, sold, & rented. No stallions overnight. Call for reservations. **Rates:** $15 per day; $40 per week; $75 monthly. **Accommodations:** RV hook-ups. Holiday Inn, Super 8 and many motels within 3-4 miles.

"D" Horse Boarding **Phone: 480-982-7778**
Wes Diekman
1650 N Vista [85219] Directions:U.S. 60, 3 miles N on Tomahawk at Tepee. **Facilities:** 30 stalls, 16'x16' and 12'x20'. Pasture/turnout available, 1 acre+ round pen, feed/hay available, trailer parking. **Rates:** $10 overnight, $50 weekly. **Accommodations:** Several motels within 3-4 miles.

BENSON

J-Six Equestrian Center / Desert Breeze Arabians **Phone: 888-668-9088**
Joyce & Jim Hostetter
3036 Williams Road [85602] Directions: 5 minutes off freeway, all paved roads. I-10 to J-Six/Mescal Exit South, 50 yds. turn left onto Williams Road, 8/ 10 mile on left side. **Facilities:** 15' x 30' pens with roof, 2 pastures 250' x 200', lighted 250' x 150' arena, feed/hay available, trailer parking. Kartchner Caverns 10 minutes from facility. **Rates:** $10 per day per horse. **Accommodations:** Holiday Inn, Best Western and others 5 to 10 minutes from facility, special rates available.

Circle R Ranch **Phone: 520-586-7377**
Bobby Joe & Elly McFadden
2850 W. Drilling Road [85602] Directions: Call for details. 1 mile off I-10, Exit 299 - Easy off and on. **Facilities:** 12 acres pasture, 5 large pipe stalls with shelter and pasture; Full size arena available at extra charge. Dog kennels; 14 spaces for vehicles and horse trailers - $8.50 per night per vehicle (includes electric and water hookup). Local Vet and Farrier available. Tack and Feed storage. 10 minutes from Kartchner Caverns. 20 min. from historic Tombstone. 30 min. from Tucson, Sierra Vista, Bisbee and Willcox. 60 min. from Chiracuaha National Monument and the Mexican border. **Rates:** $8.50 per night per horse. U-feed and care. Accommodations: Private room PLUS 3 bedroom Bed & Breakfast available with large living room and full kitchen. $17.50 per person. Several motels within 3 miles.

ALL OF OUR STABLES REQUIRE CURRENT NEG. COGGINS, CURRENT HEALTH PAPERS & OWNERSHIP PAPERS.

CASA GRANDE

Pinal County Fairgrounds **Phone: 520-723-5242**
Terry Haifley
512 S. 11 Mile Corner Road [85222] Directions: Call for directions. Exit 194 off I-10, 7 miles to 11 Mile Corner Road, turn right, stable 1/4 mile on the right. **Facilities:** 80 covered 8' x 8' stalls, trailer parking, pasture/turnout available, no feed/hay. 122-acre fairgrounds, practice arena, rodeo arena, open all year. Call for reservations. **Rates:** $10 per day. **Accommodations:** Motels within 7 miles.

CAVE CREEK

Horse Camp **Phone: 602-585-4628**
Curt Laird
27426 North 42nd Street [85331] Directions: Call for directions. 8 miles north of Phoenix, 35 miles from airport, 15 miles east of I-17. **Facilities:** 65 - 16' x 16' and 24' x 24' stalls, trailer parking, feed/hay, turnout arena, RV hook-ups, monthly boarding. **Rates:** $10 per day includes feed and water. **Accommodations:** Motels within 10 minutes.

Kukui Driving Center **Phone: 480-585-0234**
Joan M. Stearns **Barn: 480-585-0686**
4150 East Dynamite Blvd. (85331) Directions: Call for Directions. **Facilities:** 6 12X13 barn stalls with runs, 7 16X16 covered outside pens, Hay/Grass available, Trailer Parking, 2 large turnouts, Lighted Arena, Lighted 60, Bull Pen, Clean Facilities, Negative Coggins, Current Health Papers. **Rates:** $10 to $12 per night. **Accommodations:** Motels in Cave Creek, or nearby Scottsdale and Phoenix.

CHANDLER

Central Arizona Riding Academy, Inc. **Phone: 602-963-1310**
Ulrich & Dorie Schmitz
720 N. McQueen Road [85225] Directions: I-10 to Ray Road Exit. East on Ray Road. Turn right onto McQueen Road (1 mile east of Arizona Ave.). 1/2 mile on the west side. **Facilities:** Up to 36 indoor 12' x 12' stalls, 3 paddocks, alfalfa & grass. **Rates:** $15 per horse. **Accommodations:** Holiday Inn (602-964-7000) and Chandler Inn (602-963-6361).

DEWEY

Horse Breakers Unlimited **Phone: 602-632-5728**
John & Marywade Gilbert
P.O. Box 687 [86327] Directions: I-17 to Rt. 169 exit. Call for directions. **Facilities:** Up to 75 - 18' x 28' covered stalls, large show barn, wash rack, 40' x 80' pens, feed/hay. **Rates:** $15 per day; $75 per week. **Accommodations:** Days Inn, Prescott Valley, 8.5 miles from stable.

ALL OF OUR STABLES REQUIRE CURRENT NEG. COGGINS, CURRENT HEALTH PAPERS & OWNERSHIP PAPERS.

FLAGSTAFF

Flying Heart Barn **Phone: 602-526-2788**
Frank & Maxie Davies
8400 N. Hwy 89 [86004] Directions: Exit 201 off I-40. 3.5 miles north on Hwy 89N. **Facilities:** 18 indoor stalls, 4 holding pens, riding ring, hot walker, mountain trails for riding, horse rental available, feed/hay. Please call for reservations. **Rates:** $15 + tax; $75 per week. **Accommodations:** Next door to Horselodge Steak House, 20% discount.

Hitchin' Post Stables, **Phone: 520-774-1719**
Roger L. Hartman **520-774-7131**
4848 Lake Mary Road [86001] Directions: Call for directions. **Facilities:** 10 - 12' x 12' enclosed stalls connected to a 12' x 24' run, 65'diameter covered round arena, grass/alfalfa mix hay, trailer parking. Vet and Farrier on call. No studs. **Rates:** $30 per day, $75 per week, $250 per month. **Accommodations:** Many motels in Flagstaff center, 4.5 miles west of stable.

Pinon Country Cottage **Phone: 520-526-4797**
Deb Conlee
5339 Parsons Ranch Road [86004] Directions: From I-40 take Winona Exit; turn left from west, turn right from east. Drive 1-1/4 miles, turn right on Parsons Ranch Road. Drive back about 100 yds; right in driveway behind large tan house with brown trip. Honk when you arrive. **Facilities:** 10 - 12' x 10' stalls with 12' x 12' run-outs, turnout available 1/4 to 1 acre, round pen, hot walker, feed/hay extra, trailer parking. T Touch offered; farrier and vet on call. **Rates:** $15 per day, $80 per week. **Accommodations:** B&B on premises; $50 first person, $15 each additional, kids under 6 free. Weekly and monthly rates. Many motels in Flagstaff, 10 miles west of stable.

MCS Stables **Phone: 520-774-5835**
Oak Creek (89A) HC 30, Box 16 [86001] Directions: Exit 337 (Airport Exit) off I-17: 2.5 miles south on 89A. I-40 & Route 66 lead directly to I-17 & 89A. **Facilities:** 60 horse capacity, 12 - 16' x 20' outdoor pens, 8 horse barn, 8 pipe-enclosed pastures up to an acre in size, mix pellets & cubes. Horse boarding facility with manager on premises and vets and farriers on call. **Rates:** $20 for barn or $15 per night. **Accommodations:** Motel 6, Ramada Inn & Fairfield Inn located 4 miles away in Flagstaff.

ALL OF OUR STABLES REQUIRE CURRENT NEG. COGGINS, CURRENT HEALTH PAPERS. & OWNERSHIP PAPERS.

GOLD CANYON
Coyote Canyon Ranch, LLC **Phone: 480-982-8464**
Elizabeth Magee **Cell: 480-510-8290** **Fax: 480-671-6529**
2377 S. Coyote Canyon Drive [85219] Directions: From I-60 E: left on Kings Ranch Road, right on Baseline, left on Mohican, left on Valley View, right on Cloudview; first driveway on left. Mailbox says "2391." **Facilities:** 2 stalls near casita, main boarding facility offers 12' x 21' stalls if needed, 150' x 200' arena, 80′ x 80′ lighted arena, 3 round pens, 171/2 acre perimeter trial, a western theme obstacle course, feed/hay, trailer parking, shower rack, hoof soaking racking. Spectacular trails into the Superstition Mountains just a 1/4 mile away. Hiking, golf, boating, jeep tours, casino nearby. **Rates:** $10 per night. **Accommodations:** Casita with hot tub, fireplace, satelite dish, kitchen, and king bed, bath w/shower,tub and phone on premises; $75 per night. summer, $150 per night winter.

HEREFORD
Equi-Sands Training Center **Phone: 520-378-1540**
Vicki Trout
9595 Kings Ranch Road [85615] Directions: Off State Hwy 92, call for directions. **Facilities:** 400 acres, 11 box stalls, 20 pens, 5/8 mile race track, 2-250' x 150' arenas, dressage arena, surrounded by 10,000 acres of state land. 2 miles from Coronado National Monument. **Rates:** $15 per day; $75 per week. **Accommodations:** Motels in Sierra Vista, 15 miles from stable.

HOLBROOK
Navajo County Fairgrounds **Phone: 520-524-6407**
R. J. "Bob" Gates **Fax: 520-524-6824**
P.O. Box 309 [86025] Directions: Eastbound- 1st exit puts you onto Hopi go east directly onto fairgrounds. Westbound-1st exit puts you onto Navajo blvd. Go south to Hopi then east to fairgrounds. Follow racetrack fence to stalls. Your choice of block, metal or chain link. **Facilities:** 100 outside stalls, 1/2 mile racetrack, rodeo arena & 24 hr. groundskeeper. Water & electric hook-up for self-contained campers. **Rates:** $10 per night, $30 per month; $5 per night for campers. **Accommodations:** Motels within 1 mile.

KINGMAN
Mohave County Fairgrounds **Phone: 520-753-2636**
Errol Pherigo
2600 Fairgrounds Blvd. [86402] Directions: Stockton Hill Road Exit off of I-40. Call for directions. **Facilities:** 250 - 10' x 10' outside covered box stalls. Open 24 hours. Water available year round. No feed/hay available. Open door policy. RV hook-ups. Night watch person. No boarding 2nd week in May or 2nd or 3rd week in September. **Rates:** $10 per night. **Accommodations:** Motels within 1 mile. RV space available $8.00 per day.

ALL OF OUR STABLES REQUIRE CURRENT NEG. COGGINS, CURRENT HEALTH PAPERS. & OWNERSHIP PAPERS.

LAKE HAVASU

Lake Havasu Quarterhorses **Phone: 602-453-2481**
Kathy Woodall
797 Turquoise Drive [86404] Facilities: 2 stalls, outside ring, feed/hay available. **Rates:** $15 per day. **Accommodations:** Kingman Motel less than 1/2 mile from stable.

Lidstrom Farms **Phone: 602-453-9007**
Cheri Lidstrom
4033 Goldspring Road [86406] Directions: From I-40 take Rt. 95 to Lake Havasu City. Located off South McCulloch. **Facilities:** 2 stalls in open air barn, 80' x 80' working outside arena, wash rack, trails, feed/hay. Call for reservations. **Rates:** $25 per day; ask for weekly rate. **Accommodations:** Motels 3 miles from stable.

MARICOPA/HIDDEN VALLEY

Rainfire Stables **Phone: 520-424-3037**
Maryeileen Flanagan **Phoenix: 602-802-3037**
4990 North Appaloosa Road [85239] Directions: Call for directions. Located approx. 8 miles north of intersection of SR 84 and I-8, 45 miles south of Phoenix, 30 miles west of Casa Grande. **Facilities:** Eight 16′ x 16′ outdoor stalls including two stallion stalls, five large turnout pens, trailer parking, 60′ round pen, large arena, wash rack (extra fee), feed, pellets and/or hay (if available), 10 acres and room to ride. Casino nearby. Please call in advance. **Rates:** $15 per horse per night (includes two feedings); $80 per week. **Accommodations:** RV park 7 miles away; Francisco Grande Resort 23 miles away.

MOHAVE VALLEY

Double "E" Ranch **Phone: 520-768-9319**
Shawn T. Evans
9566 Evans Lane [86440] Directions: From I-40: Take exit for River R. West Broadway-Needles, Calif. Call from Carles Jr. Will meet you there. Stable is 10 min. from there. **Facilities:** 6 - 14' x 24' 1/2 covered, 1/2 open stalls, 2 - 110' x 80' turnouts, parking for self-contained RVs . Located on the edge of AZ, CA, & NV. **Rates:** $20 per day includes 2 feedings; $80 per week: $125 per month. **Accommodations:** Motel 6, Days Inn at Overland; 10 minutes from Needles, CA & 20 miles from Laughlin, NV.

McNEAL

England's Quarter Horse Ranch **Phone: 520-805-1753**
Wendall England, Owner **Toll Free: 866-845-2853**
7645 North Castlebury Lane. (85617) Directions: 1 hour South of I-10 on Routes 191 and 80, Call for Directions. **Facilities:** 2 hay feedings daily, Trailer Parking and Cleaning, Short term stay for Motor Homes and Travel Trailers, Located in the triangle of Tombstone, Douglas, and Bisbee. **Rates:** $5 per night. **Accommodations:** Campground and motels nearby.

THE SOUTHWEST'S PREMIER EQUINE FACILITY

OVERNIGHT STABLING

- 480 Box Stalls & 24 Covered Pens
- Wash Racks
- Easy Turnaround for Large Rigs
- Feed & Bedding for Sale
- Reservations Recommended

OUR FACILITY

- 360 Western Acres in Scottsdale City Limits
- Close Proximity to Night-life and Daytime Activities
- 100+ RV Hookups
- Full Restroom and Shower Facilities
- Dump Station
- 11 Show-Quality Riding Arenas
- "The Equidome" — 135' x 200' Covered Arena
- Polo & Other Great Events

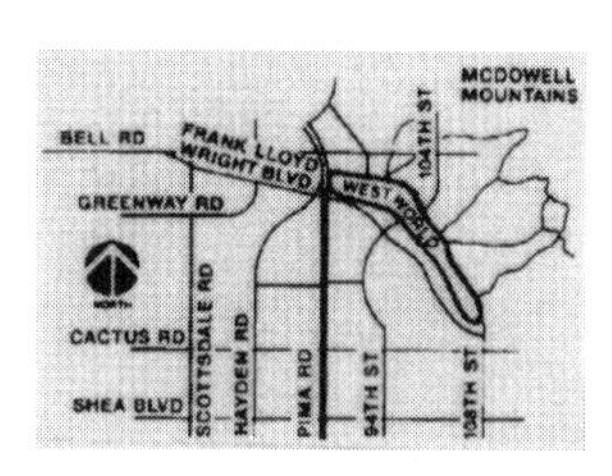

Take Bell Road Exit off I-17. East to Pima Road. North on Pima to WestWorld. Conveniently located within minutes of Scottsdale's most popular resorts.

16601 N. Pima Road
Scottsdale, Arizona 85260
1-800-488-4887
(602) 585-6416 Fax

ALL OF OUR STABLES REQUIRE CURRENT NEG. COGGINS, CURRENT HEALTH PAPERS. & OWNERSHIP PAPERS.

PAULDEN

Prescott Ranch **Phone: 800-684-7433**
Frank Marciante **or 520-636-9737**
P.O. Box 579 [80334]Directions: On Hwy 89, 20 miles north of Prescott. Next to scenic Prescott National Forest. **Facilities:** Can board up to 100 horses, 12' x 12' stalls with run in, 12' x 24' stalls with runs, covered 12' x 24' pipe corrals. Feed/hay, pastures, wash racks, restrooms, showers. Campers welcome. Home of custom-made Marciante "Lightweight Saddles" for both trail and endurance riders. **Rates:** $5 per night, $35 per week. **Accommodations:** Motels within 8 miles.

PHOENIX

Stage Line Ranch **Phone: 602-465-7492**
Michael & Sue Ewens
102 W. Desert Hills Drive [85027] Directions: Off of I-17 & Carefree Hwy. Call for directions. **Facilities:** 16 indoor stalls, 14 stalls with covered shades, 2 - 75' x 75' pens with shades, 2 arenas, wash racks, round pen, riding trails, pellets & hay, RV hookup. Breeds & sells Morgans. Trainer & vet. available. Advance reservations required. **Rates:** $15 per day; $75 per week.
Accommodations: Premier Motel-Phoenix is 10 minutes away.

SCOTTSDALE

WestWorld (see ad on opposite page) **Phone: 800-488-4887 (Press 6)**
16601 North Pima Road [85260] **or 602-585-4392 Fax: 602-585 6416**
Directions: Take Bell Rd. exit off I-17. East to Pima Rd. North on Pima. East into entrance. Only minutes from downtown Scottsdale. **Facilities:** 480 indoor "show" stalls, 24 outdoor covered pens (12' x 24'), turnout available, feed/hay available, 100+ RV hook-ups. Trail riding, livery, and cook-outs. Many major equestrian events. Please call for reservations and/or information. **Rates:** $10-$20. **Accommodations:** Several motels/resorts in area. Call for list.

SEDONA

Greyfire Farm **Phone: 602-284-2340**
David J. & Elaine Ross Payne
1240 Jacks Canyon Road [86351] Directions: I-17 to AZ 179, 7 miles to Jacks Canyon Rd., take right. 1.6 miles to Greyfire Farm. **Facilities:** 2 indoor stalls with outdoor runs, 75' x 150' turnout, alfalfa, trailer parking. Can only take horses that are accompanying the Bed & Breakfast guests. Many beautiful red rock trails accessible from farm. **Rates:** $12 per night; $80 per week.
Accommodations: Bed & Breakfast has private baths and full breakfast.

ALL OF OUR STABLES REQUIRE CURRENT NEG. COGGINS, CURRENT HEALTH PAPERS. & OWNERSHIP PAPERS.

SIERRA VISTA

Equine Inn School of Horsemanship **Phone: 520-378-1078**
Richard Zerbel
5706 S. Kino Road [85615] Directions: South of Sierra Vista on Hwy 92. Left on Ramsey Road. Go 3 miles, right on Kino. 2nd on right. **Facilities:** 15 barn stalls with runs, 16 outside pens with shades. Large riding arena (jumps, barrels, etc. available), full size dressage arena, round pen, trail course. Miles of state land and back roads for riding. Feed included. Call for availability. **Rates:** $15 per day; $75 per week. **Accommodations:** Motels in Sierra Vista, 6 miles away.

SKULL VALLEY

The Oasis Ranch **Phone: 602-442-9559**
Bruce & Bonnie Jackson
P.O. Box 256 [86338] Directions: Located off U.S. Hwy 89. At Kirkland Junction take County Rd. 15 west to Kirkland. Turn right on County Rd. 10 (Iron Springs Rd.) Go approx. 7 miles to Skull Valley. Call for further directions. **Facilities:** 6 stalls, 90' x 120' arena turnout, alfalfa or grass 3x daily. **Rates:** $15 per day; $55 per week. **Accommodations:** Oasis Ranch is an inn with priority stabling given to guests. 2 self-contained B&B cottages. $125 per night. $550-$650 per week.

SONOITA

Rainbow's End Bed & Breakfast and Gaited Horses **Phone: 520-455-0202**
Charlie & Elen Kentnor **E-mail: ElenKentnor@compuserve.com**
P.O. Box 717 (85637) **Web: www.gaitedmountainhorses.com**
Directions: Take I-10 exit #281 (Sonoita, Patagonia Scenic Hwy 83) 24 miles south, crossroads Hwy 82, continue on Hwy 83, 7/10 mile, south side (right) look for our sign. **Facilities:** 25 stalls in two large barns. Spacious comfortable and safe. 17 stalls in large barn with 12' x 12' indoor with 100' runs and automatic waterers. Feed/hay available, Colorado grass for additional fee. Horse trailer parking only. Historic breeding ranch, (Rocky Mtn, Kentucky Mtn, Saddle horses). Trainer on premises to help with training needs or guide you on scenic tours for an additonal fee. **Rates:** $15 per night per horse. **Accommodations:** 4 bedroom / 4 bath B&B on premesis, stabling for B&B guest a priority, no RVs or camping permitted.

TUCSON

Cactus Country RV Resort **Phone: 602-574-3000**
Nancy or Bob Iverson **800-777-8799**
10195 S. Houghton Road [85747] **Web: www.arizonaguide.com/cactusco**
Directions: I-10, exit 275 (Houghton Rd.), North 1/4 mile. East 1/2 mile to office. **Facilities:** Fenced pen will hold 12 "friendly" horses, no feed or hay. Short term stay only. Must have reservation during winter months. **Rates:** No cost for horse. **Accommodations:** Owner must be staying at RV Resort and have own RV or tent and pay for a full hook-up site.

ALL OF OUR STABLES REQUIRE CURRENT NEG. COGGINS, CURRENT HEALTH PAPERS. & OWNERSHIP PAPERS.

TUCSON

Catalina State Park **Phone: 520-628-5798**
PO Box 36986, 11570 N. Oracle Road [85740] Directions: I-10, exit 240 (Tangerine Road), east 12 miles to Hwy 77, turn south and go 1 mile. Park is in foothills of Catalina Mtns. **Facilities:** 8 - 12' x 12' pipe corrals, no feed/hay but feed store within 5 miles. Must stay with horse, trailer parking available, showers; open 24 hrs/day, ranger security 24 hrs/day. Access to Coronado National Forest, known for birdwatching, wildlife, variety of trails for riding. **Rates:** $10 per vehicle. **Accommodations:** Motels within 6 miles.

Rocking M Ranch, Bed & Breakfast **Phone: 520-744-2457 or 888-588-2457**
Louis & Pam Mindes **Mobile: 520-444-0308 or 520-444-0306**
6265 N. Camino Verde [85743] **520-906-3233** **Fax: 520-744-0824**
Web: www.rockingmranch.net **E-mail: lou@pamlou.com**
Directions: From Ina and I-10, west on Ina Road approximately 2 miles, south (left) on Camino Verde exactly 1 mile, road makes hard right west, continue just a few hundred feet and take first left. Follow to the end, can't miss the horse facility. Located just outside the Saguaro National Park in the Tucson Mountains. **Facilities:** Five covered 13' x 12' 3 rail pipe corrals with auto water. Lighted arena, round pen and walker and miles of great riding. Reservations preferred, visits by appointment only, this is not a commercial facility. Boarding only for B&B guests. Veterinarians and farriers on call. **Accommodations:** B&B on premises. **Rates:** October 1 to April 30, $100 per night; May 1 to Sept. 30, $75 per night.Rates include breakfast. No charge for horse corrals. Guests responsible for animal feeding and cleanup. Cash or check no credit cards.

WICKENBURG

Horspitality RV Park & Boarding Stable **Phone: 520-684-2519**
Craig & Pam Dyer
51802 Hwy. 60, Milepost 112 1/2 [85390] Directions: 2 miles south of Wickenburg on Hwy. 60. **Facilities:** 60 outdoor stalls, 20 w/shade, outdoor arena, trails, feed available. RV Park with 85 full hook-ups, restrooms with shower. Call for reservations. **Rates:** $10 per day per horse. Call for RV rate. **Accommodations:** Onsite or motels within 1/2 mile.

WIKIEUP

Bar 5 Cattle & Guest Ranch **Phone: 520-718-0000**
Ron Robach; Pete & Molly Meyer **520-753-5285**
10000 Chicken Springs Road [85360] Directions: Call for directions. **Facilities:** 6 50'x100' corrals w/turnout, 4 12'x14' pens, outdoor stalls, some covered. Alfalfa, hay and grain available, ample trailer parking. Trail riding, ranch work, cookouts. Dogs and cats not allowed unless strictly confined to a camper/ motorhome. **Rates:** $15 per night, $75 per week. **Accommodations:** Trading Post Motel in Wikieup; 9 miles. RV hookups & two guest rooms at ranch.

ALL OF OUR STABLES REQUIRE CURRENT NEG. COGGINS, CURRENT HEALTH PAPERS. & OWNERSHIP PAPERS.

WILCOX

Wilcox Livestock Auction **Phone: 520-384-2206**
Sonny Shores, Scott McDaniel
Haskell & Patti Road [85643] Directions: Take the Wilcox Exit off of I-10. Call for further directions. **Facilities:** 600 outside pens, water & feed/hay available. Office open 8 A.M. to 5 P.M. On-call night manager's number posted on door. **Rates:** $10 per horse per night plus feed. **Accommodations:** Motels within 1 mile.

B Lazy J Ranch **Phone: 520-826-3045**
William & Joanne Garrett
P.O. Box 59 (85652) Directions: From the east, exit 344 off interstate 10; from the west exit 336. Proceed to stop light at Hwy 186. Go south 5.8 miles to Kansas Settlement Rd, south 17.8 miles to Fawn Ranch Rd. Stables on right. **Facilities:** 2 stalls, 10' x 12' covered stalls. Trailer parking available. 20 acres of pasture/turnout available along with scenic riding trials. Available November through March. **Accommodations:** Best Western in Willcox (1800-262-2645) Cochise Stronghold B & B (520-826-4141) and Moonglow B & B (520-826-3448).

Dry Dock Horse Ranch **Phone: 520-824-3359**
Craig & Adel Lawson **Fax: 520-824-3579**
HCR 3 Box 5111 [85643] Directions: Call for directions. **Facilities:** 6 12'x12' indoor stalls, 6 12'x12' outdoor stalls, 1/4 acre turnout, feed/hay available, trailer parking. Located in the heart of Apacheria where Cochise, Geronimo, and Wyatt Earp rode. Guided trail rides available. Thousands of open acres to ride. **Rates:** $15 overnight, $10 per horse per night weekly. **Accommodations:** Guest Quarters available. RV hookup & dump. Bed & Breakfast 5 miles.

YUMA

Horse Haven Stables **Phone: 520-344-1450**
Audrey Hughes **or: 520-344-1030**
13750 S. Avenue 5E [85365] Directions: Exit 5E off Business 8 (Phoenix to San Diego). Call for further directions. **Facilities:** 31- 20' x 30' pens, round pen, 150' x 250' lighted arena, feed/hay & trailer parking available. Call ahead for availability. **Rates:** $12 per day. **Accommodations:** Motels within 5 miles.

Yuma-Mesa Boarding Stable **Phone: 602-726-8673**
Ord & Claudia Cox
5511 E. Hwy 80 [85365] Directions: Located 1/2 mile off Hwy. 80. Call for further directions. **Facilities:** 36 outdoor stalls, 2 arenas, feed/hay available. **Rates:** $10 per day **Accommodations:** Motels within 5 miles.

SURE-FOOTED CUSSES AIN'T THEY!
FRED LEARY

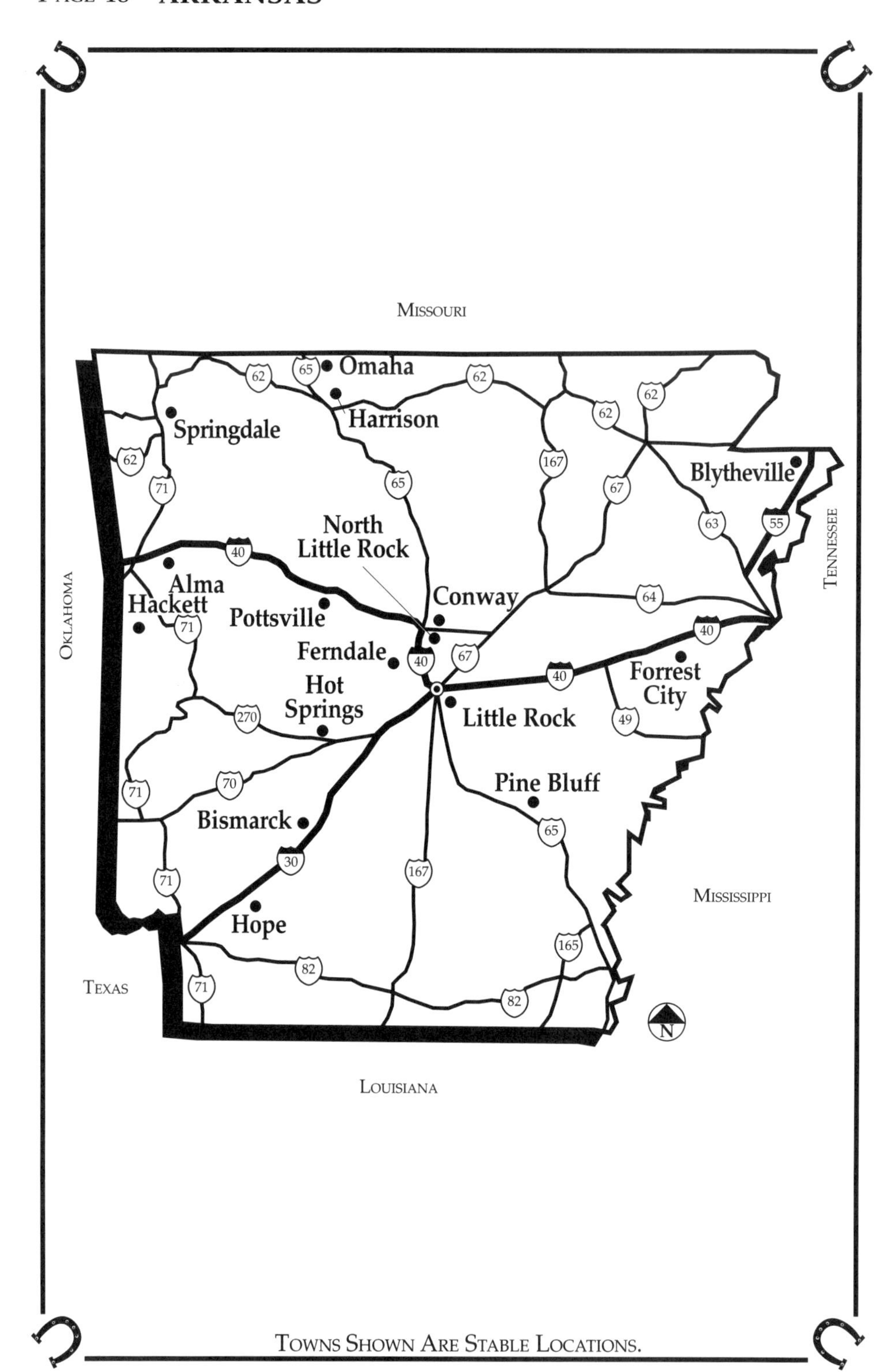

TOWNS SHOWN ARE STABLE LOCATIONS.

ALL OF OUR STABLES REQUIRE CURRENT NEG. COGGINS, CURRENT HEALTH PAPERS & OWNERSHIP PAPERS.

ALMA

Alma Inn Stables **Phone: 501-632-4501**

John Ballentine

Box 2329 [72921] Directions: 100 feet from I-40 and US Hwy 71. **Facilities:** 26 stalls, 3 - 40' x 40' paddocks, farrier and veterinarian on call. Feed store 1/2 mile from motel. **Rates:** $10 per night; $7 per night weekly rate. **Accommodations:** Alma Inn, Alma, 50 ft. from stable.

BISMARCK (Hot Springs area)

Bar Fifty Ranch, **Phone: 888-829-9570**

Bette Clay & Julian McKinney

18044 Hwy 84 [71929] Directions: Exit 91 off I-30, turn left, 9 miles on right. **Facilities:** 18 stalls with shavings, paddocks/turnout, feed/hay, trailer parking, hook-ups w/elec., sewer & water, nice trail riding. 20 min. to hot springs with mineral baths/fishing/boating/racetrack. Reservations required. **Rates:** $15 per night. **Accommodations:** On premises, lodge with 2 bedrooms suites, sleeping 2-6; pool, hot tub & sauna, full country breakfast. $80 for double occupancy -. MC/Visa.

BLYTHEVILLE

Circle S. Horse Motel **Phone: 870-763-9203**

Ronnie Self

3344 N. US Hwy 61 [72315] Directions: Exit 63 S off I-55 onto Hwy 61. Go 1 mile south. **Facilities:** 16 indoor box stalls, pasture area, camper hook-up. Stud facilities available. Tack shop. Feed/hay & trailer parking available. Call for reservations. Cash only. Check-out time 11 A.M. **Rates:** $20 per head/per night. **Accommodations:** Best Western 1 miles from stable; Comfort Inn & Days Inn in Blytheville, 4 miles from stable. No reservations accepted after 10PM.

CONWAY

Miss Toby's Horseback Riding Academy **Phone: 501-329-2233**

Toby Hart

255 E. German Lane [72032] Directions: North of Little Rock off of I-40. Call for further directions. **Facilities:** Barn with 10 stalls on 20 acres; several pastures, 150' x 200' lighted outdoor arena, 8 holding pens, round pen, walker, feed/hay. Western pleasure and rodeo lessons. Call for reservations. **Rates:** $15 per day. **Accommodations:** Motels 1 mile from stable.

FERNDALE

The Barns **Phone: 501-821-4422**

William & Leslie Barns

2 Witness Tree Lane [72122] Directions: Located 13 miles from Little Rock. Please call for directions. **Facilities:** 3 indoor stalls, 4-acre pasture, large corral, feed/hay, camper parking if self-contained. **Rates:** $15 per night; $75 per week. **Accommodations:** Motels 13 miles away.

ALL OF OUR STABLES REQUIRE CURRENT NEG. COGGINS, CURRENT HEALTH PAPERS & OWNERSHIP PAPERS.

FORREST CITY

Roberts' Racing Stable **Phone: 870-633-9041**

Stanley & Delia Roberts **870-633-3427**

2685 Hwy 1 South [72335] Directions: Exit 241 on interstate 40 between Meniphis TN and Littlerock AR. Take Hwy 1 south off I 40 , go 5 miles south. Barn 5 miles from I 40 on Hwy 1. **Facilities:** 75 indoor stalls, 12 x 12 indoor, 100 x 200 paddocks, 1-mile training track with starting gates, open runs & paddocks. Trainer of race-horses and breeder of thoroughbreds. Reservations required. **Rates:** $15 per day; ask for weekly rate. **Accommodations:** Motels within 3 miles of stable.

HACKETT

Daystar Arabians **Phone: 501-639-2401**

Tom & Annetta Tinsman

26110 Hwy 45 South [72937] Directions: I-540, exit 14, south 16 miles, stable on right. **Facilities:** At least 6 - 10' x 10' and 10' x 12' indoor stalls, 40 acres of pasture, 60' x 100' indoor arena, feed/hay and trailer parking available. Mountain trail rides, complete training and horse education. Call for reservations. **Rates:** $10 per night; weekly rates on request. **Accommodations:** Motel and restaurant within 20 miles.

HARRISON

Hilltop Stables **Phone: 501-420-3998**

Glenn & Corrin Britland

HC 33, Box 24 [72601] Directions: Located on Rt. 43 S, 20 minutes from Harrison. Call for further directions. **Facilities:** 8 indoor stalls, 3 paddocks of pasture, feed/hay and trailer parking available. Near Buffalo River National Forest and 45 min. from Branson, Missouri. **Rates:** $15 per night; $75 per week. **Accommodations:** Motel 20 minutes from stable.

HOPE

Hope Fair Park **Phone: 501-777-7500**

P.O. Box 596 [71802-0596] **Hours: 8-12 & 1-5, M-F**

Directions: From I-30: Exit 30 for Hope, take right and another right at second traffic light. Go 2 blocks to Hwy 174 and take left. Go 5 blocks to Park Drive and take right. **Facilities:** 60 indoor stalls, paddock area, 100' x 180' outside covered arena, RV hook-ups with water & electric. Open 24 hours. Hope Feed Co. nearby. Rodeos & festivals held at Park. Temporary stalls set up during shows. **Rates:** $5 per night. Please put payment in drop box.
Accommodations: Days Inn and Holiday Inn 2 miles from stable.

HOT SPRINGS

Harmony Star Riding Stables, Inc. **Phone: 501-525-0176**

Dr. George Christy

6631 Central Ave. [71913] Directions: 5 miles south of Hot Springs. Call for further directions. **Facilities:** Will take overnight boarders in an emergency situation only. **Rates:** $20 per night. **Accommodations:** Motels in Hot Springs 5 miles from stable.

ALL OF OUR STABLES REQUIRE CURRENT NEG. COGGINS, CURRENT HEALTH PAPERS & OWNERSHIP PAPERS.

Panther Valley Ranch **Phone: 501-623-5556**
Roger Stanage
1942 Millcreek Road [71901] Directions: US 70 E from Hot Springs. Take Exit 3. Go 3 miles on Millcreek Rd. Blue & white stable signs. **Facilities:** 20 indoor stalls, soft paddocks, feed/hay at add'l charge. 2,000 acres of secluded beauty. Easily accessible & well lighted. **Rates:** $10 per night. **Accommodations:** 2 fully equipped units on premises that sleep 8 or 10: $80-$100 per night.

LITTLE ROCK
AM Ranch **Phone: 501-312-2818**
Allen McKnight
13111 Colonel Glenn Road [72210] Directions: Exit 4 off I-430. Go west 1 mile on Colonel Glenn Rd. The facility is on the left. **Facilities:** 20 -12' x 12' indoor stalls; feed/hay and trailer parking are available. One acre pasture/turnout. **Rates:** $25 per night per horse. Negative coggins required. **Accommodations:** Room at facility is normally available or LaQuinta Inn on Shackleford Road 3 miles away.

Fox Creek Farms **Phone: 501-225-9384**
Kari Barber
4100 Bowman [72210] Directions: I-430 to Exit 4. Go 1/2 mile west on Colonel Glenn to Bowman Rd. Turn right (north) on Bowman. Entrance approx. 1/2 mile on left. Drive to barn. **Facilities:** 23 indoor wood stalls, 8 indoor concrete block stalls, 3 outdoor stalls, many pasture/turnout areas, no feed/hay. Farm is closed and gates locked at 9 P.M. Must call and leave message if arrival after 9 P.M. No campers, tents, etc. allowed for sleeping on the grounds. **Rates:** $15 per night. **Accommodations:** Motel 6, LaQuinta, Holiday Inn 3-5 miles from farm.

N. LITTLE ROCK
McAdams Family Stables **Phone: 501-835-2205**
Kenneth McAdams
10431 West Mine Rd. [72120] Directions: Call for directions. **Facilities:** 4 indoor stalls, arena, feed/hay available, trailer parking. Some notice if possible. **Rates:** $15 per night. **Accommodations:** Days Inn 4 miles from stable.

OMAHA
South Branson Stables **Phone: 501-426-2473**
David Bird
Rt. 1, Box 226 [72662] Directions: Located on US Hwy 65. Call for directions. **Facilities:** 10 indoor 10' x 14' stalls, 4 acres of fenced pasture, feed/hay & trailer parking for any size trailer available. Preferably no stallions. Call for reservations. Omaha is just 20 minutes away from Branson, Missouri, where there are many activities and sites. **Rates:** $20 per night; discounted weekly rate. **Accommodations:** Big Oaks Motel in Omaha 3 miles from stable.

ALL OF OUR STABLES REQUIRE CURRENT NEG. COGGINS, CURRENT HEALTH PAPERS & OWNERSHIP PAPERS.

PINE BLUFF
K-S Stables **Phone: 501-247-1517**
Kevin & Sue Medlock
4201 German Springs Road [71602] Directions: Hwy 65 South to Exit #27 W. Call for directions. 27 miles from Little Rock. **Facilities:** 4 outdoor stalls, 24' x 60' & 40' x 60' pasture/turnout, feed/hay available at additional charge. 2 weeks advance notice to board stallion. Western riding lessons for children and adults on gentle, older horses. Trail rides and monthly boarding available. **Rates:** $18 per night; $100 per week. **Accommodations:** Comfort Inn and Best Western in Pine Bluff 6 miles away.

POTTSVILLE
Galla Creek Arena & Stables **Phone: 501-967-4526**
Tom Jones
700 Mountain Base Road [72858] Directions: Exit 88 off I-40. The Arena is located northeast of Exit 88, approximately 3/4 mile on Mountain Base Road. **Facilities:** 200 - 10' x 12' indoor stalls with solid walls, two 1-acre holding pens, feed/hay, RV hookups. **Rates:** $15 per night. **Accommodations:** Motels in Russellville, 7-8 miles west of Arena.

SPRINGDALE
Ryn-Char Horse Center **Phone: 501-750-3552**
Karen Cook
4644 Hylton Rd, WC 559 [72764] Directions: Call for directions. **Facilities:** 11 indoor stalls, 10 acres of pasture/turnout, indoor arena, outdoor lighted arena, roping facilities, feed/hay & trailer parking available. Vet. on premises & 2 farriers on call. **Rates:** $10 per night. **Accommodations:** Holiday Inn and Executive Inn 3 miles from stable.

Trouble On A Long Cattle Drive

Towns Shown Are Stable Locations.

ALL OF OUR STABLES REQUIRE CURRENT NEG. COGGINS, CURRENT HEALTH PAPERS & OWNERSHIP PAPERS.

ACTON
Broken Spoke Ranch **Phone: 805-269-3653**
Greg Sachen & Jackie Tichenor
5525 Braeloch Street [93510] Directions: Call for directions. Easy access to Hwy 14. **Facilities:** 8 stalls, new 12′ x 12′ barn with stall mats & auto waterers, 12′ x 24′ run, 15′ x 60′ turnout, hay available, trailer parking available. **Rates:** $10 per day. **Accommodations:** Holiday Inn, Days Inn, E-Z 8 Motel, Palmdale Inn, Super 8 Motel all within 15-20 minutes.

ALTURAS
Lazy J Bar J Ranch **Phone: 916-233-3713**
Hugh & Jilda Comisky
Sara Lane [96101] Directions: 2 blocks west of Hwy 395 on Hwy 299 turn onto Oak St., left at first stop sign onto 19th then immediately right at Sara Lane. Ranch is approx. 1/4 mile on left. **Facilities:** 10 indoor stalls w/ runs, 2-25 acre pastures plus several paddocks, large riding arena, round pen, alfalfa & grass hay available. **Rates:** $20 per night; $90 - $130 per week. Payment expected when horses arrive. **Accommodations:** Essex Motel, Alturas, Frontier Motel, and many others within 1 mile of stable.

ANDERSON
Tony Pochop Training Stables **Phone: 916-365-7759**
Tony Pochop **916-378-2525**
5729 N. Balls Ferry Road [96007] Directions: Take Deschutes Rd. exit off I-5 for 2.2 miles. Turn left on Balls Ferry Road. Go .4 miles. Stable is on right side of road. **Facilities:** 10 indoor stalls, 2 arenas, alfalfa hay, overnight parking for campers & motor homes available on grounds. Reservations appreciated but not required. **Rates:** $15 per night for stall, $10 for arena; weekly rates negotiable. **Accommodations:** Valley Inn & Knights Inn, Amerihost Inn only 2 miles from stable.

AUBURN
Circle RD Ranch **Phone: 916-888-8396**
Ray & Diane Uebner
5645 Fawnridge Road, P.O. Box 5220 [95604] Directions: From I-80: take Bell Road. exit, turn left; go to 4th signal and take right on 49N; go 3 miles and left on Cramer Road. Go to Fawnridge Road (dirt). Ranch is. 4 miles on left - red barn w/ white fence. **Facilities:** 3 stalls, pasture/turnout available, feed/hay, camper hook-ups. Please call for reservations. **Rates:** $15 per night; $75 per week. **Accommodations:** Motels off of I-80 Forest Hill exit, 1 mile away.

ALL OF OUR STABLES REQUIRE CURRENT NEG. COGGINS, CURRENT HEALTH PAPERS & OWNERSHIP PAPERS.

AHWAHNEE (YOSEMITE)
Black Bear Ranch **Phone: 209-641-7487**
44738 Rd 628 [93601] **Toll FREE: 877-24HORSE (244-6773)**
Directions: From Hwy 41 in Oakhurst, go 6 miles North on Hwy 49 to Rd 628. Turn right on Rd 628 -- go 2.5 miles past School Bus sign. Ranch on left, take 1st left. **Facilities:** 3 corrals in trees, 2 barn stalls, 65′ X 100′ arena on 8 acres. Hay available and trailer parking. Easy access to dirt roads to Wawona horse camp in Yosmite. Reservations required. **Rates:** $18 per night, $99 per week. **Accommodations:** Partial hookup for RVs and Rustic 1 bedroom house available on Ranch. Major hotels in Oakhurst, RV Park in Ahwahnee.

BAKERSFIELD
Triple C Ranch
Cathy P. Splonick **Phone: 805-845-6937**
Rene Williams **Phone: 805-831-8143**
Holli Marshall **Phone: 805-366-5021**
5818 S. Fairfax Road [93307] Directions: 3.5 miles south of Hwy 58 on Fairfax. 5.5 miles east of 99 on Panama Lane. **Facilities:** 25 indoor stalls, 1 indoor arena, 2 outdoor arenas w/ lights, alfalfa & hay cubes & grain, parking for trailers and big rigs, camper hook-ups. **Rates:** $20 without feed. **Accommodations:** Economy Inn 5 miles away, Comfort Inn 8 miles away.

BARSTOW
Barstow Horse Motel & Ranch **Phone: 760-256-3671**
Richard & Phyllis Dye **Web: www.horsemotel.homestead.com/HorseMotel.html**
27702 Waterman St. [92311] **E-mail: horsemotelca@aol.com**
Directions: Traveling south from Las Vegas area, 6 miles west of I-15 on Old Hwy-58 or 3 miles from down town Barstow City on Old Hwy 58. From New Hwy 58, take Lenwood Road north to Old Hwy 58 and go east 4 miles. We are on the east side of Old Hwy 58. **Facilities:** 13- 20' x 20' stalls plus 24' x 24' stud pen, 11 outdoor pens, 100′ x 120′ turnout available, work area. Alfalfa hay ($2.00 per feeding), water. Electric hook-up for campers and motor homes at additional charge of $20 per night. Dogs need to be kept on leash or tied up. **Rates:** $10 per night; $35 per week; $150 per month. **Accommodations:** Many motels 3 miles form town.

BLOOMINGTON
SGS Arabians **Phone: 909-877-1084**
Sandra & Gordon Means
18382 Jurupa Avenue [92316] Directions: From I-10 take Cedar Ave. exit. Call for further directions. **Facilities:** 10 indoor box stalls, 8 outside paddocks, large paddock turnout, auto waterers, wash racks, hot walker, camper OK if self-contained. **Rates:** $15 per night; $75 per week. **Accommodations:** Best Western and others within 2 miles.

ALL OF OUR STABLES REQUIRE CURRENT NEG. COGGINS, CURRENT HEALTH PAPERS & OWNERSHIP PAPERS.

CANTIL

Rancho de Nada **Phone: 619-373-2198**
Lou & Lois Peralta **or: 619-373-2610**
End of Lake Road [93519] Directions: 6 miles off Hwy 14 north of Mohave between Los Angeles and Bakersfield; on way from Los Angeles to Las Vegas. Call for directions. **Facilities:** 6 large indoor stalls with outdoor runs, pipe railing, 400 acres of pasture/turnout, feed/hay, trailer parking. Horse motel perfect for people who want to stay with their horses or for drivers transporting horses. **Rates:** $20 per night; ask for weekly rate. **Accommodations:** 4-bedroom, 2-bath log cabin adjacent to stables, swimming pool & bar at main lodge. $50 per room per night, $25 per night for bunk space only.

CHERRY VALLEY

Gee Jay Ranch **Phone: 909-845-5859**
Gerry & Judi Brey
38660 Vineland [92223] Directions: From I-10: Beaumont Ave. exit, go north 3 miles; turn left on Vineland at stop sign. Stable is 1/2 mile on right. **Facilities:** 6 indoor stalls, 9 holding pens, pasture, camper hook-up with water & electric, veterinarian available. Morgan breeder, trainer & sales. Transports horses nationwide. **Rates:** $15 per night. **Accommodations:** Motels within 3 miles.

GOLETA

Horseman's Hangout **Phone: 805-685-4440**
Marcia Nelson **Fax: 805-685-9020**
10920 B2 Calle Road [93117] **E-mail: mnhorsin@aol.com**
Directions: Highway 101 take El Capitan Ranch Road exit, 10 miles from Santabarbara, 30 miles from Buellton. Reservations please and call if plans change. **Facilities:** 12 - 12′ x 12′ box stalls. Some outdoor pens with shade. Full size arena and outdoor round pens with shade. Natural Horsemanship/ Balanced Centered Riding. Clinics & activities, trails and beach access. Nearby summer camps. Overnight rates $10 to $25 depending on service. **Accommodations:** Full service camp ground next door (elcapitancanyon.com) and lots of motels 10 to 30 miles away.

IONE

Sunnybrook Ranch **Phone: 209-274-2680**
Jim & Lori Mote
9401 Brook Ranch Road W. [95640] Directions: Easy access off of Hwy 88. Call for further directions. **Facilities:** 9 indoor stalls, 5 outdoor paddocks, 2-4 acres pasture/turnout, wash rack, riding arenas, parking for large rigs. Tack repair and used tack for sale. **Rates:** $15 per night; $85 per week. **Accommodations:** 5 miles from Gold Country towns of Ione, Sutter Creek, and Jackson. Many motels to choose from.

ALL OF OUR STABLES REQUIRE CURRENT NEG. COGGINS, CURRENT HEALTH PAPERS & OWNERSHIP PAPERS.

LINCOLN

Magic Meadows Farm **Phone: 916-645-5521**
Michele Luna
850 S. Dowd Road [95648] Directions: Accessible from both I-80 & I-5. Call for further directions. **Facilities:** 10 stalls, paddocks, 75′ x 14′ arena, 115' x 115' arena, 2 - 50' pens, alfalfa or oat, hay & grain. Will accept stallions if not too hard to handle. Veterinarians nearby. **Rates:** $15 per night; $100 per week. **Accommodations:** First Choice Inn & Ramada Inn, Rocklin.

LOS OSOS

Barbi Breen-Gurley's Sea Horse Ranch **Phone: 805-528-0222**
2566 Sea Horse Lane [93402] Directions: 12 miles west of San Luis Obispo on the central California coast. Please call for directions. **Facilities:** Pipe corrals, large paddocks, round pens, lighted dressage area, jumps, wash rack. Ranch abuts Montana de Oro State Park vacation area & coast beach riding nearby. Please call for reservations. Must arrive by 10 P.M. **Rates:** $15 per night; $75 per week. **Accommodations:** Motels less than 5 min. away. Ms. Breen-Gurley is a AHSA Large R dressage judge & instructor/clinician.

LOST HILLS

Lost Hills KOA **Phone: 805-797-2719**
Mervin Neufeld
I-5 and Hwy 46 [93249] Directions: West on Hwy 46 from I-5. Go 1/8 mile, south at Carl's Jr. KOA is 100 yds. **Facilities:** 4 outdoor stalls with trees, each holds 4 horses: 3 - 12' x 24' and 1 - 12' x 24' for stallions, trailer parking, no feed/hay. Must clean up area before leaving. Easy walk to restaurants & service stations. By reservation only: call 1-800-562-2793. **Rates:** $10.50 per night; $50 per week. **Accommodations:** Motel 6 and Economy 8 both 200 yds from stable.

MILL VALLEY

Miwok Stables **Phone: 415-383-6953**
Gabino Saldona, mgr. **415-381-0529**
701 Tennessee Valley Road [94941] Directions: Take Hwy 101 to exit for Hwy 1 Stinson Beach; take Hwy 1 for 1/4 mile, past Holiday Inn; turn left on Tennessee Valley Road, go 2 miles to end. **Facilities:** 42 outdoor pens or pasture, alfalfa/oats, trailer parking. Public riding facility with excellent trails. Located in Golden Gate National Recreation Area, 680 acres of trails; 10 minutes from San Francisco in Marin County. **Rates:** $15 per night. **Accommodations:** Holiday Inn in Mill Valley 2.5 miles away.

ALL OF OUR STABLES REQUIRE CURRENT NEG. COGGINS, CURRENT HEALTH PAPERS & OWNERSHIP PAPERS.

MILPITAS

Diamond W Ranch **Phone: 408-262-4163**

Ernie Wool

Weller Road [95035] Directions: Take Calaveras Road off of 680. Call for further directions. **Facilities:** 100 indoor stalls, 80' x 220' indoor arena, pasture/turnout available. English & Western riding lessons, cutting & reining lessons. Call for availability. **Rates:** $15 per night, feed included; $185 per month. **Accommodations:** Motels located 2 miles from ranch.

OLEMA

Five Brooks Ranches **Phone: 415-663-1570**

Will Whitney

8001 Highway One [94950] Directions: From Hwy 101 at San Anselmo: take Sir Francis Drake Blvd. exit; continue to Olema; go south on Hwy One 3.5 miles to Five Brooks Trailhead. **Facilities:** 3 stalls, 8 outdoor paddocks, trailer parking, horses fed A.M. & P.M.. Reservations required. **Rates:** $15/paddock; $20/stall per night. **Accommodations:** Motels in Olema & Inverness plus campgrounds and unique B&Bs in area.

PORTERVILLE

Westwood Farms Equestrian Center **Phone: 559-784-9374**

DeeDee & Floyd Moore

1272 S. Westwood [93215] Directions: 50 miles NE of Bakersfield off Hwy 65. Take Hwy 99 to 190 & Westwood to Hwy 65. **Facilities:** 18 stalls available, 16 w/padded floors indoor. 150'x50'pasture/turnout area plus various smaller sizes. New barns, shavings use only, alfalfa available, trailer parking. Riding programs include mommy & me, riding camps with Teepee campouts, birthday parties. **Rates:** $15 per night, $95 per week. **Accommodations:** Best Western, Porterville (2 miles from Ranch).

QUINCY

New England Ranch **Phone: 916-283-2223**

Barbara Scott, Rick Tegeler

2571 Quincy Junction Road [95971] Directions: 3 miles from Hwy 70 along Chander Road to corner of Quincy Junction Road. **Facilities:** 12' x 12' indoor stalls, 20' x 20' outdoor stalls, 15+ acres of pasture, 50' x 100' turnout, feed/hay available, trailer parking. Riding arena, full-care boarding facility, access to Plumas National Forest riding trails. **Rates:** $15 per night. **Accommodations:** Bed & breakfast (2 rooms) and bunkhouse trailer (sleeps 5) on premises, $60-$95 per night includes gourmet breakfast.

ALL OF OUR STABLES REQUIRE CURRENT NEG. COGGINS, CURRENT HEALTH PAPERS & OWNERSHIP PAPERS.

REDDING
3 D Ranch **Phone: 530-549-3049**
Vicki Donovan
20567 Conestoga Trail [96003] Directions: Call for directions. **Facilities:** 12 indoor stalls, 1 & 2 acre pastures, round pen, lighted arena, hot walker, wash rack, electric hook-up for self-contained campers. No charge for self contained RV's. Please call ahead for reservations. **Rates:** $20 per night; $100 per week. **Accommodations:** 3 miles from I-5 and motels.

SACRAMENTO
Cracker Jack Ranch **Phone: 916-363-4309**
Lew & Jeanee Conner **Voice Mail: 916-441-8179**
10004 Jackson Road [95827] Directions: Please call for directions: **Facilities:** 8 stalls in barn , 4 stalls with outside corrals, paddocks, auto waterers, arena, turnout pens. Parking for campers if self-contained. Stallion service for thoroughbreds & quarter horses. Horse training available. Call for reservations. **Rates:** $15 per night. **Accommodations:** Howard Johnson's located 4 miles from stable.

SAN DIEGO
Clews Horse Ranch **Phone: 619-755-5022**
Christian Clew
11911 Carmel Creek Road [92130] Directions: From I-5 take Carmel Valley Road Exit. **Facilities:** 80-horse ranch with ocean views. 24' x 24' pipe corrals, auto waterers, auto feeders, lighted arena, bull pens, wash racks, miles of trails. Breeder of old-style quarter horses. Stallion is "Chubby Lobos Last." Call ahead for availability. **Rates:** $15 per night. **Accommodations:** Motels within 1 mile of stable.

SANTA BARBARA
Rancho Oso Guest Ranch & Riding Stables
Bill Krzyston, Manager **Phone: 805-683-5686**
Lil Rosen, Reservations **Fax: 805-683-5111**
3750 Paradise Road [93105] **Web: www.rancho-oso.com**
Directions & Rates: See web site (www.rancho-oso.com) or call us. **Facilities:** 24 outdoor pipe corrals, 8 indoor box stalls, arena, round pens, hay twice a day, trailer parking. Access to Los Padres National Forest trails. Open year round. Complete camping facilities plus overnight accommodations in cozy cabins or covered wagons. Weekend meals, hot showers, pool & spa. Guided trail riding on their horses.

SANTA MARGARITA
Santa Margarita K.O.A. **Phone: 805-438-5618**
Rex Jacobson, owner
4765 Santa Margarita Lake Road [93453] Directions: Santa Margarita exit from Hwy 101. Hwy 58 to Pozo Rd., follow signs to lake about 8 miles. **Facilities:** 4 outdoor pole corral stalls on 68-acre campground. Riding trails & beach riding nearby. Great fishing. Tent sites, full hookup sites, Kamping Kabins, pool, country store, & laundry. **Rates:** $7 stall (2 horses) per night, 7th night free. **Accommodations:** On premises.

ALL OF OUR STABLES REQUIRE CURRENT NEG. COGGINS, CURRENT HEALTH PAPERS & OWNERSHIP PAPERS.

SANTA ROSA
Tara Vista **Phone: 707-539-7960**
Linda & Jim Miller
246 Somerville Road [95409] Directions: Off Hwy 12. Call for directions. **Facilities:** 7 indoor stalls with two-acre turnout area, 1/2 acre paddock, dressage area, 1/4 mile track, feed/hay, trailer parking. Located across from 5,000-acre state park in middle of Sonoma wine country. Park-like setting, pool available for guest use. **Rates:** $15 per night, $100 per week; shaving for bed $6.50 per bag. **Accommodations:** Guest room for two with private bath $75 per night. Kennel available for pets.

SANTA YNEZ
Shady Creek Ranch **Phone: 805-686-1024**
Lisa Ann Childs
1475 Edison Street [93460] Directions: Hwy 101 to Hwy 154 to Edison Street. **Facilities:** 5 indoor stalls, 15 outdoor stalls, 2-1 acre pasture/turnout areas, trailer parking. Call ahead for reservations. Apartment in barn available for overnight stay. **Rates:** $10-$20 per night. **Accommodations:** Motels in Solvang, 3 miles from stable.

STOCKTON
Sperry Ranch Stables **Phone: 209-931-2961**
Marlene & George Sperry
1957 E. McAllen Road [95212] Directions: Hwy 99 from N. Central Stockton. Exit on Wilson Way. **Facilities:** 40 stalls, 2 pastures with 12-16 horse capacity, 2 riding arenas. Trains Western pleasure. **Rates:** $15 per night. **Accommodations:** Motels within 1 mile.

SUISUN
Burgundy Farms **Phone: 707-864-3520**
Sharon Cross **Evenings: 707-644-2992**
1911 Rockville Road [94585] Directions: Easy access to freeways. Close to intersection of I-680 and I-80. Call for directions. **Facilities:** 4-12' x 12' stalls, outside paddocks, 4 outdoor pole corral stalls on 68 acre campground. Riding trails and beach riding nearby. Great fishing, tent sites with full accommodations. trailer parking, 2 arenas. **Rates:** $7 per night, $42 per weekfff **Accommodations:** Motel within 5 miles of stable.

THERMAL
Deer Creek Farm
Sue Lukashevich **Phone: 619-399-1812**
or Kate Weber: **Phone: 619-399-1387**
84-393 Avenue 61 [92274] Directions: 5 miles from I-10 near Indio. Take Dillon Road exit. Call for directions. **Facilities:** 75 covered 20' x 24' pipe pens, 1/4 to 12 acres of pasture/turnout areas with covers, 5/8 mile track, 50' x 100' arena, round pen, auto waterers, full-size polo field. This is a year-round "Polo Breeding & Training Facility." Polo ponies for sale. Call for reservation. **Rates:** $20 per night; ask for weekly rate. **Accommodations:** Super 8 five miles from stable.

ALL OF OUR STABLES REQUIRE CURRENT NEG. COGGINS, CURRENT HEALTH PAPERS & OWNERSHIP PAPERS.

VACAVILLE

Ranchotel Horse Center & Tack Barn **Phone: 707-451-8225**
P.O. Box 6 [95696] **Evenings: 707-448-2435**
Directions: I-80, Pena Adobe exit, 1 mile west of Vacaville. **Facilities:** Box stalls, lighted indoor and outdoor arenas. Trail equipment, jumping equipment. Reservations required. TACK BARN on premises features everything for English/Western riders, hunt clothing, saddle seat suits, T-shirts, gifts. **Rates:** $30; rates available for extended stay. **Accommodations:** Motel on grounds. Three miles from 130 Factory Stores.

WATSONVILLE

Cando Ranch **Phone: 831-763-7702**
Iona Marsaa
1084G San Miguel Canyon Rd (96076) **Directions:** At Purnedale take San Miguel Cyn. Rd., west for 4.5 miles. The last road on left (Charmi Lane) before San Miguel Cyn. Rd., deadends (at Hill / Tarpy Rd) Take second driveway (a right turn). Go through "Cando Ranch" gate and close immediately. **Facilities:** 3 paddocks with shelters (2 are 100′ x 35′ and 1 is 100′ x 70′) plus 2 pens (40′ x 32′ and 18′ x 24′). All outdoor; 3 paddocks and 1 pen have shelters. Feed available. Large arena, round pen, and 1 acre pasture. We are in the beautiful Monterey Bay area. **Rates:** $15-20 per night. **Accommodations:** 5 room house with attached dog yard ($50-75 per night depending on number of occupants and an airstream trailor sleeps 2, $25 per night.

WESTLEY

Henderson's Walkers **Phone: 209-894-3360**
Daymond & Sue Henderson
Directions: Exit Hwy 5 at Westley. Turn right in front of McDonald's. This is Howard Road. Go one mile. Turn left at Henderson Road (large Henderson's Walkers sign). Continue 200′ past house to barn area. **Facilities:** Eight 12′ x12′ indoor stalls and 2 to 6 acre pastures plus round pen. Feed & Hay available and parking for trailers. Horseshoer on premises. Call in evenings for reservations or during day leave message about planned arrival times. **Rates:** $15 per night or $90 per week. **Accommodations:** Holiday Inn Express, Days Inn and Budget Inn all within one mile of ranch at Hwy 5 - Westley exit. We raise Tennessee Walking horses. Have two studs: one black and one palomino.

WOODLAND

Woodland Stallion Station **Phone: 530-661-1358**
Ann L. Taylor **Evenings: 530-662-1354**
34270 County Road 20 [95695] **Directions:** I-5, exit Hwy 16 west off-ramp, go 3 miles, right on Hwy 16, 4 miles to Road 94B, north 1 mile to County Road 20, turn left. **Facilities:** 12' x 16' box stalls, 60' round pen, feed/hay, trailer parking. English and Western lessons, training, miles of trails. Stallions: Arabian, Quarter Horse, Thoroughbred, Morgan. **Rates:** $25 per night. **Accommodations:** Shadow Motel in Woodland, 4 miles from stable.

<u>ALL OF OUR STABLES REQUIRE CURRENT NEG. COGGINS, CURRENT HEALTH PAPERS & OWNERSHIP PAPERS.</u>

YREKA

<u>Hawk Scry Horse Hotel</u> **Phone: 916-842-1018**

Christen Kensley

P.O. Box 1760 [96097] **Directions:** Located 4.5 miles from the nearest freeway exchange and 6.5 miles from the south Yreka exchange. Call for specific directions. **Facilities:** 25′ x 75′ outdoor paddocks, 3 acres of pasture/turnout, feed/hay available, trailer parking available. Breed and broker registered Shire draft horses and Shire cross warmbloods. **Rates:** $15 per night; $75 per week. **Accommodations:** Motel 6, Motel Orleans 6.5 miles from stable. Several others within 9 miles of stable.

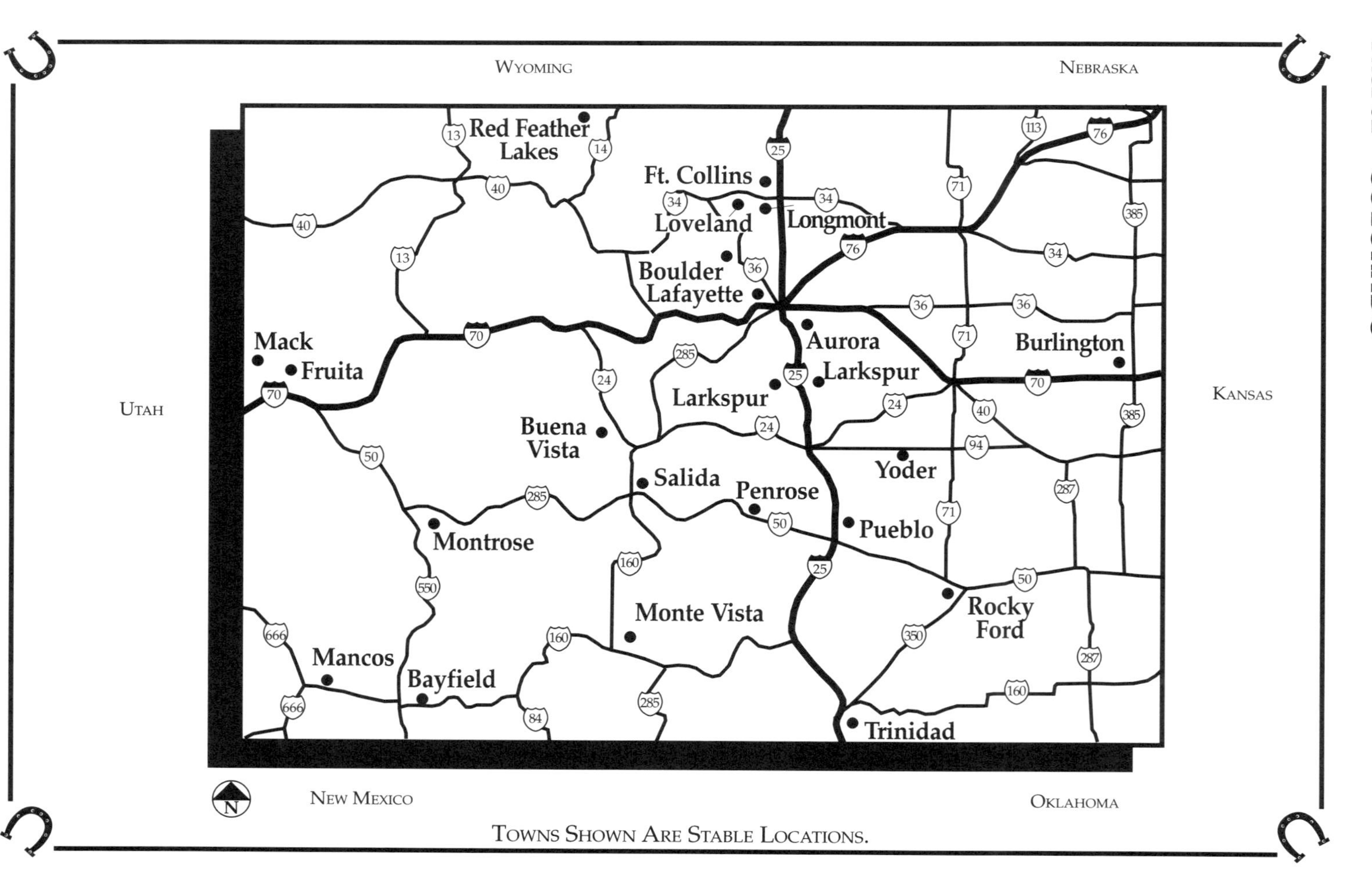

Towns Shown Are Stable Locations.

ALL OF OUR STABLES REQUIRE CURRENT NEG. COGGINS, CURRENT HEALTH PAPERS & OWNERSHIP PAPERS.

AURORA
Kenlyn Stables **Phone: 303-364-9556**
Linda Fisher **Cellular: 303-807-0062**
1000 Salida Street [80011] Directions: I-70 Airport Blvd South exit. 2 miles south to 10th Ave. East 3 blocks to Salida. We're on the corner. I-225 6th Ave east 2 miles to Airport Blvd. Left 1/4 mile to 10th Ave. Right 3 blocks to Salida. **Facilities:** 75 indoor and outdoor stalls, four 30' x 60' turnout areas, trails, indoor and outdoor arenas. Grass and alfalfa available. This is a show barn - clean and safe. Also Arabian breeding farm specializing in endurance riding, pleasure, and show. Endurance cross country horses along with halter & show winners. "Awesome Knight," Black Polish Arabian stallion standing at stud. $1000 stud fee. Lessons & sales for all ages. Call in advance. **Rates:** Varies from indoor to outdoor stalls. **Accommodations:** Holiday Inn on I-70 and Travelers Inn 2 miles from stable in Aurora.

BAYFIELD
Horseman's Lodge **Phone: 970-884-9733**
Andras & Lorraine Westwood, owners **or: 800-715-6343**
7100 County Road 501 [81122] Directions: US 160 to Bayfield. 7 miles on Rt. 501 on right side. **Facilities:** Run in shed and corral. Lodge has 9 rooms, 7 with kitchenettes with daily, weekly and monthly rental. Pets welcome. **Rates:** $10 per night per horse. **Accommodations:** On site.

BOULDER
Mary Bradley **Phone: 303-665-9247**
5610 Baseline [80026] Directions: Call for directions. **Facilities:** 2 indoor stalls, 1 to 60 acres of pasture plus one private pasture. **Rates:** $10-15 per night.

BUENA VISTA
Wapiti Run Trakehners **Phone: 719-395-8543**
Judy & Jim Moore **E-mail: dialogue@amigo.net**
17900 Vista Drive [81211] Directions: From Hwy 24: Call for directions. **Facilities:** 1/2- to 2-acre paddocks with run-in log barns & automatic heated waterers, 3 paddocks, 2 very large pipe pens, 2 outdoor stalls, grass hay available. This is a full training facility with indoor & outdoor arenas & a cross-country course and lessons and training in dressage offered. Located in the Upper Arkansas Valley at the base of 14,000 ft. Mt. Columbia with direct access to the Colorado trail. Pets welcome if controlled. **Rates:** $18 per calendar day without feed. **Accommodations:** 2 furnished studio apartments on premises available for rent.

ALL OF OUR STABLES REQUIRE CURRENT NEG. COGGINS, CURRENT HEALTH PAPERS & OWNERSHIP PAPERS.

BURLINGTON
JB Horse Motel **Phone: 719-346-8217**
Joan Chandler, owner
19790 Hwy 385 [80807] Directions: From I-70: Go west on Rose Ave. 1 1/2 blocks; Turn right on 8th St (AKA 385N) to RR tracks (approx 6 blocks) and continue north. Stable is 1.3 miles past railroad tracks on the right. **Facilities:** 7 indoor stalls, 6 stalls with pipe runs, 150' x 150' & 80' x 80' pastures, numerous corrals & pens, outdoor arena with lights. Reservations preferred. **Rates:** $15 per night. Electric and water hookups $5.00. **Accommodations:** Super 8 and Chaparral Budget Host about 2 mile away.

FORT COLLINS
Brown Quarter Horses **Phone: 303-493-0953**
Ted or Lynn K. Brown
325 E. County Rd. 56 [80524] Directions: I-25 to Exit 271 - turn left (west). Go 4 mi. to stop sign. Turn left. Go 2 miles to stop sign. Turn right & go 1 mile to County Rd. 56. Turn right & go 1/2 block to stable sign. Also accessible from Hwy 287. **Facilities:** 5 indoor stalls, 2-1 acre pastures, arena, 2 paddocks, elec/water hook-up for trailer ($10-$15). Training facility for rope & barrel horses. Horses for sale; videos available; roping lessons & clinics. **Rates:** $15 per night. **Accommodations:** Many motels in Ft. Collins, 3 miles from stable.

FRUITA
Valley View Bed & Breakfast **Phone: 970-858-9503**
Lou A. Purin
888 21 Road [81521] Directions: Exit 26 off I-70. One mile west to 21 Road, go north on 21 Road one mile. Stable on right side of road. **Facilities:** 9 outdoor stalls, 50' round pen, trailer parking. Please call in advance. **Rates:** $10 per night. **Accommodations:** Valley View B & B on premises; Westgate Motel, Grand Junction, 2 miles at Exit 26.

LAFAYETTE
Mary Bradley **Phone: 303-665-9247**
1375 N. 111th St. [80026] Directions: 10 minutes from Boulder. Call for directions. **Facilities:** 10 indoor/outdoor stalls, holding pens, pasture, arena, and riding trails. **Rates:** $12 per night. **Accommodations:** B & B on premises.

LONGMONT
CR Livestock & Animal Care, Inc. **Phone: 303-651-7193**
Rick & Chris Foster
757 Weld County Road 18 [80504] Directions: I-25 West, 3 miles on Hwy 52 to County Line Rd. Turn north - 2 miles to W.C.R. 18. **Facilities:** 3 to 4 enclosed stalls in heated barn, round pen, small indoor arena, outdoor arena, plus 1 port-a-stall barn w/access to pasture. Horse transportation - Colorado P.U.C. 24-hour reservation required. **Rates:** $15 per night. **Accommodations:** Budget Host & Super 8 in Del Camino, 8 miles away.

ALL OF OUR STABLES REQUIRE CURRENT NEG. COGGINS, CURRENT HEALTH PAPERS & OWNERSHIP PAPERS.

LARKSPUR

Spring Canyon Ranch, Camping & B&B **Phone: 303-681-3237**
Web: www.rmtc.net **Barn Phone: 303-681-2942**
E-mail: bobbi@rmtc.net **Cell Phone: 303-808-9730**
Directions: (35 miles south of Denver) Exit I25 at #172 or#173. Go west into town of Larkspur.Turn right at stop sign. Proceed 3 miles to Hwy 105 and turn right. Go 1.3 miles to Perry Park Ranch subdivision. Turn left on Red Rocks Road. Go 2 miles to split in road. Stay left. Go 1.2 miles. Pavement ends. Stay with gravel road bearing left when in doubt. Road takes you right to our gate. **Facilities:** Pens, pasture, indoor stalls and covered arena. **Rates**: Horses $15 per day. **Accommodations:** A beautiful mountain valley with hundreds of miles of America's most scenic riding trails. Sleep in authentic Sioux tipi, your camper or our beautiful ranch house bedrooms. Camper hookups & shower available.

LOVELAND

Rocky Mt. Lazy J Bar S Ranch, Inc. **Phone: 303-669-1349**
Jon Stephens, owner; Ron Webb, manager
3756 W. Country Road 16 [80537] Directions: Call for directions. From Denver, north on I-25 to first Loveland exit, Hwy. 402. **Facilities:** 100 barn stalls and runs with shelters. 350-acre horse boarding facility and working farm. Hay & alfalfa, several arenas, race track, 5-mile riding trails, round pens, large turnout, large indoor heated arena, dressage arena, jumping area , driving arena, training, lessons. Trailer parking available. Call for availability. **Rates:** $10 per night includes feed. **Accommodations:** Motels within 4 miles in town; and on I-25 and Hwy 34 (8-10 miles).

MACK

Badger Lake Ranch **Phone: 970-858-7323**
Loy & Lovell Sasser
647 R Road [81525] Directions: 5.5 miles from Exit 11 off I-70. Call for further directions and reservations. **Facilities:** Corrals, Dept. of Agriculture approved hay available, trailer parking. Located on edge of Colorado-Utah desert. "Lovell of Mack" custom products for distance riders available. **Rates:** $10 per night. **Accommodations:** Motels in Fruita, 16 miles from stable.

MANCOS

Samora's Horse Shoeing & Boarding **Phone: 970-533-7500**
Johnnie Samora
8730 CR 39 Directions: 1/4 mile off Hwy 160. 30 miles west of Durango on left or 17 miles east of Cortez. Turn right on CR 39. (Call for more info). **Facilities:** Stalls, arena & hot walker. **Rates:** $20 Per night per horse. Six miles from Mesa Verde National Park.

ALL OF OUR STABLES REQUIRE CURRENT NEG. COGGINS, CURRENT HEALTH PAPERS, & OWNERSHIP PAPERS.

MONTE VISTA
Greenie Mountain Stable **Phone: 719-852-5269**
Julie Burt
5041 Hwy 15 South [81144] Directions: From Jct. 160 & 285 in Monte Vista turn south onto Hwy 15. Follow Hwy 15 for 5 miles. Driveway is on west side, just past the 5 south road. **Facilities:** 13 12'x12' stalls, 7 w/pipe runs; runs vary in size. Grass/hay included in price, turnout available, trailer parking. Indoor and outdoor arenas. Close access to trails. **Rates:** $10 per night. **Accommodations:** Lodging available on grounds, call for availability. Other lodging 5 miles away in Monte Vista.

MONTROSE
Skyway Ranch/Stables **Phone: 970-249-6259**
Tom or Conni Hood
67909 Ogden (81401) Directions: About 2 miles East of Hwy 550, Call for Directions. **Facilities:** 17 12X12 indoor stables, outdoor pens, 2 acre turnout, feed/hay available, trailer parking, roping arena, hot walker coming soon, breeding facility (3 stallions onsite), Negative Coggins and Health Certificate, owners apartment is connected to stall barn, secured area. **Rates:** $20 per night, multiple horse discount. **Accommodations:** Holday Inn three miles from ranch.

PENROSE
Caballo Casa **Phone: 719-372-6182**
Jim & Nancy McEnulty
60921 E. Hwy 50 [81240] Directions: 28 miles west of Pueblo & 8 miles east of Canon City. Call for directions. **Facilities:** 4 box stalls, large round pen, trailer parking, hay only. **Rates:** $15 per night; $75 per week. **Accommodations:** Motels in Canon City, 8 miles from stable.

PUEBLO
Colorado State Fairgrounds **Phone: 800-876-4567**
Ask for Horse Show **After 5, call 719-561-8484**
[81004] **or 719-561-8489**
Directions: Central Avenue exit (97A) off I-25. Call for directions. **Facilities:** 400 stalls (not available if horse show in progress), feed/hay available before 5 p.m., trailer parking near stalls, camper/RV hook-ups. Call for reservations. **Rates:** $10 per night. **Accommodations:** Motels within 10-15 minutes of fairgrounds.

ALL OF OUR STABLES REQUIRE CURRENT NEG. COGGINS, CURRENT HEALTH PAPERS, & OWNERSHIP PAPERS.

PUEBLO

Five Star Ranch & Equestrian Center, LLC **Phone: 719-382-5601**
Doug Proctor, Manager **Web: www.fivestarranch.com**
18550 Midway Ranch Road (81008) Directions: From the North; take I-25 to exit 119, turn left under freeway, right on Frontage Rd. Go North 1/2 mile. From the North; Take I-25 to exit 122, turn right, then left on to Frontage Rd. Follow South. Turning right after dip in road. **Facilities:** 130- 12x12 and 12x20 foaling stalls. Feed/hay, trailer parking available. 1-5 acre pastures, 20x40 turnouts. Full breeding program, lesson program, english & western show clinics and special events. **Rates:** $15-20 per night. **Accommodations:** On-Site guest rooms, RV hookup(electric).

Fountain Valley Stable **Phone: 719-545-8350**
Lena Fox
2580 Overton Road [81008] Directions: Call for directions. **Facilities:** 9 indoor box stalls, 20 outside pens, 2 arenas, breaking pens, no feed/hay, trailer parking. **Rates:** $15 per night; $75 per week. **Accommodations:** 5 miles away.

ROCKY FORD

McComber Working Ranch **Phone: 719-980-2987**
Jack and Sylvia McComber, Owners
21777 Hwy 71 (81067) Directions: Call for Directions. **Facilities:** 4 12X16 outdoor stalls plus 2 large corrals, feed/hay available at extra cost, trailer parking, turnout arena, 25 years experience in horse training, showing, breeding and cattle ranching. Facility is functional not fancy. Close to Comanche National Grasslands for trailriding. Clean your own stalls, Negative Coggins and Health papers, If traveling with dogs or cats, our rules apply. **Rates:** $10 per night. **Accommodations:** Charming cabin onsite, or motels and and ding in nearby LaJunta.

SALIDA

The Tudor Rose **Phone: 800-379-0889 or 719-539-2002**
Jon & Terre' Terrell **Web: www.thetudorrose.com**
6720 Paradise Road [81201] Directions: Turn south onto County Rd 104 (Paradise Road) off of Hwy 50 on east side of Salida. Stable is at end of Hwy 104. **Facilities:** 4 - 10' x 10' indoor box stalls, 3 - 100' x 400' paddocks, feed/hay, trailer parking. No smoking in house or barn. Access to Rainbow Trail and Colorado Trail and BLM land for riding, guided trips available. White water rafting, mountain biking, hiking, cross country and Nordic skiing, fishing, hunting, golf available in area. **Rates:** $9.50 per night, includes hay. **Accommodations:** Beautiful Bed & Breakfast on premises.

ALL OF OUR STABLES REQUIRE CURRENT NEG. COGGINS, CURRENT HEALTH PAPERS, & OWNERSHIP PAPERS.

TRINIDAD

Happy Farm **Phone: 719-846-7093**

Kate Carlisle

101 W. Indiana [81082] Directions: Take Exit 15 off of I-25. Turn west on Goddard & proceed to first stop sign. Turn right on Arizona & go to second stop sign. HAPPY FARM is across from stop sign. **Facilities:** 2 indoor box stalls, 6-8 outdoor stalls w/ 20' x 30' paddocks, 100' x 100' & 50' x 100' turnouts. Will also accept sheep/cattle and other livestock. No feed/hay. Please call for reservations. Breeding Welsh-Arab ponies and blue-list Egyptian Arabians. **Rates:** $10/outside stall; $15/indoor stall. $20/week if available. One cold winter night a boarder arrived. Inside their large stock trailer were two horses, a pen of exotic chickens, a miniature donkey, two miniature goats, a llama named Sally and 2 great Danes. We stabled everything but the chickens and the great Danes! HAPPY FARM !

YODER

Silver Creek Ranch **Phone: 719-478-5007**

Rosalee or John Hopkins

3755 N. Ramah Hwy. [80064] Directions: Call for directions. 3 miles west and 3.25 miles north of Rush off Hwy. 94. **Facilities:** 12 stalls, 300-acre pasture, wash rack, heated barn, outside arena, round pen, special stud pen, several lean-tos and runways with tall fencing. Call for reservations. **Rates:** $15 per night; weekly and monthly boarding available. **Accommodations:** Motels in Calhan (20 miles away).

HEY, LOOK OUT UP THERE.... DO I LOOK LIKE A ROAD-RUNNER?
SORRY!
ACME ANVIL CO.
CLANG!
FRED LEARY

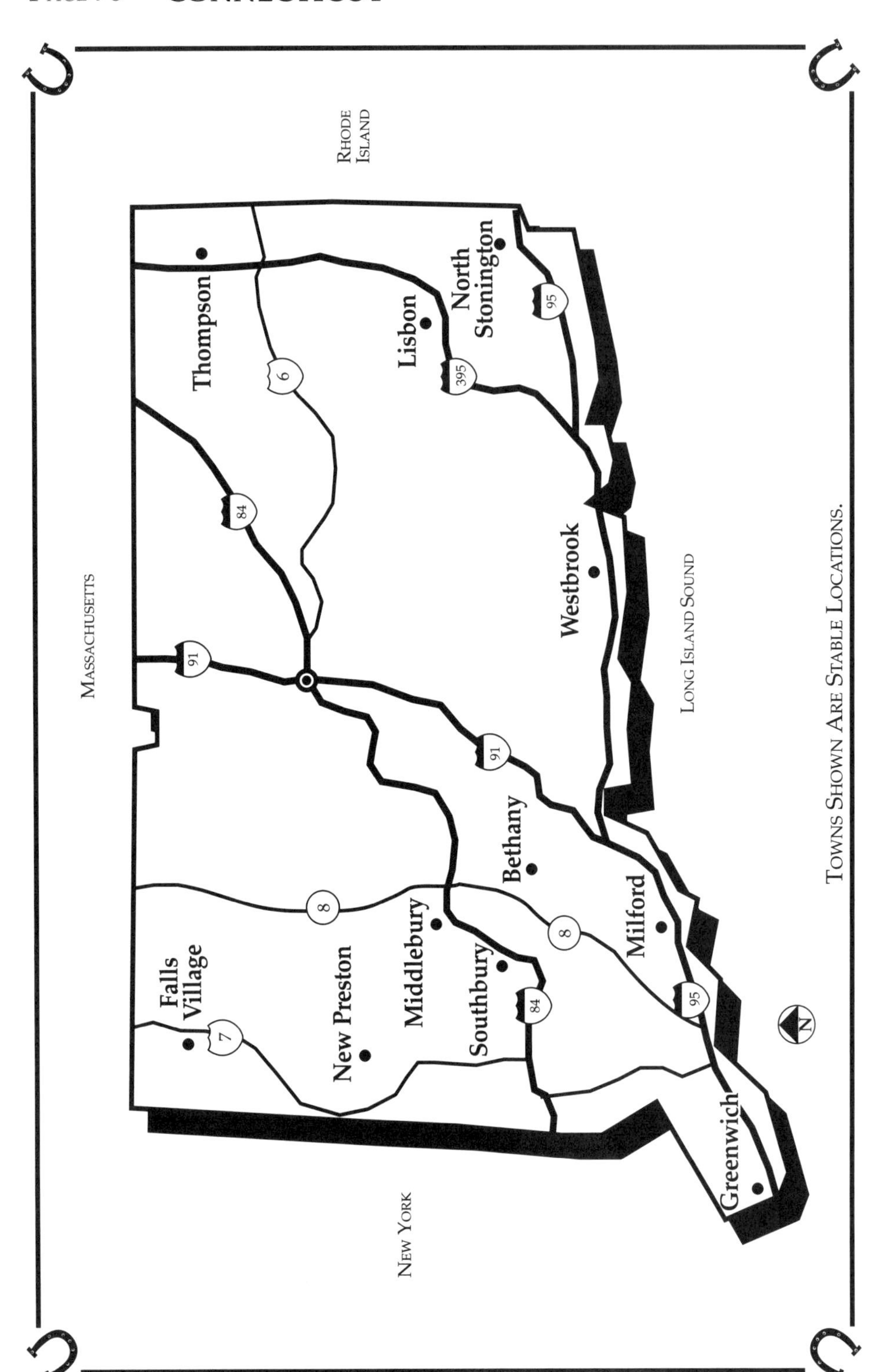
Rhode Island
Massachusetts
New York
Long Island Sound
Thompson
Lisbon
North Stonington
Westbrook
Bethany
Milford
Middlebury
Southbury
New Preston
Falls Village
Greenwich
Towns Shown Are Stable Locations.

ALL OF OUR STABLES REQUIRE CURRENT NEG. COGGINS, CURRENT HEALTH PAPERS & OWNERSHIP PAPERS.

BETHANY

Bittersweet Farm **Phone: 203-393-2586**
Lance Wetmore
325 Amity Road [06524] Directions: 7 miles from I-84. Call for directions. **Facilities:** 35 indoor stalls, 72' x 120' indoor arena, 60' x 88' outside arena, 2/10 mile track, outside paddocks, 2 large fenced pastures. Tack shop on premises. Stable specializes in Morgans & driving. Morgan stallion standing at stud: "Beta-B-Protocol." Call for reservations. **Rates:** $15-$20 per night. **Accommodations:** Holiday Inn & Sheraton in New Haven, 7 miles away.

FALLS VILLAGE (CANAAN)

Rustling Wind Stables, Inc. **Phone: 203-824-7084**
Terri Lamothe **203-824-7634**
164 Canaan Mountain Road [06031] Directions: From Rt. 7 South, go left on Undermountain Rd. (approx. 3.5 miles from Canaan Ctr.). Go 2.5 - 3 miles & take first left on Canaan Mt. Rd. Stable is up the hill on right. Also accessible from Rt. 44 & Rt. 63. **Facilities:** 10 indoor box stalls & small paddocks on 273 acres. Feed/hay & trailer parking. Grooming, exercising, etc. available. **Rates:** $12 per night; $75 per week. **Accommodations:** Super 8, Torrington, 15 miles; Weaver's House B & B, Norfolk, 5 miles from stable.

GREENWICH

On the Go Farm **Phone: 203-661-1513**
Ron Carroll
550 Riverville Road [06831] Directions: I-95 to Hwy 684. Call for further directions. Stable is 30 minutes off of I-95. **Facilities:** 20 indoor box stalls, 4 paddocks, 75' x 175' outside arena, easy access to riding trails. English lessons at all levels offered at farm and horse sales. Call for reservations. **Rates:** $22.50 per night. **Accommodations:** Ramada Inn 5 miles from stable.

LISBON

Lisbon Country Stable **Phone: 203-376-8069**
Darlene Cloutier
164 Kimball Road [06351] Directions: 4 miles off I-395. Call for directions. **Facilities:** 15 indoor stalls in 3 barns, 60' x 120' lighted indoor arena, 60' x 120' outdoor arena, 1/4 mile track, outside paddock and pastures. Stable specializes in Morgans. Call for reservations & availability. **Rates:** $15 per night. **Accommodations:** Ramada Inn in Norwich 10 miles from stable.

MIDDLEBURY

High Lonesome - Rose Hurst Farm **Phone: 203-758-9094**
John Porto & Karen Rue
Route 188 [06762] Directions: Located 2 miles off of I-84. Call for directions. **Facilities:** 4 indoor stalls, 2 small pens, feed/hay & trailer parking. Stallion stall available. Appaloosa breeder. Call for reservations. **Rates:** $15 per night. **Accommodations:** Holiday Inn, Waterbury 10 minutes away; Radisson in Southbury 2 miles away.

ALL OF OUR STABLES REQUIRE CURRENT NEG. COGGINS, CURRENT HEALTH PAPERS, & OWNERSHIP PAPERS.

MILFORD

Spring Meadow Farm **Phone: 203-877-4784**
Eleanor Malafronte **or: 203-874-1352**
918 Wheelers Farm Road [06460] Directions: Call for directions. **Facilities:** 32 indoor stalls, 80' x 200' indoor arena, large outside riding ring, fenced pasture and separate paddocks. Large heated lounge. English training of horses & riders. Also offers cross country rides & picnics and horse clinics on premises. Call for reservations. Negitive coggins required. **Rates:** $20 per night. **Accommodations:** Hampton Inn, Red Roof and Susse Chalet 5 miles away.

NEW PRESTON

Atha House Bed & Breakfast **Phone: 860-355-7387**
Ruth Pearl
Box 2015 [06777] Directions: 25 minutes from I-84 or Rte 8. Call for directions. **Facilities:** 11' x 12' and 10' x 12' in/out stalls, pasture/turnout 1/2 acre per stall, feed/hay available, trailer parking limited by size. Resident Morgan & Mini. Nearby dirt roads for hours of pleasure riding. **Rates:** $20 per night. **Accommodations:** "People" B&B on premises with continental breakfast. Pets accommodated. Advance reservations.

NORTH STONINGTON

Pickwick Farms **Phone: 203-535-1038**
Teresa Hill, manager
75A Reutemann Road [06359] Directions: Call for directions.
Facilities: 2 indoor stalls, several large & small turnout paddocks w/ water. Feed/hay, trailer parking, & blacksmith on premises. Call for reservations. **Rates:** $10 per night. **Accommodations:** Minutes from Ledyard casinos and Mystic seaport.

SOUTHBURY

Mountain Valley Equestrian Center **Phone: 203-264-4243**
Ange Chmura, manager
323 E. Flat Hill Road [06488] Directions: Exit 14 off of I-84. Call for directions. **Facilities:** 30 indoor 12' x 12' stalls, 80' x 160' heated indoor arena, large hunt course, wash room & grass turnout. Call for reservations. **Rates:** $25 per night. **Accommodations:** Radisson and Heritage Inn 7 miles away.

THOMPSON

Sunny-Croft Equestrian Center **Phone: 203-923-3060**
Dottie & Dick Bennett
415 East Thompson Road [06277] Directions: 5 minutes from I-395. Call for directions. **Facilities:** 52 stall barn on 100 acres. 60' x 144' indoor arena, 100' x 200' lighted outside ring, 3-acre fenced pasture, lots of trails for riding. Call for reservations. **Rates:** $15 per night. **Accommodations:** Motel 3 miles from stable.

ALL OF OUR STABLES REQUIRE CURRENT NEG. COGGINS, CURRENT HEALTH PAPERS, & OWNERSHIP PAPERS.

WESTBROOK
Marley Mist Farm **Phone: 203-399-8406**
Dina Sicuranza
354 Pond Meadow Road [06498] Directions: Exit 65 off of I-95. Call for directions. **Facilities:** 6 indoor 12' x 12' stalls, 150' x 225' outside paddock, outside ring on 16 acres of landscaped grounds. Hunter/jumper training facility for all levels. Call for reservations. **Rates:** $15 per night. **Accommodations:** Sandpiper and Days Inn 2 miles from stable.

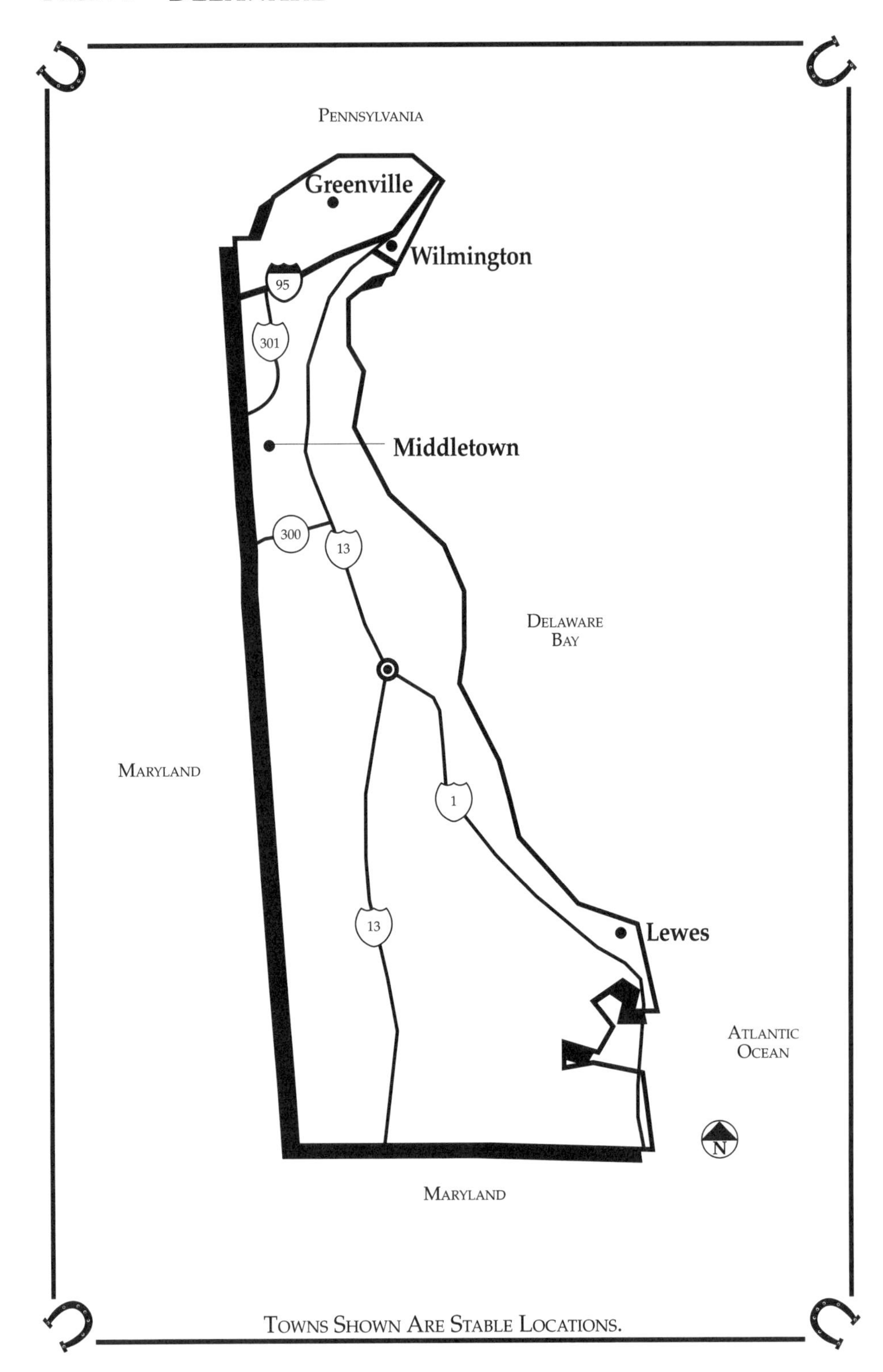

TOWNS SHOWN ARE STABLE LOCATIONS.

ALL OF OUR STABLES REQUIRE CURRENT NEG. COGGINS, CURRENT HEALTH PAPERS & OWNERSHIP PAPERS.

GREENVILLE

LRJ Enterprises **Phone: 302-655-9601**

Laurie Jakubauskas

903 Owl's Nest Road [19807] Directions: Delaware Ave. Exit off of I-95. Call for further directions. **Facilities:** 4 indoor stalls, 90' x 180' indoor arena, 50' x 150' outdoor arena, outside jump course, 50 acres of turnout & trails. **Rates:** $25 per night; $140 per week. **Accommodations:** Motels within 5 minutes.

LEWES

Winswept Stables **Phone: 302-645-1651**

Dawn & Jay Beach

Rt. 24, [19958] Directions: Call for directions. **Facilities:** 12 - 10' x 12' airy stalls, 160' x 200' sand arena, 40 acres, grass jump fields, swimming pond, riding trails, large turnouts and round pen. Located minutes from down town historic Lewes and Rehoboth beachs. Beach riding available. Reservations required. **Rates:** $25 per night, $140 per week. **Accommodations:** Many motels within minutes in Lewes, Rehoboth, & Dewey Beach.

MIDDLETOWN

Price's Public Racing Stable **Phone: 302-378-2032**

Harry Price

1239 Bunker Hill Road [19709] Directions: Located on Rt. 15 off of I-95. Call for directions. **Facilities:** 4 small and 5 large paddocks, large pasture, and riding trails. Trains and sells race horses. Call for reservations. **Rates:** $20 per night. **Accommodations:** Motels 15 miles away.

WILMINGTON

Twin Pines Farm **Phone: 302-478-9917**

Doc Talley

5700 Concord Pike [19803] Directions: 5.5 miles north of Wilmington on Rt. 202. Call for directions. **Facilities:** 28 indoor stalls, paddocks, and riding ring. Feed/hay and trailer parking available. Full boarding facility. **Rates:** $20 per night. **Accommodations:** Motels within 1 mile of stable.

ALABAMA
GEORGIA
GULF OF MEXICO
ATLANTIC OCEAN
Monticello
Milton
Marianna
Wellborn
Tallahassee
High Springs
Ponte Verda Beach
St. Augustine
Reddick
Ocklawaha
DeLeon Springs
Orlando
Fort White
Lecanto
Brooksville
Cocoa
Tampa
Plant City
Fort Pierce
Palm City
Jupiter
Port Charlotte
Immokalee
Naples
Homestead
N
TOWNS SHOWN ARE STABLE LOCATIONS.

ALL OF OUR STABLES REQUIRE CURRENT NEG. COGGINS, CURRENT HEALTH PAPERS, & OWNERSHIP PAPERS.

BROOKSVILLE

L & M Paso Fino Ranch **Phone: 352-544-0299**

Lowell & Melinda Ensinger

5114 Spring Lake Hwy [34601] Directions: From I-75: Exit 61 then west on SR 50. South on Spring Lake Hwy (541). Ranch is on left. **Facilities:** 10 indoor stalls, paddocks & large pastures. 100' x 200' riding ring with lights, round pen, 25 acres with bridle path. L & M Ranch is also a bed & breakfast with heated swimming pool & tennis court. **Rates:** $15 for first night, $10 per night thereafter; $75 per week. **Accommodations:** Bed & Breakfast on site.

COCOA

Ace of Hearts Ranch **Phone: 407-638-0104**

Dennis & Sandra Bressler **Web: www.aceofheartranch.com**

7400 Bridal Path Lane [32927] Directions: From I-95 Exit 79 (Rt #50) East to Rt 405. Turn right. Go to 3rd traffic light,Grissom and turn right. Go 3 miles south past Kings Hwy Light and turn right on Ranch Rd (dirt road). Go 1 mile on right, Bridal Path Lane. Call for U.S.#1 directions. **Facilities:** 14- 12' X 12' rubber mat stalls. Hay and feed available. 1/2 acre turn-out pasture. Trailer parking. Lots of trails. Beach Riding available Nov. - Apr. 5 minutes from Kennedy Space Center. 20 minutes from Cocoa Beach. 45 minutes from Orlando. Please call for reservations. NO STALLIONS. **Rates:** $20 per night, $110 per week. **Accommodations:** Lots of motels -- Titusville 5 miles, Cocoa 8 miles, Cocoa Beach 18 miles, Orlando 45 miles.

DeLEON SPRINGS

Spring Garden Ranch Training Center, Inc. **Phone: 904-985-5654**

Anthony & Jeanette Basile

900 Spring Garden Ranch Road [32130] Directions: From I-4: Take Exit #54. From I-95: Take Ormond Beach/Ocala exit. Call for further directions. **Facilities:** Total of 750 available stalls in 21 barns, 60 - 1/2-acre paddocks, 1 fenced arena, 2 dressage rings, stadium jumps, cross country fences. Restaurant, lounge, swimming pool, tennis, basketball, & racquetball courts & laundromat on premises. **Rates:** $10 per night w/ bedding. **Accommodations:** Holiday Inn in Deland 4 miles from ranch.

FORT PIERCE

Moon Dance Ranch **Phone: 407-466-3622**

Julia & Wayne Shewchuk

3001 Mathews Road [34945] Directions: From I-95 or FL turnpike: Drive west on SR 70 (Okeechobee Rd.) for 3 miles. Turn north (right) on Mathews Road. Go to end & take left over canal. Go straight for 3/10 mi. & stable on left hand side. **Facilities:** 10 indoor 12' x 12' stalls, 10 acres of pasture/turnout, feed/hay & trailer parking on premises, 100' x 200' lighted arena w/music, wash racks, auto waterers. Many riding trails and beach rides w/ your own horse. Call for reservation. **Rates:** $10 per night if no feed needed & you clean stall; $15 with hay, feed & cleaning inc. **Accommodations:** Motel 6, Holiday Inn, & Hampton Inn all 3 miles from stable in Fort Pierce.

ALL OF OUR STABLES REQUIRE CURRENT NEG. COGGINS, CURRENT HEALTH PAPERS & OWNERSHIP PAPERS.

FORT WHITE

Bar One Ranch **Phone: 904-497-2154**

Pat & Berney Copeland

Route 3, Box 1790 [32038] Directions: I-75 S, exit 80 or I-75 N, exit 78 to Hwy 41/441 to CR 18. **Facilities:** 8 - 12' x 12' indoor stalls & 8 outdoor shaded paddocks. Large paddocks and 30 acre pasture. Feed and hay – IF reserved. Trailer parking. **Rates:** $20 per night. We have Quarterhorses and Paints. Bordered by Florida's oldest road, "Old Bellamy Road" and we're 3 miles from Oleno State Park and 7 miles from Ichetucknee State Park. **Accommodations:** Motels 4 to 6 miles away.

HIGH SPRINGS

The Great Outdoors Inn & Cafe **Phone: 904-454-1223**

Ted Greenwald

3105 S. Main Street [32643] Directions: 5 miles west of I-75. 2.2 miles S. of High Springs on Rt. 441 & US 27. Call for further directions.

Facilities: 2 indoor stalls & two 10 acre pastures on a total of 40 acres. Bed & breakfast on premises with pool, BBQ, & 8 rooms decorated in endangered animal motif. Will board all kinds of animals. 24-hr. veterinarian. Tack store & feed store nearby. Minutes from Canterbury Equestrian Showplace. **Rates:** $20 per day; $50 per week; $100 per month. **Accommodations:** B & B on promises: $75-$85 per night.

HOMESTEAD

Bougainvillea Farms **Phone: 305-248-9177**

Margaret Edison

15460 S.W. 256th Street [33031] Directions: Call for directions. **Facilities:** 35 indoor stalls, 3 arenas & pens, feed/hay & trailer parking available on premises. Call for reservation. **Rates:** $15 per night.

Accommodations: Many major motels nearby.

Lazy Blaze Stables **Phone: 305-245-1506**

Inez Howard

24430 S.W. 157th Avenue [33031] Directions: Easily accessible from US Hwy 1. Call for directions. **Facilities:** 20 indoor stalls, 5 acres of pasture/turnout, feed/hay & trailer parking available. Call for reservation. **Rates:** $15 per night. **Accommodations:** Holiday Inn, EconoLodge, & Howard Johnson's all 4 miles from stable.

IMMOKALEE

Midge Lessor **Phone: 941-657-4569**

RR 2 Box 560 [34142] Directions: 4 miles south of I-75 on SR 29. **Facilities:** 10 indoor 10' x 12' covered box stalls, lighted regulation-size training arena, feed/hay, trailer parking. Private 160-acre farm, backs up to Fakahatchee Strand State Forest. Can handle stallions with 2 days' notice. **Rates:** $20 per night, discount for more than one ; $50 per week. **Accommodations:** B&B on premises. Motels in Everglades City, 17 miles from stable.

ALL OF OUR STABLES REQUIRE CURRENT NEG. COGGINS, CURRENT HEALTH PAPERS & OWNERSHIP PAPERS.

JUPITER

Double L Ranch **Phone: 407-744-0704**

Laurie Adams

17593 Rocky Pines Road [33478] Directions: 3.5 miles west of I-95 and Florida Turnpike. Call for further directions. **Facilities:** 4 - 10' x 12' stalls, 2 - 1/2 acre paddocks, 2 - 3/4 acre paddocks, 40' x 60' barn, feed/hay, trailer parking. Can ride on property; state park within 15 minutes. Call for reservations. **Rates:** $20 per night; $75 per week. Full board $425 a month **Accommodations:** Motels and restaurants 5 miles away in Jupiter.

LECANTO

Renab Ranch **Ranch: 352-628-9816**

Lillian Baner **Store: 352-628-2716**

5338 So. Lecanto Hwy, Rt. 491 [34461] Directions: Call for easy directions from I-75 or US 19. **Facilities:** 6 stalls, 21 acres of cross-fenced premium pasture, indoor/outdoor wash down areas, 60' x 90' exercise ring, scenic 1/4-mile bridle path access to 41,000 acres of state forest. Rolling hills and shady trails. Water/elec. hook-up for campers, shower. Reservations required. Owners live on premises and own local feed store; free delivery to your campsite. **Rates:** $10 per night. **Accommodations:** Riverside Inn, Homosassa, 8 miles from ranch.

MARIANNA

Circle D Ranch **Phone: 850-352-4882**

George E. Dryden

3121 Dryden Drive [32446] Directions: From I-10: Exit onto 231 to Cottondale. Turn right onto Hwy 90 at first light. Go 5 miles to ranch entrance on left. **Facilities:** 30 indoor stalls, trailer parking, no feed/hay, no turnout. Must sign a release. **Rates:** $10 per night. **Accommodations:** Days Inn located in Marianna.

MILTON

Coldwater Recreation Area **Phone: 850-957-6161**

Florida Division of Forestry

James Furman

11650 Munson Highway [32570] Directions: 15 miles north of Milton off Hwy 191. Facilities: 69 covered 10' x 12' stalls, 8 outdoor 15' x 15' pens, tie outs for 100+ horses, two 1-acre pasture/turnout areas, trailer parking, feed/hay arranged. Located on Coldwater Creek, Blackwater River State Forest; 35 miles of marked riding trails. Campground, showers, swimming, canoeing. Reservations required. **Rates:** $3 per night.$12 electric campsites, $7 Senior, 64 electric campsites. **Accommodations:** Quality Inn in Milton (22 miles), cabins at Adventures Unlimited canoe livery (6 miles).

ALL OF OUR STABLES REQUIRE CURRENT NEG. COGGINS, CURRENT HEALTH PAPERS, & OWNERSHIP PAPERS.

NAPLES

Lely Stable & Clinic **Phone: 813-793-3344**
Anita Martindale
11000 Tamiami Trail East [33962] Directions: Exit 15 off of I-75. Call for directions. **Facilities:** 5 indoor stalls, 3 pasture areas, indoor arena, feed/hay & trailer parking on premises. Vet. clinic on premises. Call for reservation. **Rates:** $15 per night. **Accommodations:** All major motel chains within 5 miles of stable.

M & H Stables **Phone: 941-455-8764**
Rick Morrell, Joyce Holland, & Andreas Kertscher **Fax: 941-352-1186**
2750 Newman Drive [34114] Directions: 2 miles off Exit 15 of I-75. Call for directions. **Facilities:** 51 indoor stalls with connecting paddocks, 200' x 300' riding arena. Hundreds of miles of riding trails, feed/hay & trailer parking on site. Paint stallion standing-at-stud: "Bodacious," a black Tobiano stallion. Training & breaking of horses done at stable as well as Western riding lessons. Call for reservation. **Rates:** $20 per night. **Accommodations:** Comfort Inn, Super 8, Budgetel, and Quality Inn within 3 miles of stable.

OCKLAWAHA

Wit's End Farm **Phone: 352-288-4924**
Margo Atwood-Langstaff **Phone: 352-288-8157** **FAX: 352-288-8147**
Directions: 20 miles east of I-75 & 10 miles from SR 40. Call for further directions. **Facilities:** eight 12' x 12' indoor stalls, 1/2-acre areas of turnout, feed/hay. Trailer parking and limited RV parking available. Limited smoking. **Rates:** $20 first night, $15 thereafter. **Accommodations:** Lakefront 2 Bedroom rental apartment. Swimming and canoeing. Other motels in Ocala, 20 miles away. Close to Ocala National Forest.

ORLANDO

Grand Cypress Equestrian Center **Phone: 800-835-7377**
Liz Trellue **or 407-239-1938**
One Equestrian Drive [32836] Directions: Call for directions. **Facilities:** 44 stall barn with 10 used for overnight travelers, 8 turnout paddocks, 24 hr security, automatic fly spray & watering systems, shavings for bedding, feed/hay, trailer parking. Center has received the British Horse Society's "Approval" for livery and training; instruction available, show series, clinics. Reservations required. **Rates:** $35 per night includes feeding. **Accommodations:** The Villas of Grand Cypress and Hyatt Regency Grand Cypress on same property.

ALL OF OUR STABLES REQUIRE CURRENT NEG. COGGINS, CURRENT HEALTH PAPERS, & OWNERSHIP PAPERS.

PALM CITY

Harlequin Farms **Phone: 407-286-3745**

Steve & Linda Gawlik & Debbie Schandelmeir

4101 S.W. 48th Avenue [34990] Directions: 1.5 miles from FL turnpike & 3 miles from I-95. Located on a main paved road. Call for directions. **Facilities:** 5 indoor stalls, 2 turnout paddocks, lighted 150' x 300' outdoor jump arena, feed/hay available, trailer parking. Riding trails on farm plus 30 minutes from 22,000-acre state park with trails. Mobile tack store on premises. Grain store 2 miles away. Training & lessons offered at farm. Preferably no stallions. Advance reservations required. **Rates:** $20 per night; $15 per night weekly rate. **Accommodations:** Monterey Motel in Palm City 5 miles from farm.

PLANT CITY

Turkey Creek Stables **Phone: 813-737-1312**

Peggy Womack

5534 S. Turkey Creek Road [33567] Directions: Exit 10 off of I-4 to Turkey Creek, south of State Road 60. **Facilities:** 10 indoor stalls, 3,000 acres of pasture, turnout, and trails. **Rates:** $8.00 plus .50 per feeding; weekly rate available. **Accommodations:** Holiday Inn & Best Western, Plant City, 7 miles from stable.

PONTE VEDRA BEACH

Florio Palm Valley Ranch **Phone: 904-285-2743**

Joanne Florio

505 Ranch Road [32082] Directions: Within 5 minutes of I-95. Call for directions. **Facilities:** 2 indoor stalls, several boarded corrals and paddocks, feed/hay & trailer parking available. Reservations required. **Rates:** $10 with feed; $30 per week. **Accommodations:** Marriott, Ponte Vedra Beach, 8-10 miles; Holiday Inn, 8-10 miles.

PORT CHARLOTTE

Volo Stables **Phone: 813-764-0302**

Lenora Volosin

24614 Nova Lane [33980] Directions: 5 minutes from Exit 31 off of I-75. Call for directions. **Facilities:** 8 indoor stalls, 10.5 acres of pasture/turnout, feed/hay & trailer parking available. Call for reservation. **Rates:** $15 per night. **Accommodations:** All major motels 2 miles from stable.

ALL OF OUR STABLES REQUIRE CURRENT NEG. COGGINS, CURRENT HEALTH PAPERS & OWNERSHIP PAPERS.

REDDICK

Sand Rose Run **Phone/Fax: 352-591-1711**

Cheryl Callaway

15201 NW 115 Court [32686] Directions: From I-75, Exit 72 (Irvine), go west on C318 for 1/4 mile, south on C225 for 3 miles to Fairfield; west on C316 for 3.5 miles to 115 Court. Easy drive; will fax map. **Facilities:** 5 indoor 12′ x 12′ stalls with concrete walls, straw bedding, grass pasture or paddock available, feed/hay available, trailer parking available. Small, private facility with clean, airy stalls; short-term boarding and vacation care available. Also offered: grooming, show prep, manes and braiding. **Rates:** $20 per night, $100 per week. **Accommodations:** Motel in Williston, 8 miles from stable; RV hook-ups at Irvine exit off I-75. Motels in Ocala and Gainesville.

ST. AUGUSTINE

Sunshine Acres, Inc., **Phone: 904-824-5603**

Robert W. & Kelley M. Sterner

3295 Monument Bay Road [32092] Directions: I-95 & SR 16 Exit #95. CR 208 1.3 miles. Left on Pellker Rd. for 1 mile. Left on Monument Bay Road to the end. **Facilities:** 10 indoor stalls, 4-1 acre board fence pasture/turnout areas, water & electric hook-ups for trailers & campers. Sunshine Acres is also a feed company. Dogs must be on a leash. **Rates:** $15 per night. **Accommodations:** Quality Inn & Holiday Inn 4 miles from stable.

The Irish Acre **Phone: 904-829-3771**

Margie & Francis O'Loughlin **Cellular : 904-669-7705** **Barn: 823-1952**

Directions: One mile east of I-95 on SR 207 toward St. Augustine. Stable is on the left just past the Triange Nursery. **Facilities:** 10 indoor stalls, two 50' square turnout lots, 3/4-acre pasture, & miles of wooded trails. Plenty of sightseeing in St. Augustine. **Rates:** $15 per night; $70 per week. **Accommodations:** Comfort Inn 1 mile from ranch. Exit 94 (1-800-228-5150)

TALLAHASSEE

Equine Conveyance Co. **Phone: 904-942-5928**

David E. Fabus **or 904-877-7317**

11953 Wadesboro Road [32311] Directions: From I-10: Take Hwy 90 Exit. Go east on 90 toward Monticello, 5 miles to Baum Rd. and turn right. Go to first road on left and turn. Stable is 1/4 mile on the right. **Facilities:** 6 indoor stalls, various size pasture/turnouts, feed/hay, & trailer parking. Electric hook-ups for RVs and areas to tent camp. No stallions. **Rates:** $15 per night; $12 for 2 or more. **Accommodations:** Seminole Inn in Tallahassee 5 miles from stable.

ALL OF OUR STABLES REQUIRE CURRENT NEG. COGGINS, CURRENT HEALTH PAPERS & OWNERSHIP PAPERS.

TAMPA

In the Breeze Horseback Riding Ranch Vacation **Phone: 813-264-1919**
Childrens Camp **Fax: 813-986-2265**
Woody & Lynda Fowler, owners **Web: www.breezestables.hypermart.net**
7514 Gardner Road [33625] Directions: Exit 7 off Veterians Expressway (Hwy 589) 5 minutes north of Tampa International Airport. Lovely 300 acre ranch marbled with creeks and a lovely moonlit lake. Minutes to downtown facilities. **Facilities:** Stalls $15 night or turnouts $10 includes excellent feed and care. Arena, round pen, indoor facilities, and lovely wooded shaded trails. Campsites, bonfires etc. Swimming year round. Close to beaches, Busch Gardens, State Fair Grounds, etc. **Accommodations:** Pop up or bunk house that sleeps 12 available (Aug. 15-May 15), kitchen, party dancing arena, meeting rooms and vending machines, $25 per night per person.

WELLBORN

Imperial Oaks Ranch **Phone: 904-963-2908**
Jerry & Bobbi Fenderson
16648 53rd Road [32094] Directions: I-72 to exit 92/Lake City, west on rt. 90 1/4 mile then left on Pinemount Rd (Rt. 252) 9 miles to left on 53rd Road, 1 mile on left. From I-10 take exit 41, south on 137 to rt. 252, west on 252 for 1.9 miles to left on 53rd road, 1 mile on left. **Facilities:** 4 inside 12'x12' stalls, 8 outdoor stalls, 2/ large run-ins. 2 round pens, many trails. 2 pastures. Hay/feed available. Trailer parking, self contained camper trailers and 18 wheelers O.K. Breeders of Morgan Horses, Standing Paramout Imperial@ Stud. Call ahead for reservations. **Rates:** $15 per night.

Wellborn Quarter Horses **Phone: 386-963-1556 or 386-963-1557**
Andrea & Joe Schomburg **Fax: 386-963-1557**
8660 CR-137 Directions: Exit 41 off I-10, 1/2 mile north on 137. Exit 84 off of I-75. Call for Directions. **Facilities:** 4 12x12 indoor and 2 12x12 outdoor stalls, Round pen, 11 pastures on 18 acres, Feed/Hay available, Trailer parking, Stallion-Son of Impressive (HYPP/NN) standing, Quality performance horses occasionally for sale, dogs on leash, stallions okay, semis okay. **Rates:** $20 per night, $100 per week. **Accommodations:** Motels off exit 84 on I-75, McLeran B&B in Wellborn.

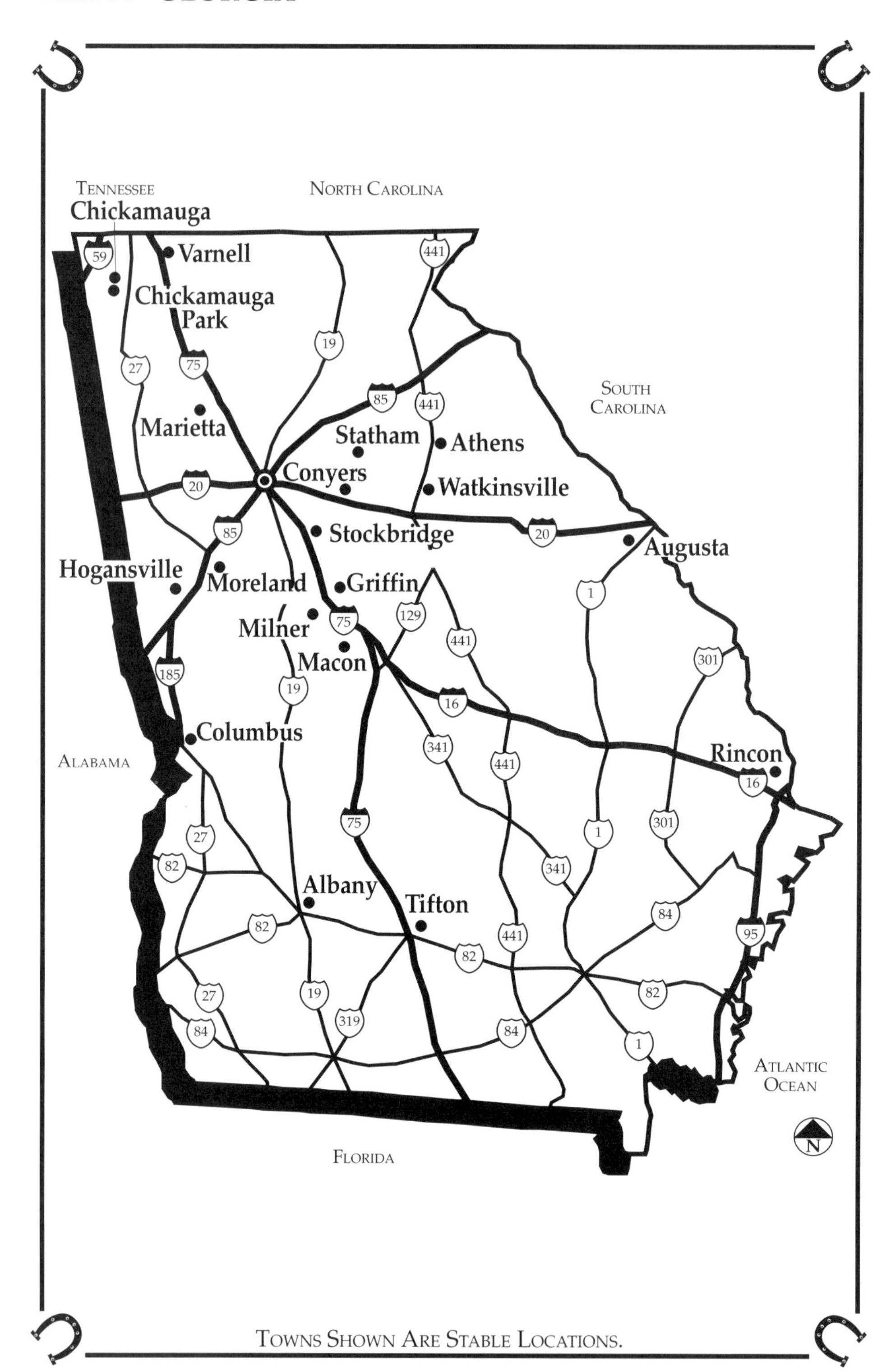

TOWNS SHOWN ARE STABLE LOCATIONS.

ALL OF OUR STABLES REQUIRE CURRENT NEG. COGGINS, CURRENT HEALTH PAPERS & OWNERSHIP PAPERS.

ALBANY

Fleming Road Stables **Phone: 912-439-4166**

Capp Council

3916 Fleming Road [31705] Directions: Hwy US 19, Moultrie Road exit. Take left at first light onto Mock Road. Next road on right is Fleming Road. **Facilities:** 60 indoor stalls, pasture/turnout areas, feed/hay at extra cost, trailer parking. Tack shops & feed stores nearby. No advance notice required. **Rates:** $10 per night. **Accommodations:** Deep South & EconoLodge 5 miles from stable.

ATHENS

Strickland Farms & Stables **Phone: 706-543-7822**

Bill Strickland

6050 Smokey Road [30601] Directions: 4.5 miles from hwy. Call for directions. **Facilities:** 10 indoor stalls, pasture/turnout area, feed/hay included. University of Georgia nearby. **Rates:** $15 per night.

AUGUSTA

Augusta Riding Center **Phone: 706-863-9044**

Ruth & Jim Jatho

1403 Flowing Wells Road [30909] Directions: From I-20: Bel-Air Road exit. Turn south & take first road on left (Frontage Rd.). Go down 1 mile & turn right on Flowing Wells Road. Center is 1/2 - 3/4 mile on right. **Facilities:** At least 5 indoor stalls, 3 large pastures & 3 paddock areas, feed/hay at extra cost, trailer parking available. Call ahead for availability. **Rates:** $20 per night. **Accommodations:** Ramada Inn & EconoLodge 2 minutes away.

CHICKAMAUGA

Chickamauga Bed & Breakfast & Stable **Phone: 706-375-3476**

Kay Red Horse

P.O. Box 81 [30307] Directions: 20 minutes SE of Chattanooga, Tennessee. Call for directions. **Facilities:** 5-stall barn, pasture turnout, trailer parking. Across the road from the Civil War Battlefield with 12 miles of trail riding. Raise and show spotted Tennessee Walkers. Nonsmoking, no alcohol. Reservations required. **Rates:** $15 per horse. **Accommodations:** Bed & Breakfast on 14-acre farm. Full country breakfast; other meals on request. $55 per couple.

CHICKAMAUGA PARK

Trail's End Ranch Riding Stables **Phone: 706-375-4346**

Sarah Clinton

Hwy 27 South, Trail End Road [30707] Directions: 18 miles from Chattanooga. Call for directions. **Facilities:** 5 indoor stalls, pasture/turnout available, feed/hay for purchase, trailer parking. 3-7 days notice. **Rates:** $10 w/o feed. **Accommodations:** Best Western 3 miles from stable.

ALL OF OUR STABLES REQUIRE CURRENT NEG. COGGINS, CURRENT HEALTH PAPERS, & OWNERSHIP PAPERS.

COLUMBUS
Shamrock Stables **Phone: 706-561-9103**
Elvin Amon
6620 Moon Circle [31909] Directions: From I-185: Take Exit 6 (Airport Thruway), go east 4 miles to Moon Circle on right. **Facilities:** 3 stalls in a pole shelter, paddocks available, feed/hay & trailer parking available. One day notice if possible. **Rates:** $10 w/o feed. **Accommodations:** Several nearby.

CONYERS
Linda's Riding School **Phone: 770-922-0184**
Linda Greene Ridley
3475 Daniels Bridge Road [30094] Directions: Take Exit 74 off of I-20E. Call for further directions. **Facilities:** 8 indoor stalls, holding pen, riding rings and trails. Hay & trailer parking available. Please call for reservations. **Rates:** $25 per night, $40 if arrival after 5 P.M. **Accommodations:** Motels within 8 miles of stable.

GRIFFIN
Silver Horseshoes Stables, Ltd. **Phone: 770-227-7681**
Lisa Goldman, manager **or: 770-227-7717**
3010 High Falls Road [30223] Directions: I-75 to Exit 205, go west, 5 miles to caution light, turn right onto High Falls Road. Stable is 6th driveway on left. **Facilities:** 40 indoor 14' x 14' box stalls, 40 acres of pasture, safe fence, indoor and outdoor arenas, feed/hay available, trailer parking; professional management. Riding lessons, boarding. **Rates:** $25 per night; $110 per week. **Accommodations:** Comfort Inn, Days Inn, Holiday Inn, Hampton Inn within 10-12 miles of stable.

HOGANSVILLE
Flat Creek Ranch, Inc. **Phone: 706-637-8920**
Joan Keegan, Manager
3564 Mountville Rd. [30230] Directions: I-85 Exit 6. West towards Hogansville. Follow campground signs. Immediate left at Waffle House (Bess Cross Road). 2 miles to 4-way stop. Left onto Mountville - Hogansville Road. Campground on right. **Facilities:** 70 - 10' X 14' indoor stalls. No feed or hay available. No pasture or turnout available. Trailer parking space provided. **Rates:** $10 per night, $200 per month. **Accommodations:** Flat Creek Campground 706-637-6001. Key West Motel 3 miles 706-637-9395. Hummingbird Inn - 3 miles 706-637-5400

MACON
Wesleyan College Equestrian Center **Phone: 912-757-5103**
Jon Conyers, Director of Riding **E-mail: JConyers@WesleyanCollege.edu**
4760 Forsyth Road [31210] Directions: Exit #3 (Zebulon Road) off I-475. See signs for Wesleyan College. **Facilities:** New facility completed summer 1999, 24 indoor stalls, 5 outside paddocks, lighted outdoor riding arena, 40 acres of wooded trails. Location of Collegiate Equestrian Team and Community Horsemanship Program. **Rates:** $30 per night. **Accommodations:** Jamison Inn (912-474-8004) and Fairfield Inn (912-474-9922) 2 miles from stable.

ALL OF OUR STABLES REQUIRE CURRENT NEG. COGGINS, CURRENT HEALTH PAPERS, & OWNERSHIP PAPERS.

MARIETTA
Hymnbrook Farm **Phone: 770-428-1065**
Barbara & Bill Dawson
150 Mt. Calvary Road [30064] Directions: Location is 25 miles of Atlanta, GA. Farm is 5 miles from I-75 and there are 7 motels at exit 269. From I-75, take Exit 269 going West. Cross Hwy 41, Hwy Old 41, Stilesboro Rd and at Burnt Hickory Rd, turn left onto Burnt Hickory and immediate right to Mt. Calvary Rd. (about 200'). Farm is about 1 mile on right. 150 Mt. Calvary Rd. **Facilities:** 10 12 x 12' indoor stalls, ring, trailer parking and electric available. Owner on premises. **Rates:** $20 per night including bedding. **Accommodations:** Six motels within 5 miles.

MILNER
Pegasus Riding School, Inc. **Phone: 770-228-3865**
Linda & Warren Abrams
392 Philip Weldon Road [30257] Directions: I-75 to Exit #198 (High Falls Road), 8 miles. **Facilities:** 7 indoor 12' x 12' stalls, 10 indoor 10' x 10' stalls, pasture/turnouts, tack shop, feed/hay, trailer parking available. Hunter/jumper, and dressage riding. **Rates:** $25 per night. **Accommodations:** Apartment with kitchenette available on premises, $25 per night per person. Days Inn, Holiday Inn within 10 miles in Griffin. Camper hookup (water & elec) $25.

MORELAND
Point to Point Farm **Phone: 404-253-0570**
Sam & Dale Bowers
83 Bear Creek Road [30259] Directions: Located 35 minutes south of Atlanta International Airport and 10 minutes from I-85, Exit 8. Call for directions. **Facilities:** 2 indoor 12' x 12' matted stalls, 1/2- to 3/4-acre pasture/turnout areas, separate paddock for stallion, 10-12% sweet feed and bermuda hay, trailer parking. **Rates:** $20 per night, special rates available for 3+ days. **Accommodations:** Days Inn, Comfort Inn, Ramada Inn in Moreland, 10 minutes from stable.

RINCON (Savannah area)
Hi Ho Hills Farm **Phone: 912-826-5808**
Robin Hughes
939 Goshen Road [31326] Directions: Exit 19 off of I-95 to Hwy 21 N. 4.5 miles to Goshen Road. Take left. Farm is 1.5 miles on left. **Facilities:** 12 indoor stalls, round pen, 3 large fields, show & dressage rings, 20 miles of trails, & 2 camper spaces available. Feed/hay & trailer parking available. 24-hr notice required. **Rates:** $15 per day; $100 per week. **Accommodations:** Sleep Inn, 5 miles away on Hwy 21 & I-95.

ALL OF OUR STABLES REQUIRE CURRENT NEG. COGGINS, CURRENT HEALTH PAPERS, & OWNERSHIP PAPERS.

STATHAM

Circle T Stables **Phone: 404-725-0637**

Doug Tate or Trish Carlyle

2010 Hwy 82 [30666] Directions: Get off Hwy 316 on 211. Go to Hwy 82 & turn right. Go 3/4 mile, stable on left. **Facilities:** 16 indoor stalls, 175' x 275' arena, round pen, H/C wash rack, pasture/turnout areas, & trailer parking. Farm store on premises. Please call in the morning the day boarding is needed. **Rates:** $12 per night; $70 weekly rate. **Accommodations:** Jameson Inn in Winder, 10 miles from stable; Days Inn in Athens, 15 miles from stable.

TIFTON

Windy Hill Stables **Phone: 229-386-0811**

Gary & Mary Anne Simmons **Phone: 229-386-8108**

4186 Hwy 82 [31794] **Fax: 229-686-7187**

Directions: Exit 18 off I-75 4.5 miles west on 82. White fence and big red barn. Number on mailbox. Facilities: New barn, steel, kickproof, Barn Master 12- 12 X 12 indoor stalls. 1 acre pasture turnout and round pen. 50 acres with lots of dirt roads to ride. 800 acres permission to ride. Golf course 1.5 miles away. **Rates:** $15 per night. **Accommodations:** 4 miles from 8 motels.

VARNELL

Cedar Creek Farm **Phone: 706-673-4040**

Judy Noel

500 Old Hwy 2 [30756] Directions: 13 miles south of Chattanooga, TN, 12 miles north of Dalton, GA. From I-75: At Exit 139, go east 1 mile, cross bridge, turn right, go 3.5 miles, sign/mailbox/gravel drive on right. **Facilities:** 15 indoor stalls, paddock, 4-board fencing, feed/hay available, trailer parking. Large tack store on site. As much notice as possible, please. **Rates:** $10 per horse per night. **Accommodations:** Many motels in area.

WATKINSVILLE

Saxon Stables **Phone: 706-769-9132**

Forrest & Denise Fulton

5011 Colham Ferry Road [30677] Directions: From I-85 S: Exit 53 (441) & turn left to Watkinsville. Call for further directions. **Facilities:** 4 indoor stalls, 120 acres of pasture/turnout, feed/hay & trailer parking available. 24-hr notice if possible. 15 minutes from Athens. **Rates:** $20 per night incl. feed; weekly rate negotiable. **Accommodations:** Motels 15 minutes from stable.

NOTES AND REMINDERS

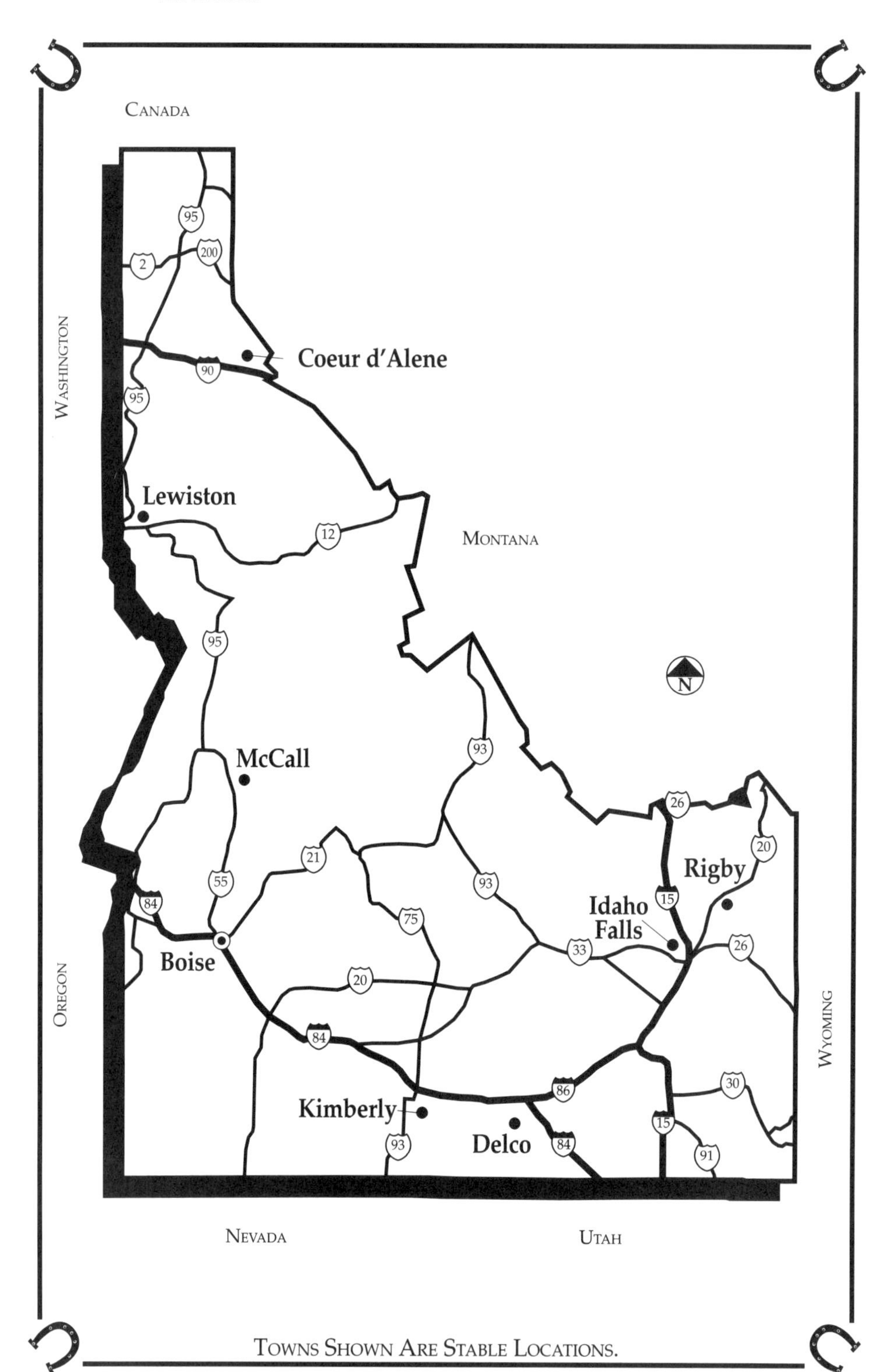

TOWNS SHOWN ARE STABLE LOCATIONS.

ALL OF OUR STABLES REQUIRE CURRENT NEG. COGGINS, CURRENT HEALTH PAPERS, & OWNERSHIP PAPERS.

BOISE/KUNA/MERIDIAN

Aspenbreak Stables **Phone: 208-922-4563**
John & Karen Vehlow
330 N. Eagle Road [83634] Directions: Exit 44 off of I-84 (about 7 miles west of Boise). Go south approx. 6 miles; turn left on Kuna Road; 2 miles to Eagle, take left. Go to first farm on right. **Facilities:** 10-16 indoor stalls and pens available for turnouts, 90' x 50' pasture/turnout, large indoor and outdoor arenas, & riding trails. Vet/farrier on call. **Rate:** $15.00; weekly rate is daily rate x 7. **Accommodations:** Available at Exit 44.

COEUR d'ALENE

Kingston 5 Ranch Bed & Breakfast **Phone (local): 208-682-4862**
Walter & Pat Gentry **Toll Free: 1-800-254-1852**
Web : www.k5ranch.com **Fax: 208-682-9445**
P.O. Box 2229, Coeur d'Alene [83816] Directions: Take I-90 to Exit 43 South to Frontage Road (1-2 blocks), turn west. Ranch is exactly 1 mile on right. 2 hrs west of Missoula, Montana & 1 hr east of Spokane, Washington. **Facilities:** 2 indoor/outdoor stalls, 4 individual pasture areas, all areas with panels or safe 2x4 horse fence. Also, round pen with 2 attached outdoor pens, 200' x 100' arena, parking, and famous 1,000-mile Silver Country Trail System. **Rates and Accommodations:** Relax and pamper yourself in one of our Jacuzzi/Fireplace suites with in-room private bath & Jacuzzi tub, in-room fireplace and personal outdoor hot tub for each room. Rates start @ $99.50 per night + 7% tax for 1or 2 guests and 1 horse. Each additional horse $10.00. Veterinarian services available. No smoking or pets indoors. Other outside pets by prior arrangement only. Overnight horse accommodations only available with B&B stay.

DECLO

Milo Erekson **Phone: 208-654-2085**
200 North Highway 77 (83323) Directions: 1/4 mile South of I-15. **Facilities:** 1 acre pasture only, feed/hay available, trailer parking. **Rates:** $5 per night. **Accommodations:** Burley 8 miles.

ALL OF OUR STABLES REQUIRE CURRENT NEG. COGGINS, CURRENT HEALTH PAPERS, & OWNERSHIP PAPERS.

IDAHO FALLS

C & D Stables **Phone: 208-522-1439**
Donna Garriott **Out of State: 1-800-837-1439**
3909 North 15 E. [83401] Directions: I-15, Exit 119 to US 20, 2 miles to Exit 15 E. Stable 1/4 mile. **Facilities:** 24 - 10' x 12' stalls, heated watering units, indoor and outdoor arenas, alfalfa/oats available, trailer parking. **Rates:** $15 per night. **Accommodations:** Motels 2 miles away in Idaho Falls.

Ellis Supreme Arabians **Phone: 208-524-7247**
Terie Ellis
1438 West 97 South [83402] Directions: 5 miles off of I-15. Call for directions. **Facilities:** 27 indoor 12' x 12' stalls, 14 with outside runs, 70' x 150' indoor arena, 60' x 100' outdoor arena, round pen, grooming stalls, year-round wash bay, heated tack room, & breeding laboratory. Sales & training of horses. Call for reservations. **Rates:** $15 per night; $70 per week. **Accommodations:** Evergreen Gables Motel 5 miles from stable, Shilo Best Western 7 miles from stable.

KIMBERLY

Western Barns **Phone: 208-423-6340**
Carol L. Sherman
3216 East, 3625 North [83341] Directions: I-84 to Twin Falls exit 182, South West on Hwy 50/30, left onto Hankins Rd, left at end, second place on left. Large white barn with green roof. **Facilities:** 8- 12' x 12' box stalls, 2-24' x x24' outdoor pipe corrals with shelter, 80' x 100' pasture/turnout, 80' round pen, and 4 automatic horse walkers. No feed/hay available. Horse must be de-wormed 1 month prior. Modern, clean & safe facility. Guide available for mountain trails. **Rates:** $20 for box stall, $10 for pipe corral, per night. **Accommodations:** Motels in area.

LEWISTON

Lewiston Roundup Association **Phone: 208-746-6324**
7000 Tammany Creek Road [83501] **Phone: 208-746-7589**
Directions: Take Hwy 12 to 21st Avenue. Call for further directions. **Facilities:** 122 indoor stalls, 150' x 240' outdoor arena, 200' x 300' indoor arena, trailer parking & RV Park with electric, water, & sewer hook-ups for 60 vehicles. Open 6 A.M. - 10 P.M. with year-round groundskeeper. Rodeos held on 80-acre facility. Call for reservations. **Rates:** $15 per night and must clean stall before leaving. **Accommodations:** Grand Plaza Hotel 5 miles from stable (208-799-1000).

MCCALL

McCall Veterinary Clinic **Phone: 208-634-8131**
Bruce Stephens, DVM
831 S. 3rd Street [83638] Directions: 1/2 mile south of McCall on Hwy 55. **Facilities:** 3 covered 12' x 12' stalls, 3 outside 12' x 12' stalls, 100' x 75' arena, feed/hay extra, trailer parking. Veterinary services available. **Rates:** $15 per night; weekly discount rate available. **Accommodations:** Campground within 1 mile; motel within 1.5 miles.

ALL OF OUR STABLES REQUIRE CURRENT NEG. COGGINS, CURRENT HEALTH PAPERS, & OWNERSHIP PAPERS.

RIGBY

Blacksmith Inn **Phone: 208-745-6208**

Mike & Karla Black

227 North 3900 East [83442] Directions: Call for directions. **Facilities:** 1 indoor 16′ x 16′ stall, 2 outdoor 16′ x 16′ stalls, 2 corrals, 8 acres of pasture/turnout, feed/hay available, trailer parking available. Raise and breed Tennessee Walkers; standing-at-stud: "Generator's Blue Boy." **Rates:** $10 per night with reservations, $20 per night without reservations. Weekly rate negotiable. **Accommodations:** Bed & breakfast on premises.

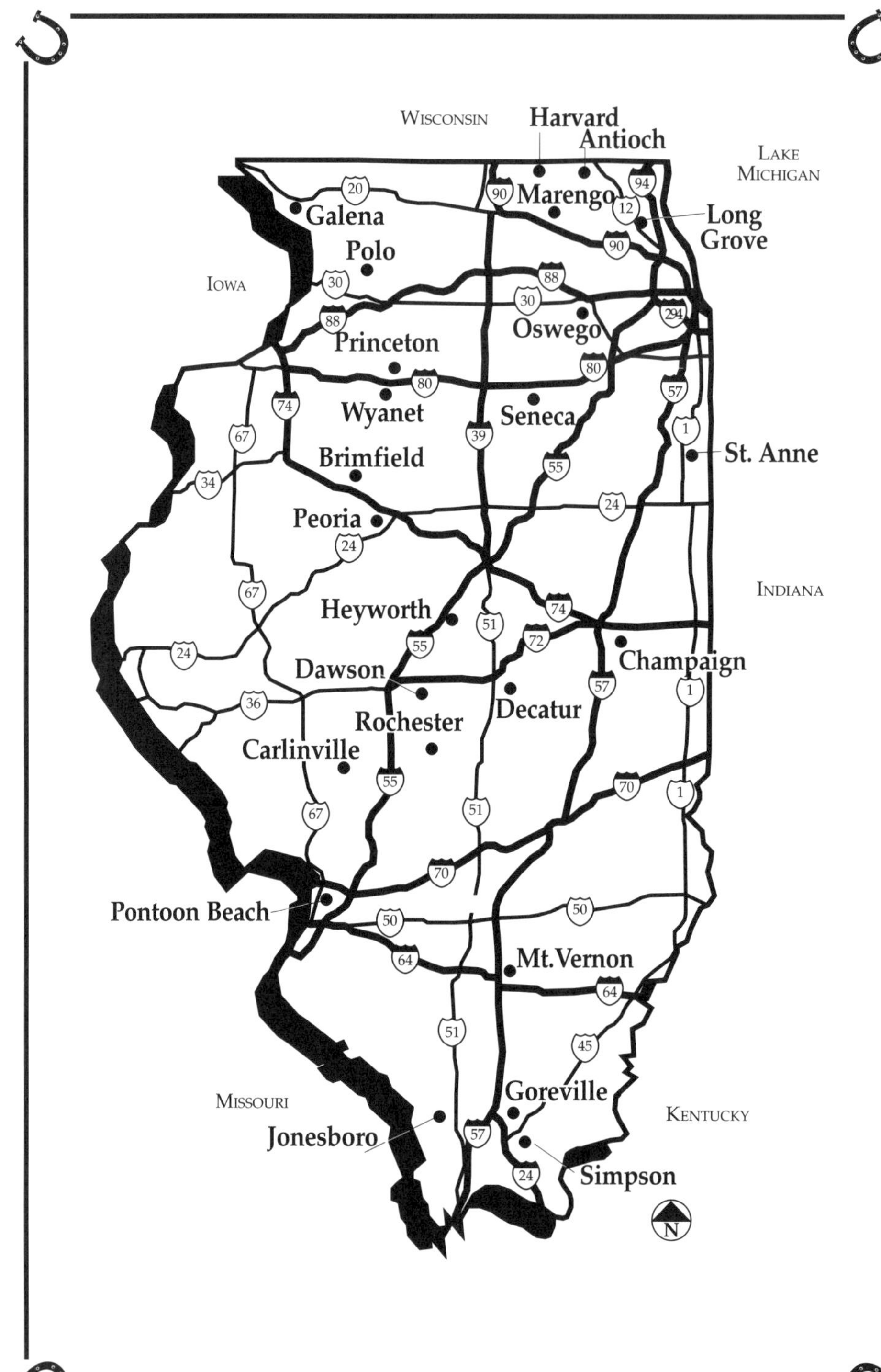

TOWNS SHOWN ARE STABLE LOCATIONS.

ALL OF OUR STABLES REQUIRE CURRENT NEG. COGGINS, CURRENT HEALTH PAPERS, & OWNERSHIP PAPERS.

ANTIOCH

Sakkara Farms, Sakkara Farms, Inc. **Phone: 708-395-0509**

19085 W. Edwards Road [60002] Directions: 3.5 miles west of I-94. Located 2 miles from Hwy 45 & 173. Call for directions. **Facilities:** 50 indoor stalls with many 1- to 5-acre pasture/turnout areas. Training & showing facility. Call for reservations. **Rates:** $15 per night. **Accommodations:** Best Western and many others in Antioch 5 miles away.

BRIMFIELD (KICKAPOO)

Mordue Quarterhorses **Phone: 309-446-9306**

Darlene & Revie Mordue

6628 Kramm Road [61517] Directions: 1.5 miles from I-74 at Exit 82. Call for directions. **Facilities:** 10 indoor stalls, 3 - 100' x 100' turnout paddocks, indoor arena, wash rack, feed/hay available. Standing-at-stud: "Charlie's Out of Oil." Call for reservations. **Rates:** $20 per night. **Accommodations:** Holiday Inn in Peoria, 15 minutes away.

CARLINVILLE

Hebron Appaloosas **Phone: 217-854-7226**

Hebron Appaloosas. Inc., Wendy Weisenberger, manager

RR 2, Box 126B [62626] Directions: 23 miles from I-55. Call for directions. **Facilities:** 3 indoor stalls, 4-acre pasture, outdoor arena. Call for reservations. **Rates:** $15 per day; $80 per week. **Accommodations:** Carlinvilla in Carlinville 8 miles away on Rt. 4.

CHAMPAIGN

Unzicker Stables **Phone: 217-359-5641**

Carolyn Unzicker

1162 County Road 900E [61821] Directions: 1.5 miles east of I-57 at Exit 229. Call for directions. **Facilities:** 40 indoor stalls, large indoor & outdoor riding arenas, & tack store on premises. Call for reservations and availability. **Rates:** $15 per night; $75 per week. **Accommodations:** Best Western Paradise Inn 3 miles away and LaQuinta 5 miles from stable.

DAWSON

Scofflaw Farm **Phone: 217-364-4350**

Donald R. Lawler

RR 1, Box 93A [62520] Directions: 8 miles from I-55 & 6 miles from I-72. Call for directions. **Facilities:** "One of the finest horse facilities in Central Illinois." 8 indoor box stalls, 11 large turnout paddocks with lean-tos, two 36' x 40' holding pens, 80' x 160' indoor arena, 200' x 247' outdoor ring with large gazebo. No stallions. Call for reservations. **Rates:** $20 per night; discounted weekly rate. **Accommodations:** Holiday Inn & Days Inn in Springfield, 8 miles away.

ALL OF OUR STABLES REQUIRE CURRENT NEG. COGGINS, CURRENT HEALTH PAPERS, & OWNERSHIP PAPERS.

DECATUR

Skyline Stables **Phone: 217-422-1051**
Linda & Ed Seaton **Barn: 217-422-7630**
4095 Rock Springs Road [62522]
Directions: 3 miles from I-72 & 1/2 mile from both Hwy 48 & 51. Call for directions. **Facilities:** 4 indoor stalls, 1/4- to 6-acre paddocks, indoor & outdoor arenas, feed/hay & trailer parking available. Breeds Arabians. Stallion services available. Call for reservations. **Rates:** $15 per night; discounted rates long term. **Accommodations:** Holiday Inn in Decatur, 3 miles from stable.

GALENA

Shenandoah Riding Center **Phone: 815-777-2373**
Galena Territory Association
200 Brodrecht Road [61036] Directions: From I-94 near Rockford, go west on Hwy 20 to entrance of Galena Territory. **Facilities:** 48 indoor stalls, lighted indoor arena, 2 outdoor arenas, cross country jumps, club room with viewing area, wash stall, riding lessons, and trails, varying sizes of pasture/turnout, feed/hay & trailer parking available. **Rates:** $25 per night. **Accommodations:** Eagle Ridge Inn & Resort.

GOREVILLE

Phoenix East **Phone: 618-995-9443**
Dee Dee Adams
1380 Sullivan Road [62939] Directions: From I-57: Take Exit 40 1.5 miles east. Turn south on gravel road at sign saying 175E & 1825N. Second house on left. Call for directions from I-24. **Facilities:** 3 outdoor covered stalls, 100' x 200' riding arena, feed/hay & horse trailer parking. 90,000 acres of horse trails in national forest & state park within 1 mile. Reservations required. **Rates:** $15 per night; $10 second night. **Accommodations:** Toupal's Country Corner 8 miles away.

HARVARD

Echo Acres **Phone: 815-943-0022**
Sue Market
19917 McGuire Road [60033] Directions: On Wisconsin border, 1 mile from town on US 14. Call for directions. **Facilities:** 15 open 10' x 12' stalls, turnouts, 60' x 190' indoor arena, 100' x 190' outdoor arena, feed/hay, trailer parking, electric outlets, campers welcome. Stallions welcome; quarterhorse, POA, APHA (paints) for stud, barrel & pole, bred for speed; horses for sale. **Rates:** $10 per night. **Accommodations:** El Rancho Motel 3/4 mile away.

HEYWORTH

Don Ellis Training Stable **Phone: 309-473-2270**
Don Ellis
RR 1 [61745] Directions: 10 miles from I-74 & I-55. Call for directions. **Facilities:** 10 indoor stalls & indoor arena with turnout area. Call for reservations. **Rates:** $10 per night. **Accommodations:** Super 8 in McLean, 10 miles away; Super 8 in Leroy, 10 miles away.

ALL OF OUR STABLES REQUIRE CURRENT NEG. COGGINS, CURRENT HEALTH PAPERS, & OWNERSHIP PAPERS.

JONESBORO
Trail of Tears Lodge & Sports Resort **Phone: 618-833-8697**
Ron & Deb Charles
Web address: www.trailoftears.com **email: ttlsr@midwest.net**
1575 Fair City Rd, Old Cape Road [62952] Directions: Between Rte. 55 in Missouri and I-57 in Illinois. Call for directions. **Facilities:** 10 under roof 10' x 12' open-air stalls, bedding, feed/hay on request, camping, trailer parking and RV space available. Trail rides, lessons. Large barn converted to lodge, miniature golf, hot tub, game room, swimming pool, and many more activities available. Reservations required. **Rates:** $15 per night, stall, bedding extra. **Accommodations:** Lodge on premises, private bedrooms with bath and shower; B&B $25 per person (2 person min.) per night; room & board, $50 per person per 24 hr.

LONG GROVE
Arrowhead Horse Enterprises **Phone: 708-438-9888**
Morrie Waud, owner Julie, manager
6697 RFD Gilmer Road [60047] Directions: 40 miles northwest of Chicago between Rts. 60, 83, & 22. **Facilities:** At least 20 indoor 10' x 12' stalls, 5 paddock areas, feed/hay included, & trailer parking. Draft horses accepted. **Rates:** $20 per night. **Accommodations:** 3 motels within 2 miles of stable.

MARENGO
Our Stable & Twin Arenas **Phone: 815-568-1736**
Michael & Bobbi Burke **Barn: 815-568-1753**
23803 Grange Road [60152] Directions: 3 blocks north of I-90. Call for directions. **Facilities:** 24 indoor stalls, private paddocks, pastures with 3 run-in sheds, 65' x 275' indoor arena, 2 outdoor lighted 200' x 300' arenas, 10' x 11' box stalls with window and rubber mats, round pens, 2 tack rooms, wash rack, concrete walkway, feed/hay, trailer parking. English & western lessons, shows, and clinics; hunt seat, barrel racing, team penning. Nearby woods and quiet roads for riding. Campers welcome, stallions welcome, horses for sale, monthly boarding. **Rates:** $15 per night. **Accommodations:** Motels within 3 miles.

MT. VERNON
Richardson Stables **Phone: 618-242-6566**
C. Wayne & Judy Richardson **Stable: 618-242-1232**
11246 N. General Tire Lane [62864] Directions: 1 mile east of I-57 on I-64 to Exit 80. Turn north on IL Rt. 37, 1.8 miles to first stop light. Turn right on IL Rt. 142, 1.5 miles to Richardson's sign. Turn right & stable is on the left. **Facilities:** 14 indoor stalls, one 3-acre pasture/turnout lot, trailer parking, & camper hook-up. Reservations requested. **Rates:** $15 per night. **Accommodations:** Many motels & restaurant in Mt. Vernon, 3 miles from stable.

ALL OF OUR STABLES REQUIRE CURRENT NEG. COGGINS, CURRENT HEALTH PAPERS, & OWNERSHIP PAPERS.

OSWEGO

Fine Meadow Gateway **Phone: 708-554-1446**

Martha Eldredge

2280 Plainfield Road [60543] Directions: Call for directions. **Facilities:** 4 indoor stalls, 70' x 180' indoor arena, 45' x 75' indoor round pen, 120' x 200' outdoor arena, 3 wash bays with warm water, 1/2-acre pasture, feed/hay & trailer parking. This is an all level training & boarding facility also offering brood mare & foal care. Reservations required; credit card confirmation if after 6 P.M. arrival. **Rates:** $20 per night. **Accommodations:** Comfort Inn, Plainfield & Aurora.

PEORIA

Black Oak Stables **Phone: 309-691-8257**

Richard & Dolores Samsel

4213 Charter Oak Road [61615] Directions: Close to I-74, I-474, and US 150. Call for directions. **Facilities:** Plenty of stalls available, pasture/turnout, indoor arena, feed/hay, wash stall, trailer parking. Jubilee State Park within 30 minutes; trails. Call for reservations. **Rates:** $10 per night. **Accommodations:** Motels and restaurants within 5 minutes.

Heart of Illinois Arena **Phone: 309-693-1805**

Ernie & Pat Frietsch **Home: 691-9161**

9201 N. Galena Road [61615] Directions: West side of Rt. 29. 1 mile south of Rt. 6. 5 miles north of Rt. 116. 7 miles from I-74. **Facilities:** 160 indoor box stalls, arena turnout, wash racks, inside & outside arenas, showers & bathrooms. Stallions welcome. Tack shop on grounds. Camper hook-ups. **Rates:** $15 per day. **Accommodations:** Red Roof Inn & Holiday Inn nearby.

POLO

Oregon Trail Stable **Phone: 815-946-2904**

Ruth Anne Spangler

2527 S. Union Road [61064] Directions: 12 miles from I-88; 6 blocks from US 52. **Facilities:** 6 indoor stalls, pasture/turnout available, holding pens, indoor arena, wash rack, feed/hay, trailer parking. Camper hook-up available. 4-hour advance notice required. **Rates:** $15 per night; cash only. **Accommodations:** Motels nearby. Bed with bathroom available in barn office.

ALL OF OUR STABLES REQUIRE CURRENT NEG. COGGINS, CURRENT HEALTH PAPERS, & OWNERSHIP PAPERS.

PONTOON BEACH

Gateway Stables **Phone: 618-931-3527**

Kelly Arnold and Rusty Wright

3514 Lake Drive [62040] Directions: Just north of Horseshoe Lake State Park. One block off Hwy 111. Convenient to Interstates 270, 70, 55, 64, 255. Call for directions. Minutes from St. Louis & Fairmount Park (T.B. Racing). Standing T.B. "Make Your Point" by Elocutionist, resident farrier and good vets on call. Please call day of arrival to confirm.

Facilities: 25 stalls, all sizes of stalls from mini to 16′ x 16′, turnout paddocks, 1- and 3-acre pastures, 1-acre dry lot, outdoor arena, 60' x 120' indoor arena, feed/hay at additional charge. **Rates:** $15 per horse. **Accommodations:** Ramada Limited, Super 8 in Pontoon Beach, Best Western in Mitchell, smaller local motels available with reservations.

PRINCETON

The Prairie Hill - Barn, Bed & Breakfast **Phone: 815-447-2487**

Janine Klayman, owner; Janet , manager **Cellular: 773-251-7044**

RR 4, Box 74 [61356] Directions: 8 miles from I-80. Call for directions.

Facilities: Country estate & horse facility. 12 - 12′ x 12′ indoor stalls, feed/hay, paddocks by reservation only. Riding school and combined training facility. Facility offers 30 acre cross-country course, 2 arenas, miles & miles of trails. Reservations in advance. Health certificate and negative coggins required.

Rates: $25 per horse per night. **Accommodations:** Historic, antique filled Farmhouse with 6 bedroom B & B including homemade breakfast.

ROCHESTER

Willow Creek Farms **Phone: 217-498-8136**

Nancy Wright **or 217-498-9859**

4818 Oak Hill Road [62563]

Directions: 4 miles from I-55. 6 miles from I-72. Call for directions. **Facilities:** 2 indoor stalls, 2 - 150' x 150' paddocks, 3-5 acres of pasture. Feed/hay and trailer parking available. Hunter/jumper facility. Call for reservations.

Rates: $15 per night. **Accommodations:** Red Roof Inn & Days Inn in Springfield, 2.5 - 3 miles away.

ST. ANNE

Sunrise Farms, Inc. **Phone: 815-932-6170**

Karen Hemza, manager **815-935-8897**

4370 E. 3500 South Road [60964]

Directions: 15 minutes east of I-57. Call for directions. **Facilities:** 10 indoor stalls, ten 1-acre paddocks, 4 with run-in shelters, indoor & outdoor arenas, stadium & cross-country jumps. Bring own buckets. Call for reservations.

Rates: $10-$20 per night. **Accommodations:** Motel 8 & Howard Johnson's in Bourbonnais, 15 minutes away.

ALL OF OUR STABLES REQUIRE CURRENT NEG. COGGINS, CURRENT HEALTH PAPERS, & OWNERSHIP PAPERS.

SENECA

The Pony Place **Phone: 815-695-5913**

Elaine Owens

2945 N. 37th [61360] Directions: Adjacent to I-55 and I-80. Call for directions (let ring 10-15 times). **Facilities:** 8 indoor 12' x 12' boxes in new barn with concrete floor, feed/hay, trailer parking, limited turnout, dry lot available. Traveling horses will be stalled at night. Specialize in large children's riding ponies. Welsh Cob at stud: "Winks Titus"; Black Morgan at stud: "CassaNovas' Cashon." Welsh Cobs & Morgan Crossbreeds for sale. **Rates:** $20 per night. **Accommodations:** Local campground nearby; onsite camping, no hook-ups. Motels within 15 miles in Ottawa and Morris.

SIMPSON

Livin' Color Farm, Inc. **Phone: 618-695-3570**

Dennis and Terri Marr **Fax: 618-695-3571**

RR 1, Box 221A [62985] **E-mail: livincolr@aol.com**

Directions: Call for directions. **Facilities:** 12 indoor 10' x 12' stalls, individual turnout pens, 10+ acres pasture, indoor & outdoor arenas, round pen, wash rack, feed/hay available, trailer parking. Extra services (lunging, grooming, etc.) provided for reasonable fee. Retirement home for horses, lay-ups welcome. Call for reservations or e-mail. **Rates:** $15 per night; $75 per week. **Accommodations:** Motels within 10 miles at Vienna.

WYANET

Huskey's Horse Motel **Phone: 309-895-2314**

Richard & Karen Huskey

RR 1, Box 67 [61379] Directions: 4 miles south of I-80. Take Exit 45, Buda, IL. Call for further directions. **Facilities:** 6 large indoor box stalls, indoor arena, feed/hay & trailer parking available. Electric & water hook-up for campers. Call for reservations - same day usually OK. **Rates:** $15 per night; weekly rate negotiable. **Accommodations:** Sheffield Motor Lodge at junction of I-80 & I-88, 7 miles from stable.

Hardships Endured By Early Settlers.

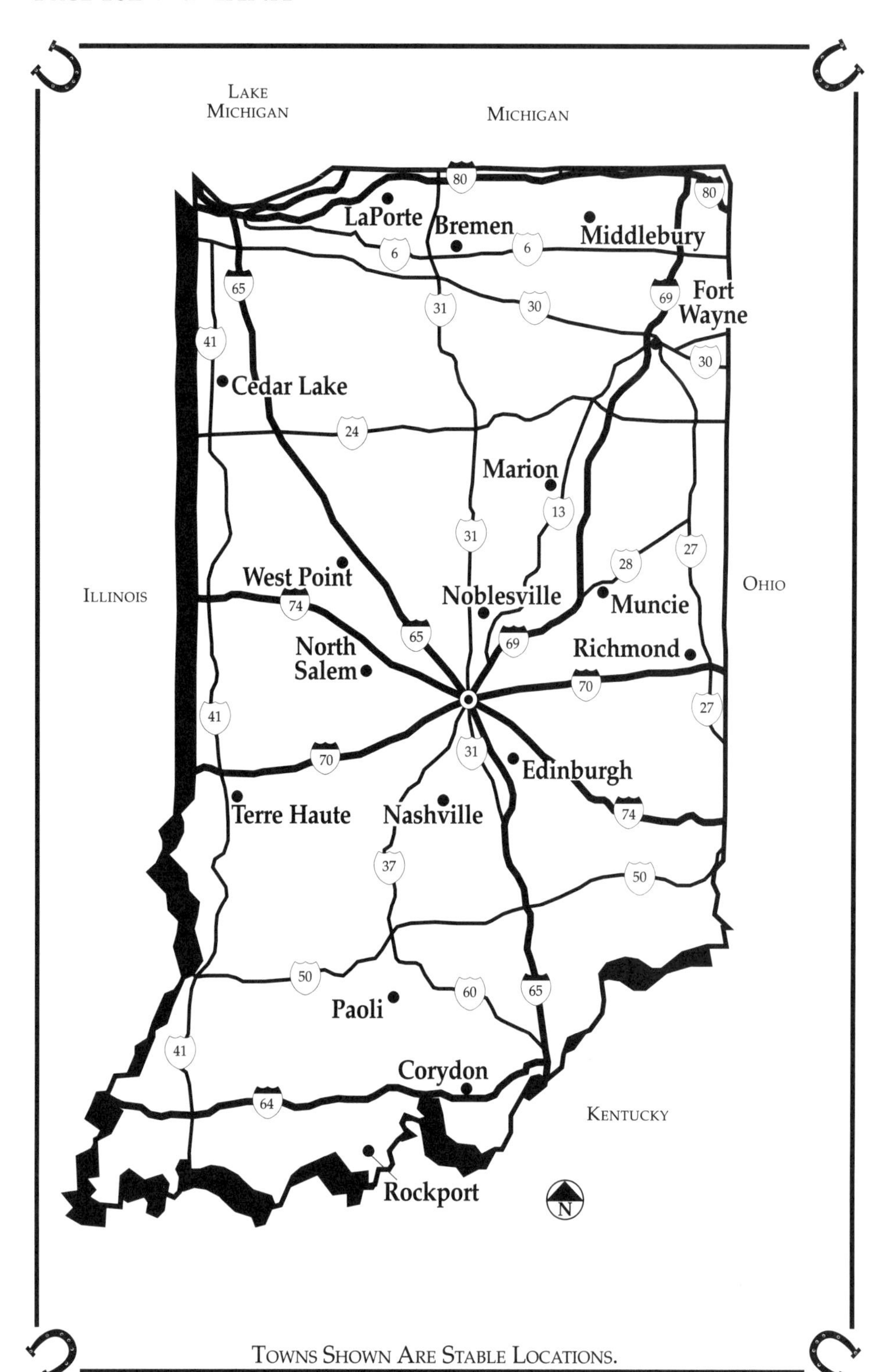

TOWNS SHOWN ARE STABLE LOCATIONS.

ALL OF OUR STABLES REQUIRE CURRENT NEG. COGGINS, CURRENT HEALTH PAPERS, & OWNERSHIP PAPERS.

BREMEN

Brown Group, Inc. **Phone: 219-546-3608**
Marilyn M. Brown
5294 E. 3rd Road [46506] Directions: Call for directions. **Facilities:** 10 indoor 10' x 12' stalls, holding pens, pasture, ample trailer parking, camper hook-up. **Rates:** $10 per night. **Accommodations:** Bremen Inn in Bremen, 2 miles from stable; Holiday Inn in Plymouth, 20 miles away; Inn at Amish Acres in Nappanee, 8 miles away.

CEDAR LAKE

Rainbow Equestrian Center, Inc. **Phone: 219-365-2127**
Ilse Vogelmann
11750 W. 117th Avenue [46303] Directions: 4 miles from Rt. 30; 1/2 mile from Rt. 41; 7 miles from I-65; and 8 miles from I-80 & I-294. Call for directions. **Facilities:** 5 indoor stalls, many size areas of pasture/turnout, feed/hay & trailer parking available. Call for reservation. **Rates:** $25 per night; $20 for 5 or more nights. **Accommodations:** Crestview Motel in Cedar Lake, 1 mile from stable.

CORYDON

Harrison County Forest Boarding Stables, Inc. **Phone: 812-738-3694**
Howard or Rosemary Saylor **Stable: 812-738-6838**
1278 Hwy 462 [47112] Directions: Hwy 62W to Hwy 462. Call for directions. **Facilities:** 24 indoor box stalls, 64 tie-downs, pasture, feed/hay & trailer parking available. Over 100 miles of trails. Camping facilities with restrooms, showers, & firerings, electric hook-up for campers in season only. Open 24 hrs. 2 miles from entrance to Wyandotte State Recreation Area. **Rates:** $10 per night. **Accommodations:** Best Western & Budgetel in Corydon; Days Inn in Carefree, 12 miles from stable.

EDINBURGH

Frost Hill Farm **Phone: 812-526-6102**
Debbie Amos Cook
8248 W. 1150 S [46124] Directions: Located just off I-65 on SR 252 East. 1/4 mile on right. 5 miles from Hoosier Horse Park. **Facilities:** 21 indoor 12' x 12' stalls including 10 foaling size, 8-10 paddocks with sheds, pasture. Can handle semi-trailers (15 horse). Electric hook-up for trailers. Dogs welcome if nice. Vets on call at all times. Can accommodate stallions. 1 hr. & 20 min. from pari-mutuel track at Anderson. **Rates:** $20 per night. **Accommodations:** Holiday Inn & Comfort Inn in Columbus, 4 miles from stable.

ALL OF OUR STABLES REQUIRE CURRENT NEG. COGGINS, CURRENT HEALTH PAPERS, & OWNERSHIP PAPERS.

EDINBURGH

Hoosier Horse Park, Johnson County **Phone: 812-526-5929**

E-mail: hhp@netdirect.net

P.O. Box 67 [46124] Directions: I-65 to Exit 80; take Hwy 252 W through Edinburgh; follow HHP signs approx. 3 mi. to US 31 intersection. Go straight 3 more miles to Schoolhouse Road; turn right; HHP is .7 miles on left. **Facilities:** 324 indoor 10' x 10' indoor box stalls, 7 arenas, including 300' x 160' covered. Call to make reservation. **Rates:** $10 per night. **Accommodations:** Hilltop Motel in Franklin; Holiday Inn Express in Taylorsville.

Pick up ad frm sixth edition page 104
Hoosier Horse Park

FORT WAYNE

Union Chapel B & B **Phone: 219-627-5663**

Larry & Jo Ann Burkhart

6336 Union Chapel Road [46845] Directions: Exit 116 on I-69. Take Dupont Road East 1 mile to light, turn left onto Tonbel Road, go north 1 mile to Union Chapel, turn right, stable is 3/4 mile down on right. **Facilities:** 4 - 10' x 10' steel pipe/corral type stalls, 30' x 30' round pen, feed/hay, trailer parking available. Peruvian horses for sale. **Rates:** $12 per night for first horse, $10 for second horse. **Accommodations:** B & B on premises, $40-$60 per night.

LA PORTE

Gillerlain Quarter Horse Farm **Phone: 219-362-1122**

Paul & Joanne Gillerlain

0102 E. 200 North (Severs Road) [46350] Directions: 4 miles from Indiana Toll Road; 6 miles from I-94. Call for directions. **Facilities:** 8 indoor 10' x 10' stalls, turnout with shed, feed/hay available. Call for reservation. **Rates:** $15 per night. **Accommodations:** Ramada, Super 8, Cassidy Motel within 1 mile in LaPorte.

ALL OF OUR STABLES REQUIRE CURRENT NEG. COGGINS, CURRENT HEALTH PAPERS, & OWNERSHIP PAPERS.

MARION

Hoosier Stables **Phone: 317-664-5188**

Wendell Donaldson

4546 E. 100th South [46953] Directions: Marion Exit off of I-69. Call for directions. **Facilities:** 8 indoor wood box stalls on a 23-acre farm. Feed/hay and parking for big rigs available. Call for reservation. **Rates:** $20 per night; ask for weekly rate. **Accommodations:** Holiday Inn 5 miles away on Hwy 18.

MIDDLEBURY

Coneygar **Phone: 219-825-5707**

Mary Dugdale Hankins

54835 County Road 33 [46540] Directions: Located south of Exits 101 and 107 (Middlebury) on Ind. Toll Road 80-90, between US 20 and SR 120. Call for specific directions to stable. **Facilities:** 10 indoor stalls, large dry barnlot and several acres of pasture. Feed/hay and trailer parking available. Boarding only for guests of B&B. **Rates:** $10 for stall per night; $5 for pasture. **Accommodations:** Bed & Breakfast in country home on 40 acres with stable facilities. Includes hearty country breakfast. Near Northern Indiana Amish country.

MUNCIE

Lions Delaware County Fairgrounds **Phone: 317-288-1854**

Jim Scott, mgr. **or: 317-285-1837**

1210 N. Wheeling [47308] Directions: I-69 to Hwy 332 to Wheeling. Turn right; fairgrounds is 1 mile on right. **Facilities:** 100 indoor 10' x 10' box stalls, 1/2-mile racetrack on 43 acres. Rodeos, public auctions, & standardbred sales held at fairgrounds. Near Ball State University. **Rates:** $7 per night. **Accommodations:** Signature Inn & Days Inn 1 mile away.

NASHVILLE

Rawhide Ranch **Phone: 812-988-8755**

Dick Pardue

1292 South State Road 135 [47448] Directions: 14 miles west of I-65, 13 miles from Columbus, & 1 mile off State Rd. 46. Call for directions. **Facilities:** 32 indoor 14' x 12' stalls, two 1-acre paddocks, 120' x 60' outdoor riding arena, hot walker, riding in nearby Brown County State Park with 100,000 acres of trails. Horses for sale. Hunters & fishermen welcome. **Rates:** $30 per night. **Accommodations:** New cabin on premises available to rent: $100 per night with horse boarded. Also, Brown County Inn, Seasons, & Salt Creek Inn 2 miles away.

NOBLESVILLE

Janet Keesling Stables **Phone: 317-773-5482**

Janet Keesling

11930 E. 211th [46060] Directions: 1/2 mile from US 37; 16 miles from I-465; 7 miles from I-69. Call for directions. **Facilities:** 30 indoor 12' x 12' stalls, 60' x 120' indoor arena, 63' round pen, feed/hay, & trailer parking. Veterinarian within 3 miles. Stallions accepted. Call for reservation. **Rates:** $15 per night. **Accommodations:** Super 8 in Noblesville, 5 miles from stable.

<u>ALL OF OUR STABLES REQUIRE CURRENT NEG. COGGINS, CURRENT HEALTH PAPERS, & OWNERSHIP PAPERS.</u>

NORTH SALEM

<u>Ken & Joyce Greene</u> **Phone: 317-676-6615**

Kirk Greene

7287 W. County Road 1000N [46165] Directions: I-74 west to Jamestown Exit; south on State Road 75 about 6 miles; at Tomahawk Hills Golf Course, turn west onto Rte 1025; turn south at T; turn right onto next road (1000N); stable is second house on left. **Facilities:** 5 indoor 10' x 12' indoor stalls, round pen, 300' x 96' lighted outdoor arena, horse walker & barn, feed/hay, trailer parking available. Caters to rodeo performers; outdoor area set up for calf roping, team roping, bull dogging; steers and calfs available for practice for a fee. **Rates:** $12 per night; $50 per week. **Accommodations:** Campground 1 mile away; bed & breakfast 2 miles away; motels in Brownsburg, 20 miles away.

PAOLI

<u>Wilstem Dude Ranch</u> **Phone: 812-936-4484**

Pam Clark, mgr. **or: 812-634-1413**

US Hwy 150 West [47454] Directions: Call for directions. **Facilities:** 55 indoor 10' x 12' stalls, paddock, large indoor & outdoor arena, 30 miles of trails on a total of 800 acres, feed/hay & trailer parking available. Call for reservation. **Rates:** $10 per night. **Accommodations:** Historic lodge on premises that sleeps 18, plus cabins.

RICHMOND

<u>Dreamland</u> **Phone: 317-935-7930**

Tony & Chris Bertsch, Grag Shears

2334 Niewoehner Road [47374] Directions: Exit 156 off of I-70. Call for directions. **Facilities:** 27 indoor stalls, 80' x 120' indoor ring, 100' x 200' outdoor ring, 2 paddocks, pasture area, jump courses, feed/hay & trailer parking available. English, Western, huntseat & jumping lessons for horses & riders. Horse sales. Call for reservation, leave message. **Rates:** $15 per night. **Accommodations:** Motels 2 miles from stable.

ROCKPORT

<u>Ramey Riding Stables</u> **Phone: 270-771-5590**

Joan Ramey **(KY office)** **Barn: 812-649-9579**

Web: www.mindspring.com/jramey **E-mail: jramey@mindspring.com**

2152 S. County Road 200W [47635] Directions: Call for directions. **Facilities:** 30 -8′ x 10′ indoor stalls, 4 acres of pasture turnout, round pen, heated barn, indoor arena, 50' x 100' outdoor arena & miles of riding trails. CHA certified instructor for English, Western & dressage. Residential riding camps, picnic area & electric RV hook-up available. Reservation required. **Rates:** $15 per night. $60 per week. **Accommodations:** Bed & Breakfast on premises with 3 BR condo cottage, finely appointed, sleeps 10: $85/night double. Tennis courts, swimming pool, workout rooms, & more in second facility 20 minutes away.

ALL OF OUR STABLES REQUIRE CURRENT NEG. COGGINS, CURRENT HEALTH PAPERS, & OWNERSHIP PAPERS.

TERRE HAUTE

Persimmon Hollow Farm **Phone: 812-299-4754**
Bill & Donna Isaacs **812-232-0542**
2966 E. Harlan Drive [47802] Directions: From I-70 and US 41: 7 miles south to Stuckey Pecan Shop; go 9/10 mile east and you will see long white board fence of farm. **Facilities:** 5 indoor 12′x12′ stalls, 3 72′x20′ holding pens, trailer parking available. Camping facilities 1/2 mile away. Call for reservation. **Rates:** $12 per night. **Accommodations:** Holiday Inn and many others 7 miles from stable.

WEST POINT

Flint Run Stables **Phone: 317-572-2803**
Dave & Nancy Shaw
6339 South 700 W [47992] Directions: 2 miles south of SR 25 at West Point on 700W. Call for specific directions. **Facilities:** 8 indoor stalls, 2 - 1/2-acre turnouts, 60' x 85' indoor riding arena, hay & trailer parking available. Electric & water RV hook-ups. Spring water for horses. 10 minutes from Purdue University Large Animal Clinic. Reservations not necessary but call in advance. **Rates:** $15 per night; $75 per week. **Accommodations:** Howard Johnson's & Ramada Inn in Lafayette, 15 minutes away.

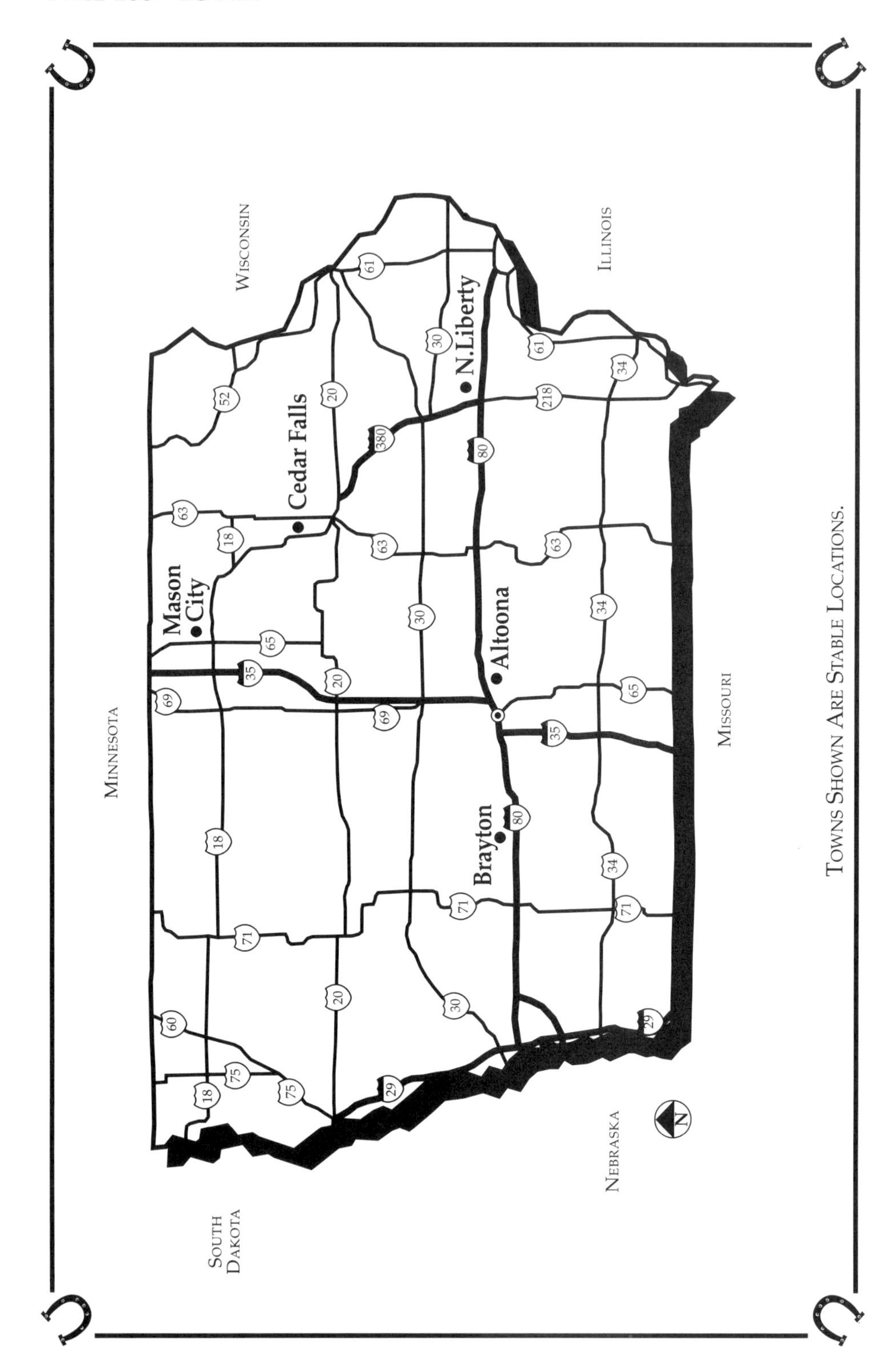
Minnesota
Wisconsin
Illinois
Missouri
Nebraska
South Dakota
Mason City
Cedar Falls
N.Liberty
Altoona
Brayton
Towns Shown Are Stable Locations.

ALL OF OUR STABLES REQUIRE CURRENT NEG. COGGINS, CURRENT HEALTH PAPERS, & OWNERSHIP PAPERS.

ALTOONA

Georgia's Arabian Stables **Phone: 515-967-5553**
Georgia Campbell
5055 NE 96th Street [50009] Directions: 5 miles from I-80: Take Exit 143. At stop sign, turn to go to Altoona then go 1/2 mile to first road on left. Take left and go 4 miles. Turn right at stop sign & go 1/2 mile. Stable is 2nd house on left. **Facilities:** 38 10x12 indoor box stalls, 90' x 100' indoor arena, 3 indoor tie stalls, 3 large turnout pens, 3 round pen, & electric camper hook-up, 20-year 4-H leader, Hay Rack rides, Horse trailer rentals, riding lessons for beginners, advanced and handicapped, Reg. Pinto stud(Black and White), Arabian stud, Stud pen with shed available. Will clean out horse trailer for free. Negative Coggins and current health papers. **Rates:** $15 per night. **Accommodations:** Eight motels within 6 miles of stable. Most will also take pets.

BRAYTON

Hallock House Bed & Breakfast **Reservations: 800-945-0663**
Guy & Ruth Barton
3265 Jay Avenue [50042]
Directions: Take Exit 60 off I-80, north to Brayton 3 miles; stable is approx. 1 mile east of Brayton. **Facilities:** 4 inside stalls with outside pens, 2 separate outside pens, 12 inside box stalls available across the road. 1 acre of pasture/turnout, trailer parking available. **Rates:** $10 per night. **Accommodations:** 2 guest rooms on premises. EconoLodge within 4 miles at I-80 and Hwy 71.

CEDAR FALLS

Beahr Ridge Legendary Stables **Phone: 319-988-3021**
Richard & Sylvia Beahr
5533 Hudson Road [50613] Directions: On Hwy 58 two miles north of Hudson & 3.5 miles off Hwy 20. Call for final directions. **Facilities:** 47 box stalls, 80' x 120' riding arena, 100' outdoor arena, 3 paddocks with covered shelters. Arabian horses for sale. Standing-at-stud: "Rho-sabee." Seven national top-ten titles. Reservations required. **Rates:** $15 per night; $75 per week. **Accommodations:** Motels within 5 miles of stable.

MASON CITY

North Iowa Fair **Phone: 515-423-3811**
Katy Elson **E-mail: office@northiowafair.org**
3700 4th Street SW [50401-1590] Directions: From I-35: Take Exit 196 (Clear Lake). Go east on Business Hwy 18 (US122). From Hwy 18 take Eisenhower exit. North 3+ miles to fairground entrance on right after stop light. Or continue through light to entrance on left. **Facilities:** Up to 209 indoor stalls plus trailer parking. Horse shows often use all of the stalls on weekends

ALL OF OUR STABLES REQUIRE CURRENT NEG. COGGINS, CURRENT HEALTH PAPERS, & OWNERSHIP PAPERS.

during the summer. Electricity & parking for campers available. **Rates:** $20 with wood shavings; $15 without shavings. **Accommodations:** Hanford Inn and Super 8 in Mason City, less than 1 mile away. Camper hook-ups on site $10 per unit per night.

NORTH LIBERTY
Whinnee Acres **Phone: 319-626-6416**
Jean Eckhoff
2237 Scales Bend Road NE [52317] **Directions:** North off Exit 240 of I-80. Call for directions. **Facilities:** 20 large indoor stalls, double-size stall available, 63' x 120' indoor arena, pasture/turnout, feed/hay, trailer parking available. **Rates:** $15 per night. **Accommodations:** Motels. 7 within 5 miles at Exit 240.

NOTES AND REMINDERS

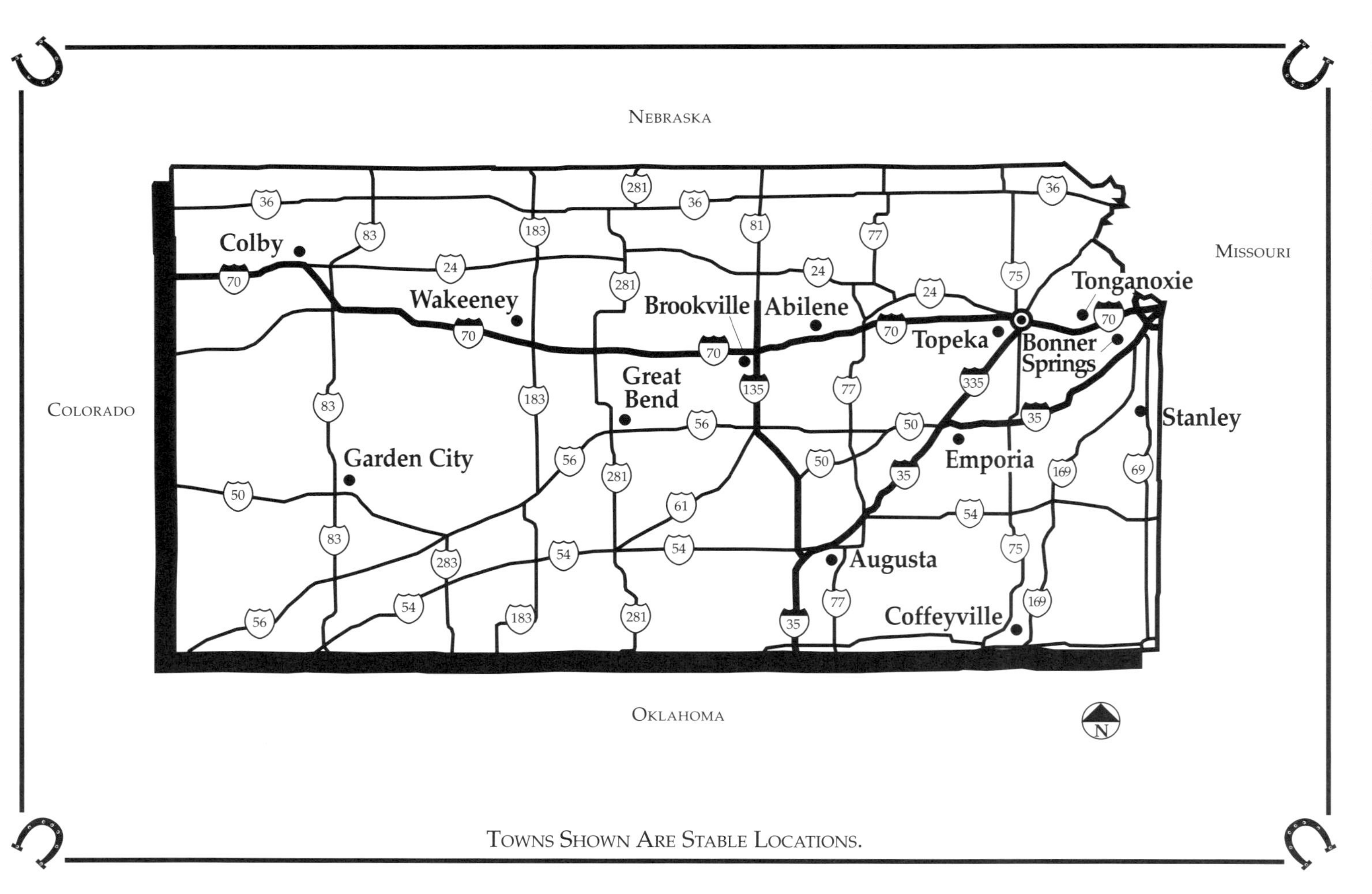
Nebraska
Missouri
Colorado
Oklahoma
Colby
Wakeeney
Brookville
Abilene
Tonganoxie
Topeka
Bonner Springs
Great Bend
Stanley
Garden City
Emporia
Augusta
Coffeyville
N
Towns Shown Are Stable Locations.

ALL OF OUR STABLES REQUIRE CURRENT NEG. COGGINS, CURRENT HEALTH PAPERS, & OWNERSHIP PAPERS.

ABILENE

B and W QuarterHorse Ranch **Phone: 913-479-2220**
Wanda West
Rt. 2, 1777 Hawk Road [67410] Directions: I-70 to Hwy 15 South on Abilene Exit. Go through town. Cross Smoky Hill River Bridge and turn on Hawk Road to the east. Stable is 2.5 miles. **Facilities:** 6-8 stalls, outdoor runs, 60' x 200' pipe arena turnout. Feed/hay and trailer parking available. Reservations in advance. Arrival by 8 P.M. **Rates:** $20 per night. **Accommodations:** Motels in Abilene, 4 miles from stable.

AUGUSTA

DKG Farms **Phone: 316-733-0445**
Dennis Gumieny
14247 Tawakoni Road [67010] Directions: Off of Hwy 54. Call for directions. **Facilities:** 5 stalls on over 100 acres. Pasture, turnout, feed/hay, and trailer parking. Reservations required. **Rates:** $15 per night. **Accommodations:** Motels 7 miles from stable.

Unity Morgans **Phone: 316-775-1554**
Jim & Patricia Michael **Fax: 316-775-1554**
12698 SW Thunder Road [67010] Directions: Three miles southwest of highway 54 & 77. Call for specific directions. **Facilities:** 2 - 12 X 12 indoor stalls. Feed and Hay available. Trailer parking available. Outdoor arena, round pen. **Rates:** $15.00 per night. **Accommodations:** Lehr's Motel and Agusta Inn near by.

White Training Stables **Phone: 316-775-3602**
Steve White
6404 Southwest Hunter Road [67010] Directions: Call for directions. **Facilities:** 20 indoor stalls, indoor arena, feed/hay, & trailer parking available. Specializes in Arabian and Quarter horses. Reservations required. **Rates:** $15 per night; ask for weekly rate. **Accommodations:** Motels 4 miles from stable.

BONNER SPRINGS

Z7 Stables Over Hill **Phone: 913-441-8860**
Hank & Linda Perrin
3601 S. 142nd Street [66012] Directions: Bonner Spring, Hwy 7, exit off of I-70. Call for further directions. **Facilities:** 6 indoor stalls, 3 turnout pens, indoor arena, jumps available. Horseshoeing available and vets nearby. Cannot accommodate tractor-trailers. Minutes away from dog race track & riverboat gambling. **Rates:** $15 per night. **Accommodations:** Motels 15 minutes from stable.

BROOKVILLE

Castle Rock Ranch **Phone: Toll Free 888-225-6865**
Judy A. Akers **785-225-6865**
1086 29th Road [67425] Directions: Call. **Facilities:** 4 - 12' X 12' indoor stalls. Soft wood chips. Hay available. 2 acre pasture surrounding barn. Trailer parking available. **Rates:** $20 per night. **Accommodations:** Overnight stabling only available to B&B guests. 8 miles from Kanopolis Lake which has 26 miles of riding trails.

ALL OF OUR STABLES REQUIRE CURRENT NEG. COGGINS, CURRENT HEALTH PAPERS, & OWNERSHIP PAPERS.

COFFEYVILLE
The Horse Center **Phone: 316-251-4511**
Elton Weeks **Home: 251-4649**
Rt. 4 [67337] Directions: 2 miles north of Hwy 166 West and 1 mile north of Coffeyville Country Club. Call for directions. **Facilities:** 10 indoor 12' x 12' box stalls, 60' x 106' arena, stud pen with run, feed/hay and trailer parking available. Trains driving horses. Reservations required. **Rates:** $6 per night. **Accommodations:** Motels 6.5 miles from stable.

COLBY
Colby Livestock Auction **Phone: 913-462-3231**
Dave Dickey
125 South Country Club Drive [67701] Directions: Country Club Exit off of I-70. Call for further directions. **Facilities:** 130 outside steel pens, water, feed/hay & trailer parking available. Reservations required. **Rates:** $2 per night. **Accommodations:** Motels 1 mile from stable.

EMPORIA
Emporia Livestock Sale **Phone: 316-342-2425**
Ask for: "Hutch"/day, "Jim"/evening
502 Albert [66801] Directions: Located on Business Hwy 50. Call for directions. **Facilities:** 200 outside pens, some with sheds, trails nearby, feed/hay and trailer parking available. Reservations required. **Rates:** $15 per night.

GARDEN CITY
Finney County Fairgrounds **Phone: 888-876-3844**
Angie Clark
601 Lake Ave [67846] Directions: Turn west off Hwy 83 South Fairgrounds 1 block west. **Facilities:** 17 indoor stalls, 21 covered stalls, adequate and easy trailer parking. **Rates:** $15 per stall, $10.00 per electrical hook up, $40 per week, reservations required. **Accommodations:** Best Western, Days Inn, Garden City Inn, Holiday Inn Express, Super 8 within 6 miles.

GREAT BEND
Riverside Stables **Phone: 316-793-6523**
Londa Combs
222 South Kiowa Road [67530] Directions: Located 1/2 mile south of Hwy 56 on east edge of town. Call for directions. **Facilities:** 6 indoor box stalls, 13 stalls with outdoor runs, 75' x 100' indoor lighted arena, 200' x 150' outdoor arena, trailer parking available. Complete tack store on premises. Reservations required. **Rates:** $20 per night. **Accommodations:** Motels 2 miles from stable. Best Western, Super 8, Holiday Inn and others.

ALL OF OUR STABLES REQUIRE CURRENT NEG. COGGINS, CURRENT HEALTH PAPERS, & OWNERSHIP PAPERS.

STANLEY

Parsons Farm Stables **Phone: 913-897-5882**
Larry Parsons
14504 Kenneth Road [66224] Directions: Call for directions. **Facilities:** 8 stalls in 36' x 96' barn, 50-acre pasture on a total of 230 acres, round pen. **Rates:** $15 per night; ask for weekly rate. **Accommodations:** Motels 10 miles from stable.

TONGANOXIE

Laurel Hill Stables **Phone: 913-845-3346**
Chris van Anne
19478 Hay Sixteen [66086] Directions: 25 miles off of I-70. Call for directions. **Facilities:** 15 indoor stalls, 150' x 200' outdoor arena, indoor arena, separate paddocks & pastures. Feed/hay and parking for big rigs available. Morgan show barn. Trains English & Western & saddleseat lessons. Reservations required. **Rates:** $15 per night; call for weekly rates. **Accommodations:** Motels 3 miles from stable.

TOPEKA

Horse O'Tel, **Phone: 785-346-2579**
Harold M. Smith
8540 NW Hwy 75 [66618] Directions: 8 miles north of I-70 and Hwy 24 on Hwy 75. Milepost #170. **Facilities:** 4 indoor 10' x 10' stalls, no pasture/turnout. Feed/hay and trailer parking available. **Rates:** $10 per night. **Accommodations:** Motel 6 & Holiday Inn in Topeka, 8 miles from stable.

WAKEENEY

Saline River Hunting Lodge & Guide Services, Inc. **Phone: 785-743-6676**
Roger & Kelli Flax **E-mail: rogflax@ruraltel.net**
R.R. 1 Box 47 (67672) **Web: www.salinelodge.com**
Directions: I-70, exit 128, 7 miles north on Hwy 283, 2 miles west on county road 422. **Facilities:** Indoor pens for up to 3 horses, 2 large and 2 small outdoor holding pens. No feed/hay, trailer and camper parking with electrical and water hookups, Dog kennels also available. **Rates:** $15/horse/night. **Accommodations:** Very nice lodge for up to 6 individuals, full home cooked breakfast included with lodging. Reservations required.

Thistle Hill Bed & Breakfast **Phone: 785-743-2644**
Dave and Mary Hendricks
Rt. 1, Box 93 [67672] Directions: 1.5 miles from I-70 halfway between Kansas City and Denver. Complete directions given at time of reservation. **Facilities:** 2 indoor pens, 4 large outdoor holding pens. No feed/hay. Trailer parking available. Pets subject to restraint. Accommodations not suitable for pets indoors. Reservations required. Smoking restricted. **Rates:** $15 per night per horse. **Accommodations:** Bed & Breakfast is a comfortable, secluded cedar farm home situated on 320 acres. Wonderful, homemade complimentary breakfast included. $59-$75 per room/double occupancy; $59 single occupancy. Deposit requested to guarantee reservation.

NOTES AND REMINDERS

Long Before There Were Cowboys and Indians...

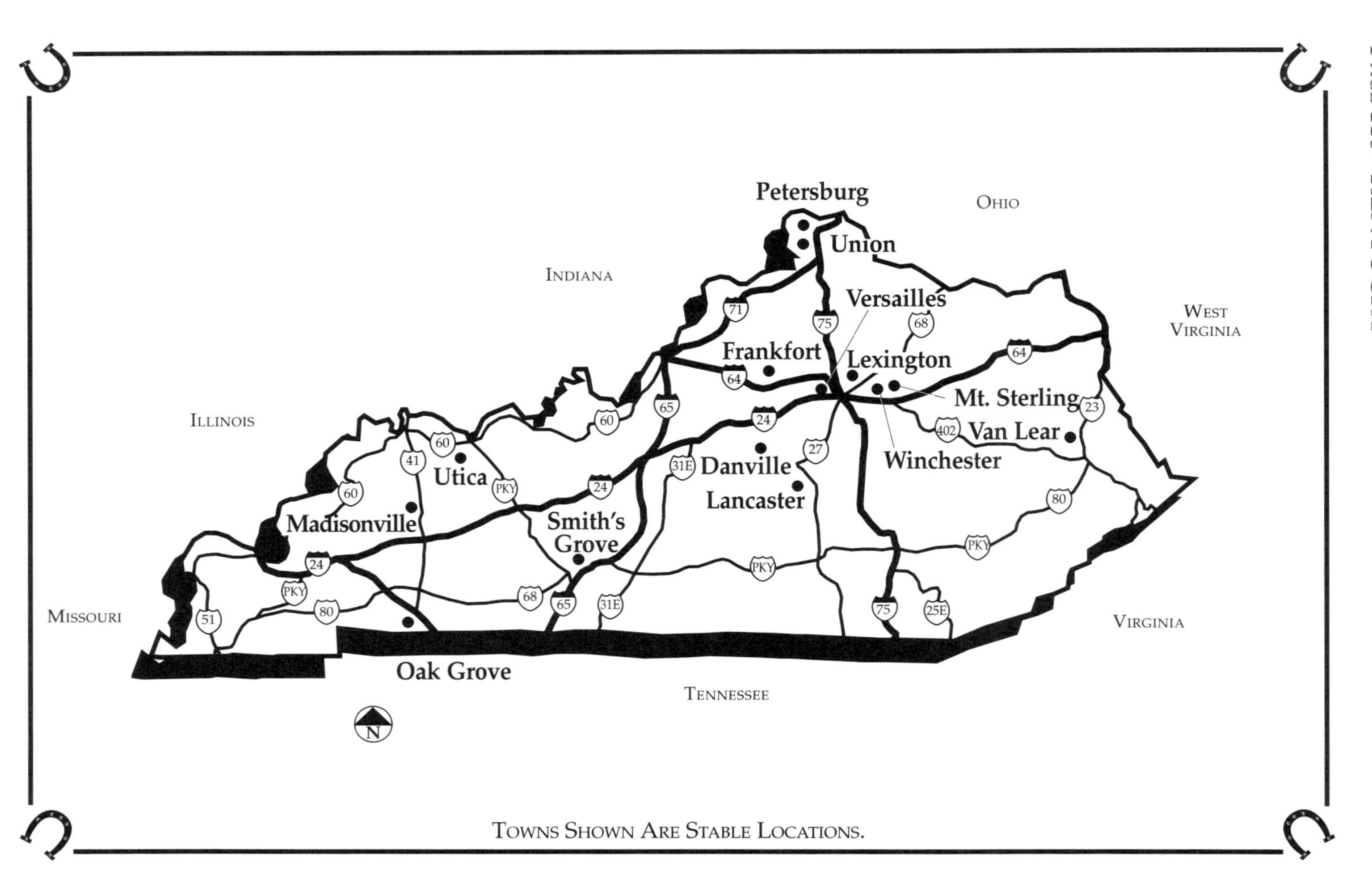

TOWNS SHOWN ARE STABLE LOCATIONS.

ALL OF OUR STABLES REQUIRE CURRENT NEG. COGGINS, CURRENT HEALTH PAPERS, & OWNERSHIP PAPERS.

DANVILLE
Stoneymeadow Farm **Phone: 606-238-7754**
Lloyd & Teri Wineland
2805 Lancaster Road [40422] Directions: Near Rtes. 27, 127, and 150; 35 miles from Lexington and I-64 and I-75. Call for directions. **Facilities:** 10 - 10' x 10' and 10' x 12' indoor stalls plus run-in shed, 50' x 80' corral, pasture/turnout, feed/hay, trailer parking. Excellent fencing, lush pastures, clean stalls. Small outdoor riding ring & access to 80+ acres to ride on. Call for reservations. **Rates:** $15 per night; $50 per week. **Accommodations:** Super 8, Holiday Inn Express, Days Inn, and RV Park within 3 miles in Danville.

FRANKFORT
Jackson Stables **Phone: 502-223-5412**
William & John Jackson **or: Wm: 502-695-2853; John: 502-695-1298**
126 Shadrick Ferry Road [40601] Directions: 10 miles off of I-64. Call for directions. **Facilities:** 8 indoor stalls, 3-acre pasture, 3-acre turnout area, 86' x 210' outdoor riding ring, indoor riding ring, feed/hay & trailer parking available. Horse training, buying & selling. Specializing in Tennessee Walkers. Call for reservation. **Rates:** $15 per night. **Accommodations:** Holiday Inn 5 miles.

LANCASTER
Oliver Stables **Phone: 606-792-4141**
Gary Oliver **(Evenings)**
Hwy US 27, Box 522 [40444]
Directions: Located 16 miles from I-75. Call for directions. **Facilities:** 37 indoor box stalls, 4 acres of fenced pasture/turnout, feed/hay & trailer parking available. Training of Tennessee Walkers & pleasure horses done at stable. Also offers riding lessons at all levels and horses for sale. Trails for riding. Call for reservation. **Rates:** $15 per night. **Accommodations:** Sunset Motel in Stanford, 2 miles from stable.

LEXINGTON
Brook Ledge Vans **Phone: 800-331-0142**
Keith Boyer
2810 Newtown Pike [40511] Directions: Call for directions. 1.5 miles off of I -75, 1.5 miles from Kentucky Horse Farm. **Facilities:** 20 stalls and turnout paddocks, feed/hay, trailer parking available. Local & long distance horse transportation available; truck & trailer maintenance available at facility. Vet on call. Call for reservations. **Rates:** $20 per night. **Accommodations:** Hotels within 1.5 miles (at interstate exit).

Meoldie's Farm **Phone: 606-255-8741**
Mike & Glenda Meadows **or: 606-293-5058**
4222 Iron Works Pk [40511] Directions: Exit 120 (Horse Park) off I-75, farm directly across from Horse Park. **Facilities:** Five 12'x12' indoor stalls, three small paddocks. Feed/ hay available, some trailer parking. Farm is located directly across from the Kentucky State Horse Park, and next to a large Equine Clinic. **Rates:** $20 per day. **Accommodations:** Marriott, Holiday Inn, Embassy Suites, in Lexington , 5 miles away.

ALL OF OUR STABLES REQUIRE CURRENT NEG. COGGINS, CURRENT HEALTH PAPERS, & OWNERSHIP PAPERS.

LEXINGTON

Moonridge Farm **Phone: 606-299-9816**
Dianne Charter
4200 Winchester Road [40509] Directions: Located 3 miles east of Exit 110 off of I-75 on Rt. 60 (Winchester Rd.). Also 3 miles from downtown Lexington & 5 miles from KY Horse Park. **Facilities:** 350-acre horse farm: 62 indoor stalls, 2 outdoor stalls, 20 paddocks, 10 fields, outdoor riding areas, excellent fencing. 24-hr security, hay & grain included & free trailer parking on premises. Vet on call 24 hours a day. **Rates:** $25 per night. **Accommodations:** 15 motels within 2 miles of farm.

Spring Bay Farm - Bed & Stall **Phone: 859-231-8702**
John S. & M. Stanley Wiggs
1249 Greendale Road [40511] Directions: From West: I-64, Exit Georgetown/ Rt. 62. From East: I-64 W, Exit Newtown Pike/Rt. 922. **Facilities:** 4 indoor stalls in a separate guest barn, 10' x 14' box stalls in wood barn with clay floors, one 5-acre pasture, several holding paddocks, feed/hay within 5 miles, trailer parking. Prefer calls for reservations between 7 & 10 P.M. but recorder on the phone at all times. Within 5 miles of 2 major horse parks. **Rates:** $20 per night includes shavings & turnout. **Accommodations:** Bed & continental Breakfast available in 1880s restored Victorian farmhouse (call for details). No pets or smoking in house. Also LaQuinta, Holiday Inn & many others within 5 miles.

Rowland Farm **Phone: 606-299-5674**
Henry Waites
3497 Cleveland Pike [40516] Directions: Located off of US 60. Call for directions. **Facilities:** 42 indoor stalls, pasture/turnout depending on season, feed/hay, & trailer parking available. Boarding & breeding of thoroughbreds done on farm. Call for reservation. **Rates:** $20 per night.
Accommodations: Many motels within 5-10 miles.

MADISONVILLE

Daybreak Horse Park **Phone: 502-825-0761**
Donna Baldridge **502-821-4182**
700 Daybreak Drive [42413] Directions: Call for directions. **Facilities:** 60-acre farm: 20 indoor stalls, 10 of which open into paddocks, fenced pasture/ turnout, 75' x 150' outdoor ring, indoor ring. Breeder of saddlebreds. Standing at stud: "Playboys Wild Hunter." Riding program, including for handicapped and local Scouts. Saddlebreds for sale. Call for reservation. **Rates:** $15 per night. **Accommodations:** Days Inn & Holiday Inn 1 mile from stable.

ALL OF OUR STABLES REQUIRE CURRENT NEG. COGGINS, CURRENT HEALTH PAPERS, & OWNERSHIP PAPERS.

MOUNT STERLING

Bittergreen Farm — **Phone: 606-498-3068**
Jack & Judy Gurnee
1441 Harper's Ridge Road [40353] Directions: 3 miles off I-64. Call for directions. **Facilities:** 5-6 indoor box stalls 12' x 12' to 12' x 16', pasture/turnout, private paddock with run-in shed for stallions, feed/hay, trailer parking. Certified instructor & school horses available; vet and farrier nearby. Close to Cave Run State Park, which has trails. 24-hr. advance reservations requested; emergencies welcome. **Rates:** $15 per night, includes bedding. **Accommodations:** B&B on premises, $80 per night for separate guest house that sleeps 4 adults; weekend package available.

OAK GROVE

Tuckaway Farms — **Phone: 502-439-6255**
Bill Shaut
Box 674 [42262] Directions: 2 miles on 41A off of I-24. Call for directions. **Facilities:** 100-acre farm, 8 indoor stalls, 5 separate pasture areas covering 15 acres, feed/hay & trailer parking. Call for reservation. **Rates:** $15 per night. **Accommodations:** Bed & Breakfast on premises: 6 rooms available with full breakfast. $50 per couple.

PETERSBURG

First Farm Inn — **Phone: 606-586-0199**
Jennifer Warner & Dana Kisor — **Fax: 606-586-0299**
2510 Stevens Road [41080] — **E-mail: firstfarm@goodnews.net**
Directions: From I-275, take Exit 11 (Ky 20 to Petersburg) southwest .9 mile. Turn right onto Stevens Road. First Farm Inn is 1 mile ahead on right. Watch for white fence. **Facilities:** 20 acres of pasture/turnout, feed/hay available, grass riding arena, 5 stalls, trailer parking; historic old barn under renovation. Easy access to Cincinnati, Louisville, Lexington, and the Kentucky Horse Park. **Rates:** $15 pasture, $25 stall per night. **Accommodations:** Bed & Breakfast on site.

SMITHS GROVE

Maple Hill Farm — **Phone: 502-563-4247**
Annemarie Bryce — **502-563-5141**
1888 Hays-Smiths Grove Road [42171] Directions: Exit 38 off of I-65. 1/4 mile to stable. **Facilities:** 4 indoor 12' x 12' stalls, paddock, 3 acres of pasture/turnout, feed/hay & trailer parking available. Rocky Mountain Horses for sale. 20 miles from Mammoth Cave National Park with 60 miles of riding trails. Camper hook-up. **Rates:** $15 per night. **Accommodations:** Bryce Motel 1/4 mile from stable.

UNION

Canterbury Hill Equestrian Center — **Phone: 606-384-3421**
Doug Miller
3383 Hathaway Road [41091] Directions: Exit 178 off of I-75. Call for directions. **Facilities:** 5 indoor and outdoor stalls, depending on availability, pasture/turnout area, feed/hay & overnight trailer parking on premises. No stallions. Call for reservation. **Rates:** Call. **Accommodations:** Many motels within 5-10 miles.

ALL OF OUR STABLES REQUIRE CURRENT NEG. COGGINS, CURRENT HEALTH PAPERS, & OWNERSHIP PAPERS.

UNION
Choco-Ridge Equestrian Center **Phone: 606-384-1600**
Nancy & Elmer Baute
2976 Long Branch [41091] Directions: US 42 off of I-75. Call for directions. **Facilities:** 32 indoor stalls, paddocks, feed/hay & trailer parking on premises. Call for reservation & availability. **Rates:** $15 per night. **Accommodations:** Many motels and restaurants within 4 miles of stable.

UTICA
Twin Ridge Quarterhorse & Paint Farm **Phone: 502-275-4005**
Jack & Gayle Leibfried
4078 Ronnie Lake Road [42376] Directions: Take Hwy 231 south from Owensboro. Call for further directions. **Facilities:** 30 indoor stalls with portable stall available, 37 acres of pasture/turnout, 75' x 100' indoor pipe arena, 150' x 300' outside paddock, hot walker, feed/hay & trailer parking available. 2 Paint stallions standing-at-stud: "Hot Cabin Jack," "Big Leaguer." Quarter horse stallion-at-stud: "Expressions By Boss." All bred for pleasure & halter. Overnight boarding in emergency situations only. **Rates:** $20 per night. **Accommodations:** Days Inn in Owensboro, 13 miles away.

VAN LEAR
Miss T Stables **Phone: 606-789-8424**
Theresa J. Sligh
4705 Kentucky Hwy 302 [41265] Directions: 1/2 mile off of Hwy 23. Call for directions. **Facilities:** 300-acre farm: 25 indoor stalls, 50-acre pasture - grazing for 50 horses, indoor training ring, 14 miles of beautiful mountain riding trails, 24-hour security in barn. 1/2 mile from Thunder Ridge Harness Track. Barn is on city water. 6 campsites available in summer. Call for reservation. **Rates:** $25 per night. **Accommodations:** The Carriage House in Paintsville, 3 miles from stable.

VERSAILLES
River Mountain Farm B&B **Res: Phone: 859-873-7216**
Elaine & Trey Schott, DVM **Office: 859-873-1604**
E-mail: RMFhorses@aol **Fax: 859-879-8019**
3085 Troy Park [40383] Directions: I-64 or I-75 (Newtown Park Route 922 exit). Follow signs to BG Parkway. BG Parkway to first exit, Versailles. Turn left, go 1 mile, farm on right. Great access to Kentucky Horse Park **Facilities:** 36 - 12′ X 12′ wood stalls. Feed, hay and trailer parking available. Total 100 acres with 15 paddocks. English/Polo riding lessons; University of Kentucky Riding Team Training Center; Hunter Jumper our specialty. Private turnout available. **Rates:** $20 per night. $125 per week. **Accommodations:** Best Western, US 60, Frankfort. Comfort Inn, Harrodsburg Rd, Lexington. Bed & Breakfast available on farm: $85 per night (sleeps 4, full kitchen, private entrance).

ALL OF OUR STABLES REQUIRE CURRENT NEG. COGGINS, CURRENT HEALTH PAPERS, & OWNERSHIP PAPERS.

WINCHESTER

Mint Lick Lodge B & B & Horse Layover **Phone: 859-737-1408**
Sarah E. Fralix **Web: www.mintlick.com**
5497 Bybee Rd (40391) Directions: Call for directions. 3-1/2 miles from Point Breeze tack shop, 55 miles from Red River Gorge, 30 miles from Ky. Horse Park. 28 miles from Red Mile, 33 miles form Keeneland and 5 miles to Ky. River. **Facilites:** 4 private paddocks, with planked fences. 10 double box stalls, 11' x 12' with mats and shavings /feed. 4 -1/2 to 3/4 acre paddock. Health cert. and neg coggins. **Rates:** $25 per horse $18 mulitiple or weekly. **Accommodation:** 3 units. B & B on site. Equestrian suite has complete kitchenette, laundry service. Trail riding on site. 110 trailer hook-up.

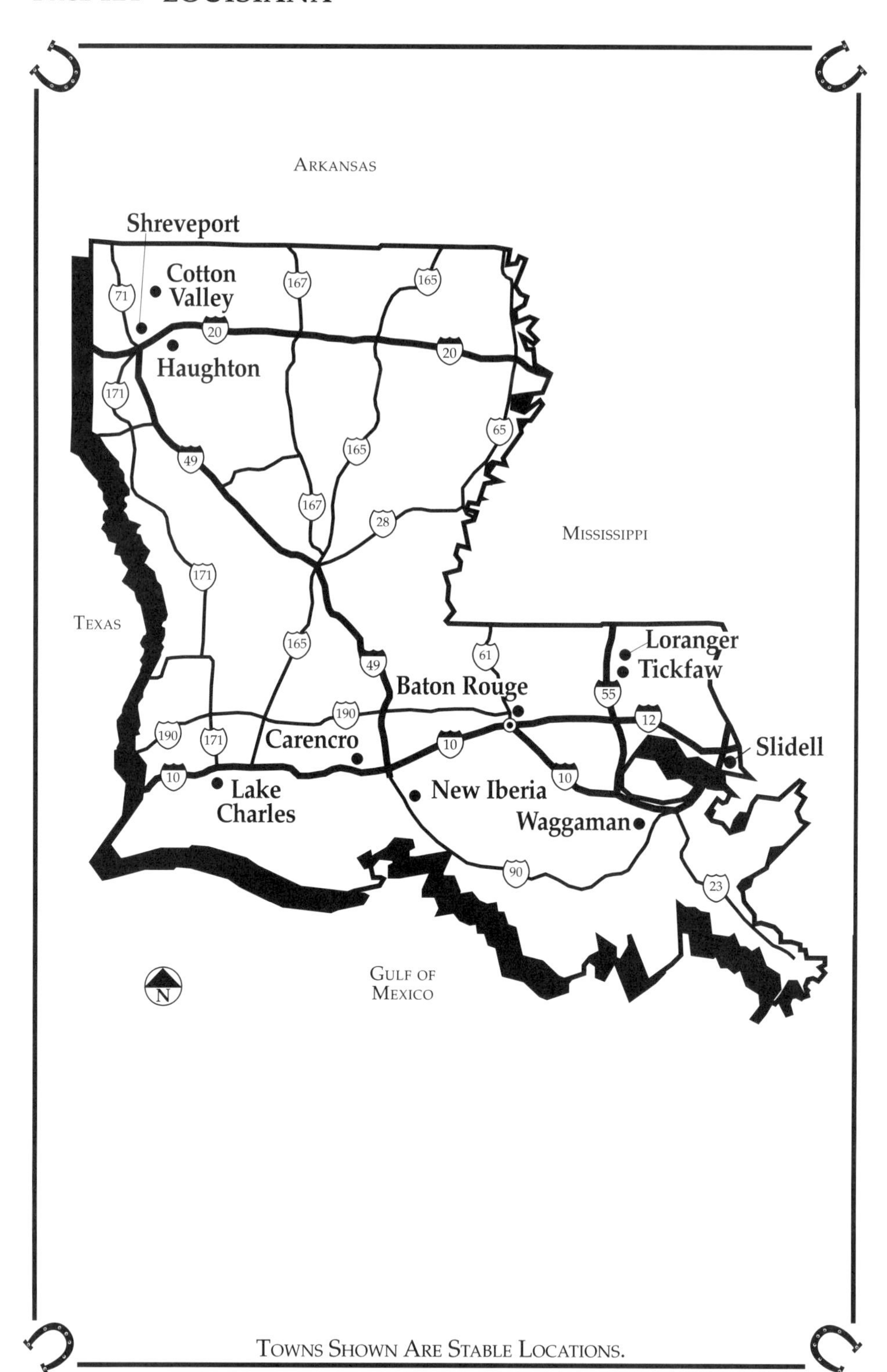

TOWNS SHOWN ARE STABLE LOCATIONS.

ALL OF OUR STABLES REQUIRE CURRENT NEG. COGGINS, CURRENT HEALTH PAPERS, & OWNERSHIP PAPERS.

BATON ROUGE

Farr Park Horse Activity Center   **Phone: 225-769-7805**

Gretchen Morgn, Director

6402 River Road [70820] Directions: 2.5 miles south of LSU campus on River Road; 6 miles south of downtown Baton Rouge; 5.5 miles south of I-10 College Drive Exit. Call for specific directions. **Facilities:** 190 outdoor covered stalls, 121' x 244' indoor arena with grandstand seating for 1,500, wash racks, 140' x 300' outdoor arena, polo field, 24-hr security on the 335-acre facility. 151-unit RV area with electric, water & sewer hook-ups for $10 per night, plus bathrooms, showers, & laundry. Center is site of rodeos, team-penning, dressage & jumping events. Boarding arrangements must be made in advance. **Rates:** RV $12 per night, stalls $15 per night **Accommodations:** Ramada Inn 4 miles from Center.

CARENCRO

Traders Rest Farm   **Phone: 337-234-2382**

Don Stemmans   Toll Free Reservations: 800-544-6773

P.O. Box 156 [70520] Directions: 2 miles from I-49 near Lafayette. 1 mile north of I-10 at Ambassador Caffrey. **Facilities:** 105 indoor stalls, pasture/turnout available, no feed/hay, trailer parking on premises. Call 1-800-544-6773 to make reservations. **Rates:** $15 per night. **Accommodations:** Red Roof Inn & Holiday Inn North nearby.

COTTON VALLEY

Tullett Farm   **Phone: 318-326-4087**

John Tullett   Fax: 318-326-7487 Home: 326-4851

Route 1, Box 132 [71018] Directions: 26 miles north I-20 & 10 miles off of Hwy 3. Call for directions. **Facilities:** 6 large indoor box stalls, 120-acre pasture with pond, feed/hay & trailer parking on premises. Riding trails on property. Security system on farm. Riverboat casinos & Louisiana Downs nearby. **Rates:** $15 per night; weekly rate negotiable. **Accommodations:** Sheraton & LaQuinta 26 miles away.

HAUGHTON

Double Rainbow Equestrian Center   **Phone: 318-949-9133**

Sig North   or: 318-949-9948

1860 Adner Road [71037] Directions: 3 miles from I-20 and 220 Loop. Also 3 miles from Louisiana Downs. Call for directions. **Facilities:** 50 indoor stalls in 6 barns on 146-acre Center, numerous 1/2 to 30-acre paddocks, 10 wash racks with H/C water, racetrack, electric walking wheel at each barn, 2 outdoor arenas, 2 round pens, feed/hay & buckets available, trailer parking on site. 24-hr security. Farrier on premises. Multi-disciplinary riding & training programs offered at Center including therapeutic riding for the handicapped. Call for reservation. **Rates:** $15 per night; $75 per week. **Accommodations:** leBossier Motel & Grande Isle Motel both 4 miles from Center.

ALL OF OUR STABLES REQUIRE CURRENT NEG. COGGINS, CURRENT HEALTH PAPERS, & OWNERSHIP PAPERS.

LAKE CHARLES

Grass Roots Stable **Phone: 318-474-2606**
Kurt Courville **or: 318-478-9436**
6065 Tom Hebert Road [70605] Directions: 5 miles from I-10 & 2 miles from I-210 Loop. Call for directions. **Facilities:** 14 indoor 12' x 14' stalls with automatic waterers and insect sprayers, 5 large outdoor paddocks with shelters, automatic waterers. All wood fencing. 100' x 200' lighted outdoor arena. Hot walker and washracks. Trailer parking on premises. **Rates:** $20 per night. **Accommodations:** EconoLodge and Motel 8, 5 miles from stable.

LORANGER

Live Oak Boarding Stables **Barn Phone: 504-878-3300**
Debi Brown **Home: 504-878-6553**
53384 Hwy. 40 [70446] Directions: Easy access from I-10, I -12, and I-55. Call for directions. **Facilities:** 40 - 12' x 12' stalls, 3 large foaling stalls, round pen, 120' x 220' arena, covered round pen, riding area, pasture/turnout, feed/hay, trailer parking. Inside wash room, tack & feed rooms. Lessons & trainer available by appointment; picnic area. Call for reservation. **Rates:** $15 per night. **Accommodations:** Rooms available on premises with laundry & cooking facilities. 10 RV hook-ups, primitive camping. Motels within 15 miles.

NEW IBERIA

Butcher Farms **Phone: 318-365-1883**
Michael Butcher
5411 Norris Road [70560] Directions: About 14 miles off I-10 on Hwy 90 East. Call for directions. **Facilities:** 13 indoor stalls, pasture/turnout, feed/hay, trailer parking and RV hook-ups available. Vet on call. Breeding farm for quarter horses, training; dog kennel available. Right in the heart of Cajun country. Call for reservation. **Rates:** $13 per night. **Accommodations:** Motels within 5 miles of stable.

SHREVEPORT

State Fair of Louisiana **Phone: 318-635-1361**
Tommy Lacy, Livestock Director
3701 Hudson Street [71109] Directions: Call for directions. Off I-20. **Facilities:** 132 permanent stalls, 40 in cattle barn, trailer parking available ($15.00 per night), feed store close by. Reservations required; gates locked at 5 P.M. Fair dates October 23 -- November 8; very full at that time. **Rates:** $15 per stall per night. **Accommodations:** Cluster of motels near airport (1 mile).

ALL OF OUR STABLES REQUIRE CURRENT NEG. COGGINS, CURRENT HEALTH PAPERS, & OWNERSHIP PAPERS.

SLIDELL
Lewis Stables **Phone: 504-643-8025**
Bob Lewis **E-mail: lewistrl@cmg.net** **Fax: 504-643-6791**
80 Tortoise Street [70461] Directions: From I-10: Take Exit 266; go east 1 mile to first traffic light; take left (north) & go to 3rd stop sign; take right & go 2 blocks to stable. **Facilities:** 130 indoor stalls, 6-8 holding pens, Purina dealer on site. Can handle large transports. Trailer sales, service, & rentals on site as well as a tack store. Horse transportation business from this location primarily serving the Southeast but will transport nationwide on request. Personal & speedy service their specialty. Advance reservations required. **Rates:** $15 per night; weekly rate negotiable. **Accommodations:** Ramada Inn, LaQuinta, within 1.5 miles.

TICKFAW
R & L Farm **Phone: 504-542-7262**
Richard Pisciotta & Toni Wilner
Rt. 2, 49273 Fedele Road [70466] Directions: Tickfaw Exit off of I-55; go east 1 block to Fedele Road; go 3/4 mile and farm is on left. **Facilities:** 10 indoor stalls, paddocks, 10-acre pasture, hot walker, round pen, feed/hay, limited trailer parking. **Rates:** $15 per night; $100 per week. **Accommodations:** Motels in Hammond, 4 miles from stable.

WAGGAMAN
Emerald Fox Farm **Phone: 504-431-0831**
Lynne Kurilovitch
7601 River Road [70094] Directions: Exit I-10 to I-310 South: Take Hahnville/Luling exit from Hale Boggs Bridge; turn right on to River Road, 9.5 miles from I-310. Or Exit I-10 in Metairie at Clearview Pkwy South to Huey Long Bridge; take Hwy 90 to Hwy 18W (1/2 mile from bridge); 18W is Historic River Road - 3.5 miles to farm. **Facilities:** 4 indoor stalls, 2 outdoor stalls, 40' x 100' pasture/turnout available, ring riding or levee riding available, small amounts of feed/hay. Close to New Orleans and other vacation spots. Phone ahead for reservations. **Rates:** $15 per night; $75 per week. **Accommodations:** Room available on premises with simple breakfast included: $40 single; $55 double. Also, motels 3 miles from farm.

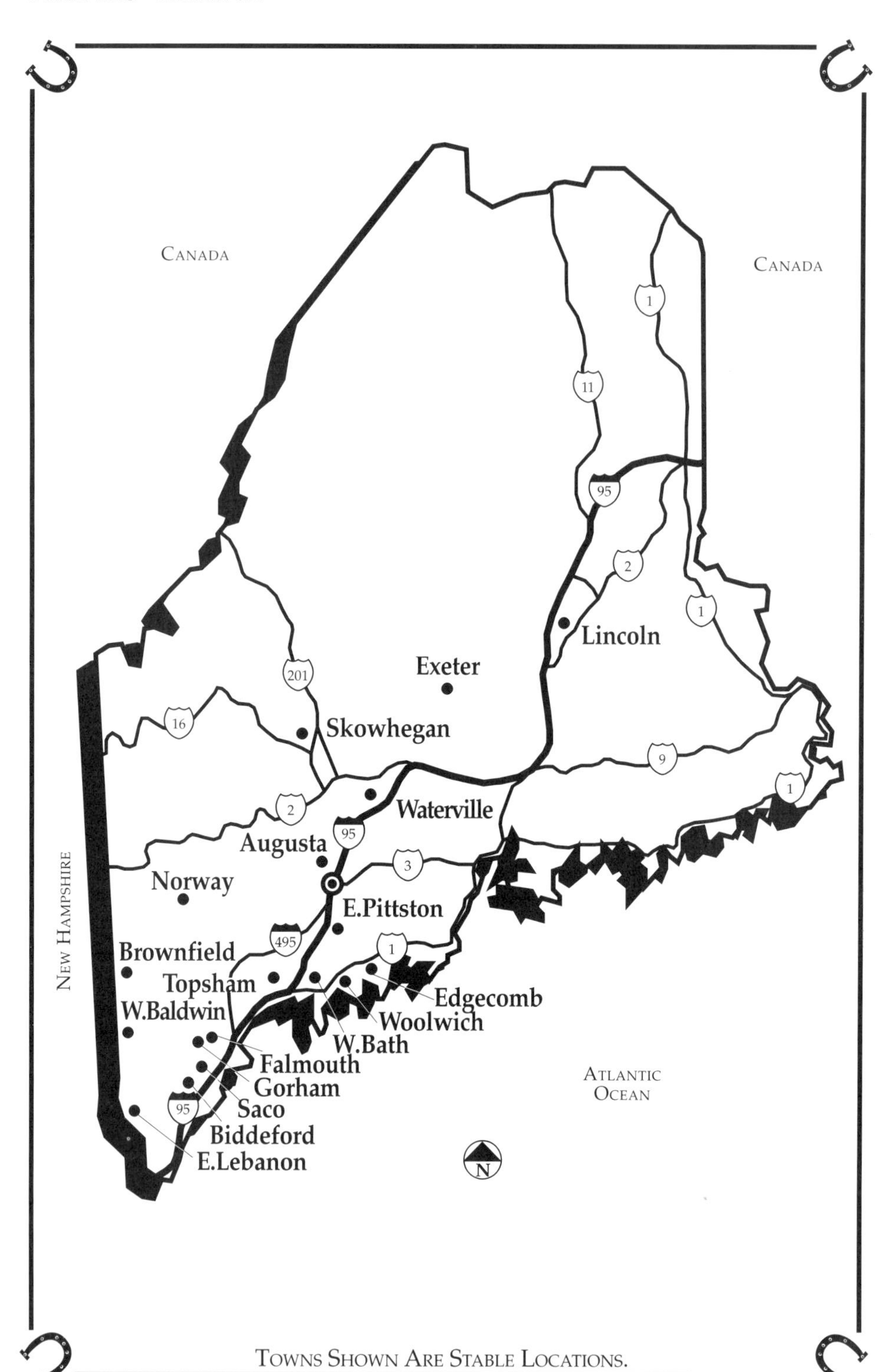

TOWNS SHOWN ARE STABLE LOCATIONS.

ALL OF OUR STABLES REQUIRE CURRENT NEG. COGGINS, CURRENT HEALTH PAPERS, & OWNERSHIP PAPERS.

AUGUSTA

Friendship Stables **Phone: 207-622-5047**

Jackie McConkie/Sarah Wright

Eight Rod Road, Box 1180 [04330] Directions: Civic Center Exit off of I-95. Call for directions. **Facilities:** 20 indoor stalls, indoor arena, 2 outside paddocks, large outside arena. Saddle seat instructions. Paddocks only for overnight boarding. Call for reservations. **Rates:** $15 per night. **Accommodations:** Motels within 2 miles of stable.

BIDDEFORD

Bush Brook Farm **Phone: 207-284-7721**

Mona Jerome **Home: 207-284-8311**

4463 West Street [04005] Directions: Exit 4 (Biddeford) off of I-95. Call for directions. **Facilities:** 4 indoor stalls, feed/hay, trailer parking available. Jumps & trails for riding. Near 3 beaches. **Rates:** $20 per night. Call for weekly rate. **Accommodations:** RV park 1/2 mile and motels 3 miles from stable.

BROWNFIELD

The Foothills Farm B & B **Phone: 207-935-3799**

Kevin Early & Theresa Rovere

RR 1, Box 598 [04010] Directions: 9 miles from Route 302 & 10 miles from Route 16. Call for directions. **Facilities:** 3 indoor, 2 outdoor lean-to stalls, pasture, and riding trails available. No smoking preferred. Reservations required. 10 miles from Conway, NH and 9 miles from Fryeburg. Must stay at B&B. **Rates:** $15 per night. **Accommodations:** B&B: $42-$48 double occupancy with shared bath.

E. LEBANON

Menomin Meadow **Phone: 207-457-1774**

Corine Crossmon

RR 1, Box 1319D [04027] Directions: Call for directions. **Facilities:** Up to 10 indoor stalls in a new barn, some varied size paddocks, feed/hay at additional cost. Vaccinations for E&WEE, tetanus, flu, rhino, & Potomac in last 3 months. Payment upon arrival by cash or travelers check only. **Rates:** $20 for 10' x 12' stall; larger stalls available; call for weekly rate. **Accommodations:** Cardinal Ranch Motel in Rochester, NH, about 7 miles from stable.

E. PITTSTON

Woodrose Farm **Phone: 207-582-6315**

Karen Lyons

Mast Road [04345] Directions: Rt. 1 to Rt. 27. 5 miles off Rt. 27. Call for further directions. **Facilities:** 3 indoor stalls, indoor arena, pasture, paddocks, & pens. Dog boarding OK. Please call ahead. Dressage lessons available. **Rates:** $10 per night. **Accommodations:** Motels 10 miles from stable.

ALL OF OUR STABLES REQUIRE CURRENT NEG. COGGINS, CURRENT HEALTH PAPERS, & OWNERSHIP PAPERS.

WISCASSET

Ledgewood Riding Stable **Phone: 207-882-6346**
Carol Reed
432 Lowell Town Rd [04578] Directions: Located on Rt. 27 on way to Boothbay Harbor. Call for directions. **Facilities:** 15 box stalls, 2 stalls with turnout shelter, paddocks, two 10-acre pastures, riding ring. English & Western lessons. Please call ahead. **Rates:** $15 per night; $75 per week. **Accommodations:** Motels 2 miles away.

RC Quarterhorses **Phone: 207-882-7661**
Carol Heaberlin
Middle Road [04556] Directions: Exit 10 off of I-95. Call for directions. **Facilities:** 4 indoor stalls, outdoor paddocks, 60' x 120' indoor arena. Trail riding. Breeds, shows, and sells quarter horses. Standing-at-stud: "Exelans," who produced a World Champion Halter Class Horse. Please call ahead for reservations. **Rates:** $15 per night. **Accommodations:** Motels within 5 miles.

EXETER

Sirsarg Stables **Phone: 207-379-2776**
Noel Sirabella
Champion Road [04435] Directions: Newport Exit off I-95 to Rte 11; approx. 13 miles off exit. **Facilities:** 5 indoor 10' x 10' box stalls, 3 indoor 5' x 8' walk-in stalls, 16' x 40' lean-to, 50' x 70' paddock, round pen, 150' x 100' outdoor riding ring, barrel course, 14- & 10-acre fenced pastures, feed/hay, trailer parking available. Trail riding available. Farrier, grain and feed store, vet in town. Buys and sells horses, generally quarter horses. **Rates:** $12 per night; weekly rate available. **Accommodations:** Spare bedrooms on premises; motels in Newport (13 miles away) and Bangor (18 miles away).

FALMOUTH

Norton Farms **Phone: 207-797-7577**
Lori Graffam
613 Blackstrap Road [04105] Directions: Exit 10 off of Maine Turnpike (I-95). Call for directions. **Facilities:** 28 indoor stalls, 4 paddocks, 4 pastures, walker. Horse trailer parking available. Breeds & trains Standardbreds. Please call ahead for reservations. **Rates:** $15 per night. **Accommodations:** Motel 8 miles from stable.

GORHAM

Kents' Stables **Phone: 207-839-5351**
Lisa Kents **207-839-6428**
726 Fort Hill Road [04038]
Directions: Exit 8 off of I-95. Call for directions. **Facilities:** 5 indoor stalls, paddock area, indoor & outdoor arenas, outside stadium arena, outside dressage arena, cross-country course, miles of trail riding including beach riding. Lessons available. Call for reservations. **Rates:** $15 per night. **Accommodations:** Motels within 5 miles of stable.

ALL OF OUR STABLES REQUIRE CURRENT NEG. COGGINS, CURRENT HEALTH PAPERS, & OWNERSHIP PAPERS.

LINCOLN

Trail's End Morgans **Phone: 207-794-6273**

Richard Tolman

RR 1, Box 462 [04458] Directions: 3 miles outside of Lincoln off Rt. 2. Six miles off I-95. Call for directions. **Facilities:** 6 indoor 12' x 12' stalls, 100' x 150' outside paddock, grazing pasture, 100' x 150' outside arena. Breeds & sells Morgans. Two stallions standing-at-stud: "T.E.M. Firemaker" and TEM's Pride," registered Morgans. **Rates:** $15 per night; ask for weekly rate. **Accommodations:** Motels 3 miles from stable.

NORWAY

Hidden Brook Farm **Phone: 207-743-6546**

Beth & Paul Brainerd

RFD 1, Box 968 Howe Road [04268] Directions: 8 miles from Rt. 26 & 30 minutes from Fryeburg. Call for directions. **Facilities:** 20 indoor 12' x 12' stalls, 20m x 60m wood-fenced grass pasture areas. Farrier on premises. Dressage & combined training. Call for reservations. **Rates:** $15 per night; $75 weekly rate. **Accommodations:** Inn Town Motel in Norway, 6 miles away.

SACO

Breezy Meadow Horse Farm **Phone: 207-284-9409**

Doreen Metcalf **207-284-4074**

184 Buxton Road [04072] Directions: 2 miles off of I-95. Call for directions. **Facilities:** 6 indoor stalls, indoor ring, 2 outdoor rings, 4 small paddocks, 5- & 10-acre paddocks. Horse training & sales. Call for reservations. **Rates:** $15 per night. **Accommodations:** Motel 2 miles from stable.

SKOWHEGAN

North Star Horse Center **Phone: 207-474-7314**

Judy & Jaymee Dore

RFD #1, Box 3307 [04976] Directions: From I-95: Route 201 Exit for Skowhegan. Located on Rt. 150. Call for further directions. **Facilities:** 37 indoor 10' x 12' stalls, 70' x 140' indoor ring, indoor round pen, outdoor paddock. Barn partially heated and heated observation lounge. Guided trail rides. Full riding instruction & summer day camp. **Rates:** $20 per night.

TOPSHAM

Cathance Morgan Farm **Phone: 207-729-0478**

Lynda Bernier

Box 3468, Cathance Road [04086] Directions: Exit 25, Bowdoinham, off of I-95. Call for directions. **Facilities:** 3 indoor stalls, outdoor riding ring, feed/hay & trailer parking available. Please call ahead for reservations. **Rates:** $15 per night; $75 per week. **Accommodations:** Motels 5 miles from stable.

ALL OF OUR STABLES REQUIRE CURRENT NEG. COGGINS, CURRENT HEALTH PAPERS, & OWNERSHIP PAPERS.

WATERVILLE
Serendipitous Stables **Phone: 207-872-7616**
Tom Millett
Marston Road [04901] Directions: Exit 33 off of I-95. Call for directions. **Facilities:** 20 indoor stalls, outdoor ring, indoor arena, trailer parking available. Please call ahead for reservations. One mile from Colby College. **Rates:** $15 per night. **Accommodations:** Motels 1 mile from stable.

WEST BATH
Whorff Stables **Phone: 207-443-3965**
Rhonda Whorff
Foster's Point Road [04530] Directions: Call for directions. **Facilities:** 6 indoor stalls, 80' x 100' riding ring, large pasture, feed/hay available. Trail rides and lessons offered at stable. Call ahead for reservations. **Rates:** $20 per night. **Accommodations:** Motels 3 miles from stable.

WOOLWICH
Gallant Morgan Horse Farm **Phone: 207-443-4170**
Margarite Gallant
Days Ferry Road [04579] Directions: Call for directions. **Facilities:** 4 indoor stalls, indoor & outdoor rings, farrier available. Tack shop on premises. Morgans for sale. Farm offers a summer horse camp for girls. Please call for reservations. **Rates:** $15 per night. **Accommodations:** Motels 6 miles from stable.

I TOLD YOU OLD LIGHTNIN' WAS A SLIPPERY ONE!
FRED LEARY

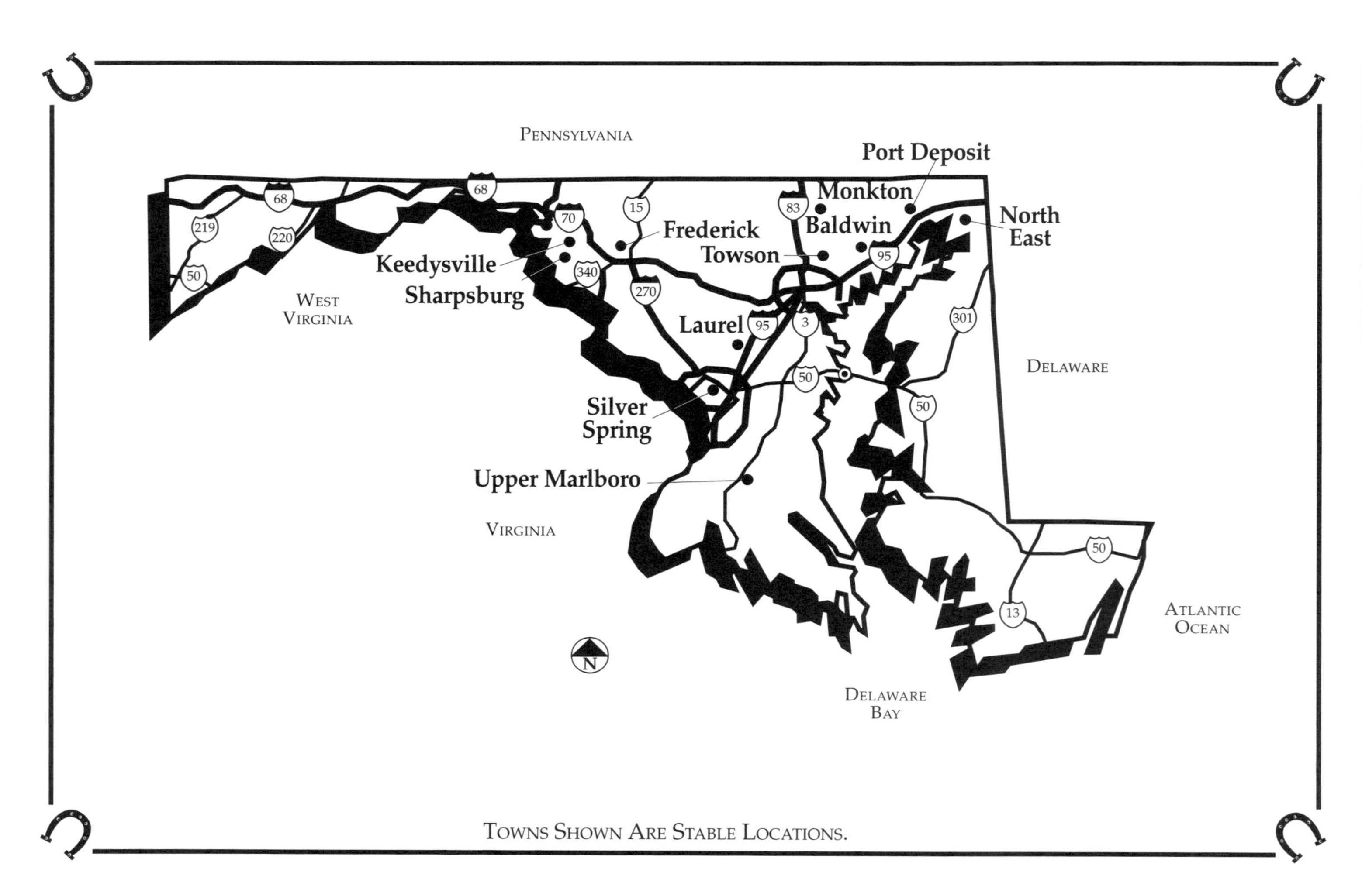

TOWNS SHOWN ARE STABLE LOCATIONS.

ALL OF OUR STABLES REQUIRE CURRENT NEG. COGGINS, CURRENT HEALTH PAPERS, & OWNERSHIP PAPERS.

BALDWIN

Wintergreen Farms **Phone: 410-592-5806**
Kevin Jones, manager
14252 Lynch's Lane [21013] Directions: Bel-Aire Exit, Rt. 165, off of I-95. Call for directions. **Facilities:** 21 indoor stalls, indoor & outdoor rings, paddocks, wash rack, overnight trailer parking. English riding lessons given. Call for reservation. **Rates:** $20 per night. **Accommodations:** Motels 10 miles from stable.

FREDERICK

Rushing Winds Farm **Phone: 301-898-9133**
Angela Klinger
9912 Masser Road [21702] Directions: Off Rte. 15. Call for directions. 70 minutes from Baltimore; 70 minutes from Washington, D.C.; 40 minutes from Hagerstown, MD; 90 minutes from Harrisburg, PA. **Facilities:** 6 - 12' x 12' box stalls, 100' x 200' and 140' x 240' outside arenas, several pastures & paddocks, wash stall, feed/hay, RV hook-ups, trailer parking available. RV shop and feed and tack stores nearby. **Rates:** $15 per night. **Accommodations:** Bed and meals provided at a very reasonable cost. Several motels within 10 miles.

KEEDYSVILLE

Laura J. Parrish **Phone: 301-416-2706 or: 301-432-6054**
18929 Shepherdstown Pike [21756] Directions: From Hagerstown, at intersection of Rt. 70 & I-181 follow Rt. 70 east to Sharpsbury Pike exit, Rt. 65 south and follow for 12 miles. Left at Rt. 34 east, go 2.4 miles; farm on right. From Baltimore, take Rt, 70 west to Middletown/Braddock exit onto Alternate Rt. 40 west follow for 12 miles, left onto Rt. 34 west, 4 miles to farm on left. **Facilities:** Eight 12'x12' indoor stalls w/outdoor pastures. Hay, bedding, grain and separated pastures available. Antietam Battlefield and the C&O canal trail riding, trailer parking. **Rates:** $20 night, $100 weekly. **Accommodations:** Bed & Breakfast 2.5 miles.

LAUREL

Columbia Horse Center **Phone: 301-776-5850**
Susan Wentzel
10400 Gorman Road [20723] Directions: Exit 35B off of I-95. Call for directions. **Facilities:** 20 indoor stalls, 2 large outdoor rings, 2 indoor arenas, 12 paddocks on 88 acres. Lessons & training for huntseat only. Call for reservation. **Rates:** $25 per night. **Accommodations:** Motels 5 miles from stable.

MONKTON

Upper Crondall Farm **Phone: 800-447-1782**
Jack & Betsy Ensor **After 6 p.m.: 410-472-4528**
16909 Gerting Road [21111] Directions: I-83 to Exit 27, east 1 mile to stop light, right onto York Road, 1 block to left onto Monkton Road, go 3 miles to Sheppard Road, follow Sheppard Road for 2 miles, left at first stop sign (Gerting Road), farm is first house on right. **Facilities:** 12' x 12' indoor stall, pasture/turnout, hay & trailer parking available. Call for reservation. **Rates:** $15 per night. **Accommodations:** Bed & breakfast on premises, $85 per night, fireplace, private bath, air conditioning.

ALL OF OUR STABLES REQUIRE CURRENT NEG. COGGINS, CURRENT HEALTH PAPERS, & OWNERSHIP PAPERS.

NORTH EAST

Tailwinds Farm **Phone: 410-658-8187**

Ted & JoAnn Dawson

41 Tailwinds Lane [21901] Directions: From I-95 take exit 100 (MD), this will be Rt. 272. Take 272 north 4 miles, farm is on the right. **Facilities:** 25 12'x12' indoor stalls, 25 acres of separated pastures, feed/hay available, ample trailer parking. Indoor and outdoor rings, trail rides, riding lessons and carriage rides available. Farm is located 10 minutes from Fair Hill State Park, featuring 5,000 acres with riding trails. **Rates:** $25 per night box stall. $15 per night turnout only. **Accommodations:** Tailwinds is a Bed & Breakfast, 2 rooms available at $75 per night. Crystal Inn in North East just 4 miles away.

PORT DEPOSIT

Anchor and Hope Farm **Phone: 410-378-4081**

Edwin Merryman

P.O. Box 342 [21904] Directions: Located 4 miles off of I-95: Please call ahead for directions. **Facilities:** 3 indoor stalls, paddocks, pasture/turnout, feed/hay & trailer parking available. **Rates:** $30 per night. **Accommodations:** Comfort Inn at I-95, 4 miles from farm.

SHARPSBURG

Poor Boy Stables **Phone: 301-223-9089**

Raymond Ramsey

16419 Woburn Road [21782] Directions: 5 miles off I-81 & I-70. Call for directions. **Facilities:** 22 indoor box stalls; 6-, 2-, & 1-acre pasture/turnout areas; 85' x 165' indoor arena; 6 outside paddocks; rings; and riding trails. Large rigs OK. Lessons & carriage rides, sleigh & hay rides, etc. Also buying & selling of horses, specializing in child-proof horses. Located 4 miles from Antietam Battlefield. Call for reservation. **Rates:** $15 per night. **Accommodations:** Days Inn in Williamsport, 6 miles from stable.

SILVER SPRING

Woodland Horse Center **Phone: 301-421-9156**

Michael Smith

16301 New Hampshire Avenue [20905] Directions: 8 miles from Beltway. Call for directions. **Facilities:** 5 indoor stalls, pasture & small turnout paddock, 5 - 2- to 5-acre pastures. No stallions. OVERNIGHT BOARDING IN AN EMERGENCY SITUATION ONLY. **Rates:** $15 per night. **Accommodations:** Many motels 15 minutes away in Silver Spring.

ALL OF OUR STABLES REQUIRE CURRENT NEG. COGGINS, CURRENT HEALTH PAPERS, & OWNERSHIP PAPERS.

TOWSON
Dunroman Stables **Phone: 410-296-5945**
Robin & Paul Farace
1102 Hart Road [21286] Directions: Exit 28N (Providence Rd) off of I-695. Call for directions. **Facilities:** 25 indoor box stalls, 100' x 150' outdoor arena, feed/hay & trailer parking available. Riding trails. Vet on call 24 hours. Truck & trailer repairs on site. Campers OK. Monthly boarding. Call for reservation. **Rates:** $15 per night; $75 per week. **Accommodations:** Motels 3 minutes from stable.

UPPER MARLBORO
Prince George's Equestrian Center **Phone: 301-952-7900**
Maryland National Capital Park and Planning Commission
14900 Pennsylvania Avenue
(MD Rte. 4) [20778] Directions: Call for directions. **Facilities:** 240 - 12' x 12' block barns. Stabling available by appointment only, Availability limited on weekends between mid March & November due to horse shows. Ship-ins to reserve stalls in advance, feed & bedding can be delivered if arranged in advance. Trailer parking available. Negative coggins required. **Rates:** $12 per night. **Accommodations:** Call for information. Hampton Inn/Largo

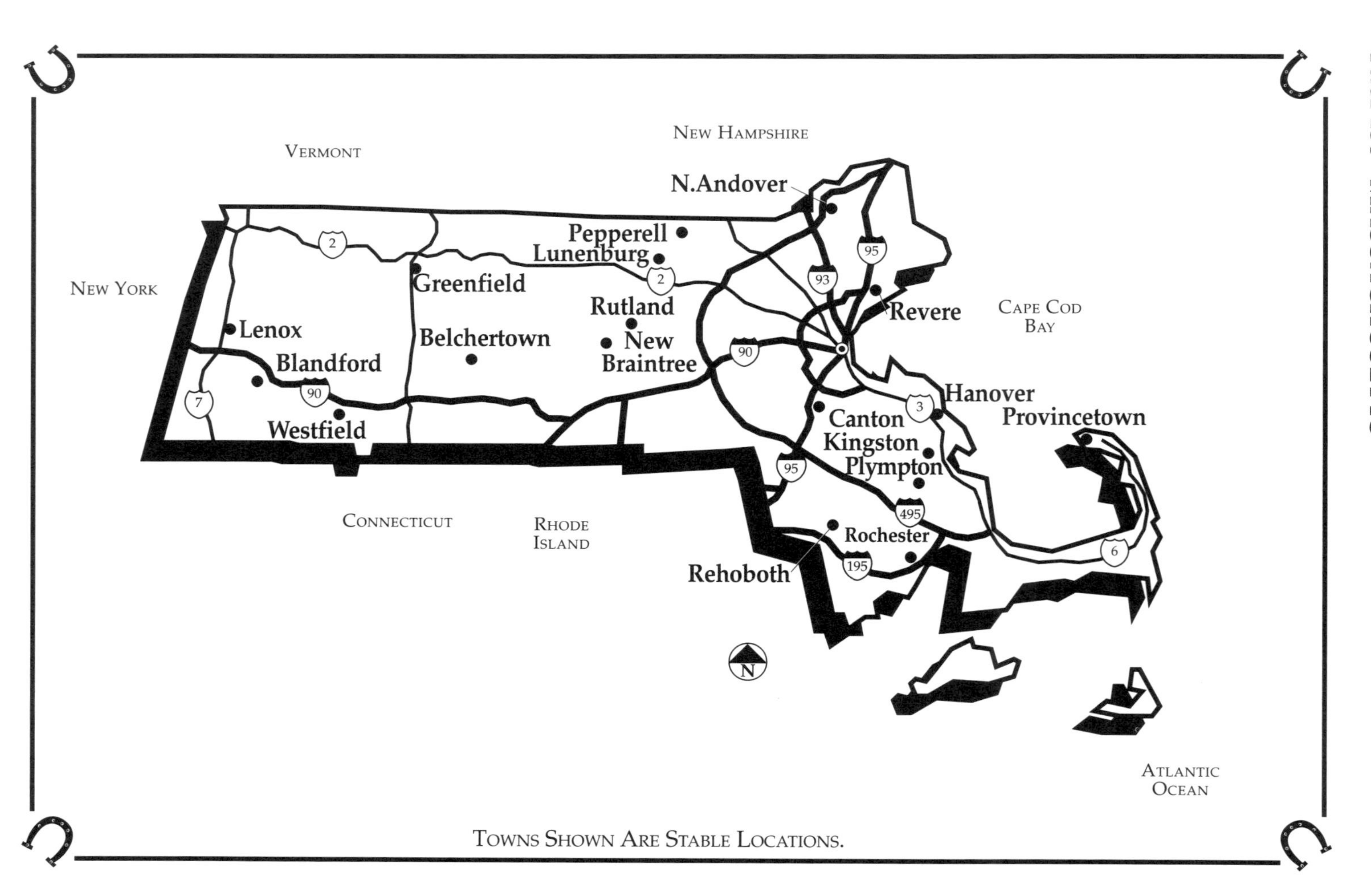

Towns Shown Are Stable Locations.

ALL OF OUR STABLES REQUIRE CURRENT NEG. COGGINS, CURRENT HEALTH PAPERS, & OWNERSHIP PAPERS.

BELCHERTOWN

Ingate Farms **Phone: 413-253-0440**
William (Bill) W. McCormick **or: 253-9856**
60 Lamson Road [01007] Directions: East on Rt. 9 off of I-91. Go to Rt. 116 in Amherst; take right (south) to Bay Rd. at Atkins Farm; take left on Bay Rd. and go 3.8 miles to Lamson; take right into Ingate property. **Facilities:** 20 indoor stalls, five 1/4-acre pasture/turnout areas, indoor ring, 2 outdoor rings. 500 acres adjacent to state lands - miles of riding trails. Trailer parking with easy turnaround. **Rates:** $15 per night; $90 per week. **Accommodations:** Ingate Bed & Breakfast on property: 5 BRs with breakfast included. Outdoor Olympic pool, 8 units with kitchenettes. (Publisher's note: This is a beautiful home and horse facility located in a very scenic and cultural area.)

BLANDFORD

Pleasure Horse Paso Fino Barn **Phone: 413-848-2214**
Ingrid Gureckis
43 Russell Road [01008] Directions: Call for directions. **Facilities:** 4 wood indoor 10' x 10' stalls, 2-acre pasture, 40' x 40' turnout, feed/hay, trailer parking available, riding ring, trails, saddlery, gift shop. Paso Fino training, breeding, lessons. Standing-at-stud: "Rascelleo de Vez," grandson of "Coral Lace." **Rates:** $15 per night. **Accommodations:** Bed & breakfast on premises.

CANTON

Canton Equestrian Center **Phone: 617-821-5527**
Mary Hughes **617-828-0335**
1095 Randolph Street [02021] Directions: Take Rt. 138 off of Rt. 128 (I-95). Call for further directions. **Facilities:** 35 indoor stalls, indoor arena, outdoor arena, jump course, 3 fenced turnouts, miles of riding trails nearby. English & Western, & hunter/jumper lessons at all levels. Call for reservations. **Rates:** $15 per night. **Accommodations:** Holiday Inn 2 miles from stable.

GREENFIELD

Meadowcrest Farm **Phone: 413-773-7842**
Jim & Rita Adams
290 Leyden Road [01301] Directions: 2 miles off of I-91 at Exit 26. Call for directions. **Facilities:** 12 indoor 12' x 12' box stalls, outdoor lighted ring, pasture/turnout on 64 acres. Meadowcrest is also a Christmas tree farm with hay rides & sleigh rides drawn by a team of Percherons. Many activities offered at this scenic location. Call for reservation. **Rates:** $20 per night. **Accommodations:** Howard Johnson's 2 miles from stable at I-91.

ALL OF OUR STABLES REQUIRE CURRENT NEG. COGGINS, CURRENT HEALTH PAPERS, & OWNERSHIP PAPERS.

HANOVER
Briggs Stable **Phone: 781-826-3191**
John D. & Vicki F. Dougherty
623 Hanover Street, Rt. 139 [02339] Directions: Convenient to Rt. 3. Call for directions. **Facilities:** 70 indoor stalls, turnout, 80' x 210' indoor ring, 2 outdoor riding rings, hunt course, riding trails. Tack shop & trailer sales on premises. Standing-at-stud: "Boston Pops," Westphalian stallion - Jumping 4' Dressage 4th Level Champion. Private to group lessons in all seats. **Rates:** $20 per night. **Accommodations:** Motels in nearby Rockland.

KINGSTON
Dusty-J Farm **Phone: 781-585-6258**
Rick & Karen Johnson **Barn: 585-1940**
199 Grove Street [02364] Directions: Off Rt. 3 & 3A. Call for directions. **Facilities:** 13 indoor stalls, 4 outside rings (1 lighted), feed/hay & trailer parking available. National Champion for Hunter Under Saddle: "Tardee Straw" - registered Appaloosa horse. Call for reservation. **Rates:** $15 per night. **Accommodations:** Motels 2 miles from stable.

LENOX
RCR Stables **Phone: 413-637-0613**
Charlene McAteer **800-840-0613**
430 East Street [01240] Directions: Exit 2 off of Mass. Pike (I-90). Take Rt. 20 East. Call for further directions. **Facilities:** 13 indoor box stalls, 3 paddocks, 40 acres fenced pasture, 1/3-mile race track, open ring, trail rides. Located at "Eastover" resort facility. Call for reservation. **Rates:** $15 per night. **Accommodations:** At "Eastover" or motels 2 miles away.

LUNENBURG
Pine Fall Farm **Phone: 508-582-7748**
Tammy McAlpine **Evenings: 508-582-7601**
271 Elmwood Road [01462] Directions: Located off of Rt. 2A. Call for directions. **Facilities:** 24 indoor stalls, 60' x 120' indoor arena, fenced grass paddocks, 70' x 140' outdoor arena. English & Western training for horses & riders. Also horse sales. Call for reservation. **Rates:** $20 per night. **Accommodations:** Motels 15 miles away.

NEW BRAINTREE
Ash Lane Farm **Phone: 508-867-9927**
Mary Kay Newton
Havens Road [01531-0192] Directions: Sturbridge Exit off I-90 , Rte. 20 east, left onto Rte. 49, left onto Rte. 9, right onto Rte. 67, left at Reeds Country Store. Second farm on right. **Facilities:** 4 indoor 12' x 12' stalls, 2- and 3-acre fields, indoor arena, outdoor sand ring, feed/hay, trailer parking available. Arabian breeding farm, foals for sale. Arabian stallion standing-at-stud: "Ganesh." **Rates:** $25 per night. **Accommodations:** Bed & breakfast on premises, $40 per person, $75 per couple. Motels available near Sturbridge Village.

ALL OF OUR STABLES REQUIRE CURRENT NEG. COGGINS, CURRENT HEALTH PAPERS, & OWNERSHIP PAPERS.

NORTH ANDOVER
Andover Riding Academy **Phone: 508-683-6552**
Frank Fiore **Evenings: 508-683-9387**
16 Berry Street, Rt. 114 [01845]
Directions: Centrally located near I-93, I-495, & I-95. Call for directions.
Facilities: 80 indoor stalls at a complete horse facility. Feed/hay & trailer parking available. Call for reservation. **Rates:** $20 per night.
Accommodations: Candlelight Motel within 1 mile of stable.

PEPPERELL
Twin Pine Farm **Phone: 508-433-5252**
Cathy & Toby Tyler
34 Jewett Street [01463] Directions: 1 mile from Rt. 113 & Rt. 111. Call for directions.
Facilities: 25 indoor stalls, 6 outside paddocks, 70' x 200' indoor arena, 150' x 225' outdoor arena, wash racks, riding trails, jumping arena. Training for horses & riders in English, Western, jumping, dressage, & Western reining. Overnight boarding in emergencies only. **Rates:** $25 per night. **Accommodations:** Motels 8 miles from stable.

PLYMPTON
Southfield Farm **Phone: 781-582-2710**
Joanne Heath
11 Oak Street [02367] Directions: Exit 10 off of Rt. 3 to Rt. 106. Call for directions. **Facilities:** 20 indoor box stalls, 4 turnout areas on 8.5 acres, outside riding ring, round pen, heated tack room, riding trails. Training and showing of horses and riders at all levels. Morgan breeder. Call for reservation. **Rates:** $15 per night. **Accommodations:** Motels 4 miles from stable.

PROVINCETOWN
Bayberry Hollow Farm **Phone: 508-487-6584**
Chris Lorenz
P.O. Box 1427 [02657] Directions: Rt. 6A east and take 3rd Provincetown exit on Shankpanter Road; take right onto Bradford St.; go 1/2 mile & take right onto West Vine St. extension. See Farm sign. **Facilities:** 6 indoor straight stalls, 4 paddocks, trailer parking available. Massage therapy available. Hay provided. Pony rides. Call for reservation. **Rates:** $15 per night; $75 per week. **Accommodations:** Motels nearby.

REHOBOTH
Gilbert's Tree Farm Bed & Breakfast **Phone: 508-252-6416**
Jeanne D. Gilbert
30 Spring Street [02769] Directions: Call for directions. Located 3.5 miles from Rte. 195.
Facilities: 5 indoor box stalls, 50' x 50' pasture, feed/hay, trailer parking available. Riding trails through 100 acres. Jeanne is licensed riding instructor. **Rates:** $15 per night.
Accommodations: Non-smoking bed & breakfast in 150-year-old home with in-ground pool and full breakfast. Also a secluded cabin on premises.

ALL OF OUR STABLES REQUIRE CURRENT NEG. COGGINS, CURRENT HEALTH PAPERS, & OWNERSHIP PAPERS.

REVERE

Revere-Saugus Riding Academy **Phone: 617-322-7788**
Mary Ward **617-324-1594**
122 Morris Street [02151] Directions: Located right off of Rt. 1. Call for directions. **Facilities:** 15 indoor box stalls, 4 turnout paddocks, large turnout area, outside hunt course, riding trails, feed/hay & trailer parking. Training & riding lessons and sales of horses. Call for reservation. **Rates:** $15 per night; $100 per week with group discount. **Accommodations:** Many motels on Rt. 1 within 1 mile.

ROCHESTER

Bowen Lane Stable **Phone: 508-763-1741**
Joseph, Diane & Curry DeLowery
EXPERT FARRIER SERVICES AVAILABLE -- SEE ARTICLE ON PAGE 332
68 Bowen's Lane [02770] Directions: Rochester exit off 195 onto Route 105. Bowen Lane on left 1/2 miles from Rochester Town Hall/Plum Corner. **Facilities:** 20 full size indoor stalls. Feed & hay available. Trailer parking available. 2 - 600' paddocks; smaller areas will be available. Indoor arena, outdoor dressage area. Acres of trail riding available plus lots of woodlands and corn fields. **Rates:** $16.95 per night. **Accommodations:** 5 miles away in Marion, Ma.

RUTLAND

Holiday Acres Equestrian Center **Phone: 508-886-6896**
Deborah & Clifton Hunt
331 Main Street [01543] Directions: Located on Rt. 122A. Call for directions. **Facilities:** 6 indoor stalls, nine - 1/8-acre paddocks, boarded riding ring, indoor riding arena. No stallions. Hunter shows held at Center. Call ahead to check availability. **Rates:** $15 per night; $75 per week. **Accommodations:** Motels in Worcester, 20 minutes away.

WESTFIELD

AJ Stables **Phone: 413-562-5974**
Tammy & Irene Lowe
1040 E. Mountain Road [01086] Directions: Exit 3 off of Mass. Turnpike (I-90). Also easy access off of I-91. Call for directions. **Facilities:** 10 indoor box stalls, large rodeo riding arena, pasture, paddocks. Blacksmith on premises. Training and breaking of horses at all levels. Trains racehorses. Also, horse transportation available serving New England and New York. Call for information and reservations. **Rates:** $25 per night. **Accommodations:** Motels 2 miles from stable.

NOTES AND REMINDERS

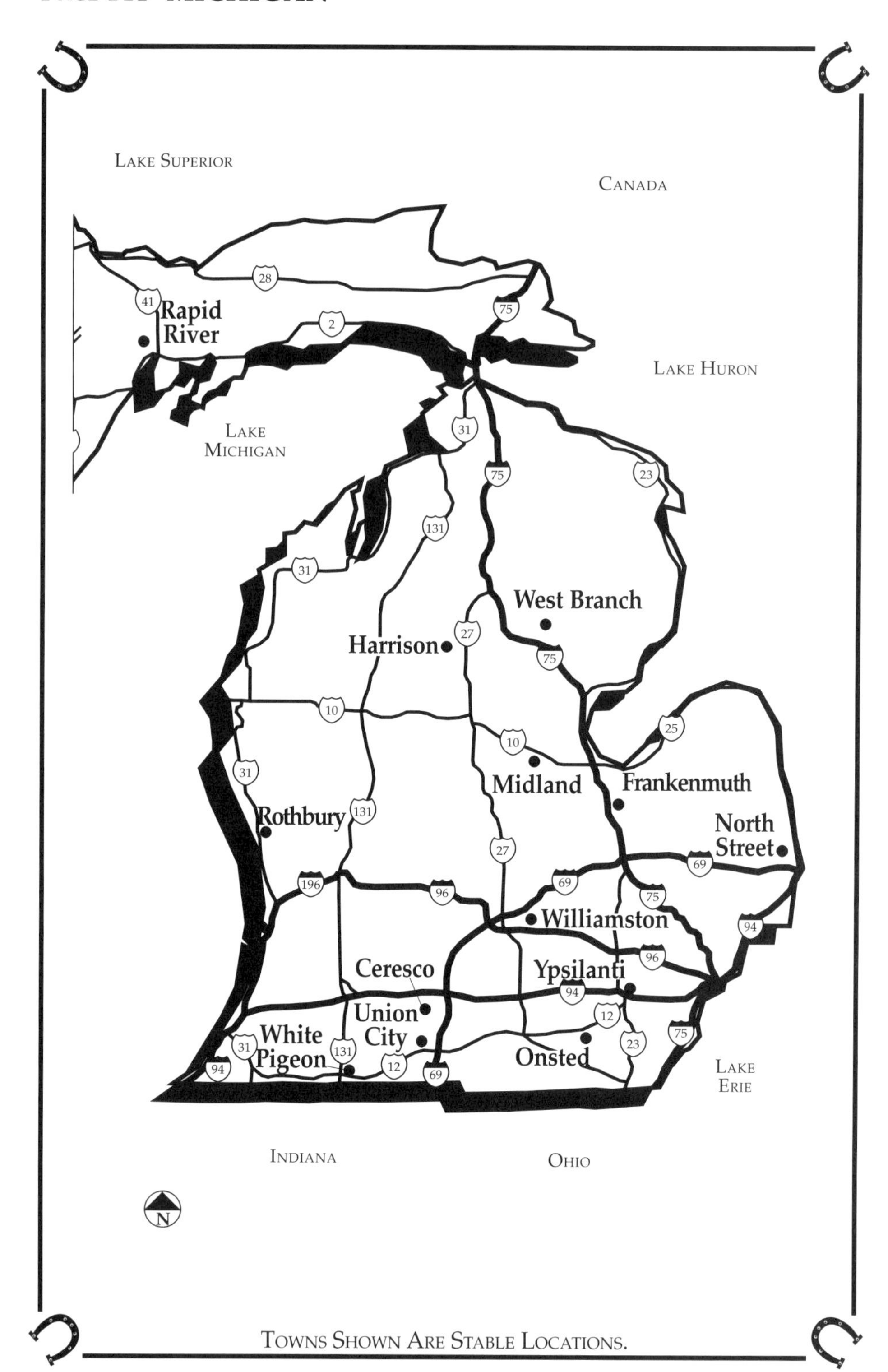
Lake Superior
Canada
Rapid River
Lake Huron
Lake Michigan
West Branch
Harrison
Midland
Frankenmuth
Rothbury
North Street
Williamston
Ceresco
Ypsilanti
Union City
White Pigeon
City
Onsted
Lake Erie
Indiana
Ohio
N
28
41
2
75
31
23
131
27
10
25
69
196
96
94
12
Towns Shown Are Stable Locations.

ALL OF OUR STABLES REQUIRE CURRENT NEG. COGGINS, CURRENT HEALTH PAPERS, & OWNERSHIP PAPERS.

CERESCO
Harper Creek Stables **Phone: 616-979-1554**
Deenna Hamilton
11432 8 1/2 Mile Road [49033] Directions: Exit 100 off of I-94. Call for directions. **Facilities:** 45 indoor box stalls, 3 outdoor paddocks, 60' x 120' indoor arena, 12 acres of fenced pasture at this boarding facility. Call for reservation. **Rates:** $20 per night. **Accommodations:** Comfort Inn 4 miles.

FRANKENMUTH
Marigold Stable **Phone: 517-652-8761**
Teri Cox
7670 E. Curtis Road [48734] Directions: Frankenmuth/Birch Run Exit off of I-75. About 7.5 miles on Dixie Hwy. Call for directions. **Facilities:** 65 indoor & outdoor stalls, paddock areas, all levels of lessons in all seats. Feed/hay at extra charge, trailer parking in summer but limited in winter. Reservations required. **Rates:** $10 per night. **Accommodations:** Many motels within 5 miles of stable.

HARRISON
Horse'n Around Tack Shop & Equestrian Center **Phone: 517-539-8500**
Joyce Hamsher
8400 N. Bass Lake Road [48625] Directions: Go from US 27 to Old 27: 6 miles north of Harrison on Old 27; left on Long Lake Road; go 1 mile to Bass Lake; turn right. Stable is on right side of road. **Facilities:** 22 indoor stalls, 7 pastures & lots, 60' x 120' indoor arena, 110' x 200' outdoor arena. Tack shop & feed store on premises, trailer parking. Mobile tack shop goes to rodeos, shows, etc. Western & English training for horses & riders available. **Rates:** $15 for indoor stall, $10 for outdoor per night. **Accommodations:** Deer Trail Motel 6 miles from Center & rustic camping on site if self-contained.

MIDLAND
Triple L Training Stable **Phone: 517-835-4673**
Linda Town & Lanny Warner
184 North 9 Mile Road [48640] Directions: Bay City Exit off of I-75. Call for directions. **Facilities:** 14 indoor stalls, indoor arena, 3 large outdoor runs, 1/2-mile conditioning track, large outdoor paddock, large trucks welcome. Training of quarter horses & thoroughbreds plus barrel racing, pleasure riding, and sale conditioning. Call for reservation. **Rates:** $15 per night. **Accommodations:** Many motels nearby.

NORTHSTREET
Rio Vista **Phone: 810-987-4840**
Doug Martinez
3945 Vincent Road [48049] Directions: From M-69, exit at Wadhams Road. Call for further directions. **Facilities:** 6 indoor stalls, turnout available, indoor & outdoor arenas, feed/hay included, trailer parking. 2 days notice if possible. Beautiful riding facilities. **Rates:** $10 per night. **Accommodations:** Several motels within 5 miles.

ALL OF OUR STABLES REQUIRE CURRENT NEG. COGGINS, CURRENT HEALTH PAPERS, & OWNERSHIP PAPERS.

ONSTED
Green Hills Stables **Phone: 517-467-7614**
J. Scott Schultz
10500 Stephenson Road [49265] Directions: Located within 5 miles of US 12, US 223, & M-50. Call for further directions. **Facilities:** 6 indoor 10' x12' stalls, 2 outbarns with paddocks, 4 pasture/turnout areas, 150' x 300' lighted arena used for roping, nature trails for riding, 12% sweet feed plus hay available and lots of trailer parking. **Rates:** $20 per night. Call for weekly rate. **Accommodations:** Located in the heart of the Irish hills with many motels and things to do. Call for more information.

RAPID RIVER
Spruce Winds Farm **Phone: 906-474-9701**
Deb and Craig Olsen, Pat and GaryDeGrave **906-474-6520**
9811 Y.25 Lane (49878)
Directions: Call for Reservations and Directions **Facilities:** 6 tie stalls, 3 indoor 10x12 Box stalls. Feed/hay, trailer parking available. 1.5 acre turnout, 1 150′x150′ pen. Adjacent to Hiawatha Forest and numerous trails. **Rates:** $10 per night. **Accommodations:** Several small motels within 6 miles.

ROTHBURY
Double JJ Resort Ranch **Phone: 231-894-4444**
Bob Lipsitz **800-DoubleJ**
6886 Water Road [49452] Directions: Located 20 miles north of Muskegon. Exit off U.S. 31 at the Winston Road, Rothbury Exit. Turn east on Winston Road for .5 miles, north on Water Road. Ranch office is one mile on right. **Facilities:** 10 indoor 12' x 12' stalls, many 12' x 30' turnout pastures on 1000 acres of riding trails, feed/hay, trailer parking available. Dude ranch, full-service resort, championship golf course, weekly rodeo, steak & breakfast rides. **Rates:** $10 per night; $25 minimum. **Accommodations:** RV Park, Hotel, Cabins & Condominiums on premises. Call for more information.

UNION
Camp Bellowood Recreation World **Phone: 616-641-7792**
Dan Galbraith
14260 East US 12 [49130] Directions: Located off of Indiana Toll Road (I-80/90) at Elkhart or Bristol exits. Call for directions. **Facilities:** 20 indoor stalls, indoor and outdoor arenas, 300 acres of riding trails, feed/hay & trailer parking available. Campground facilities. 24-48 hours notice required. **Rates:** Negotiable. **Accommodations:** Several motels on US 12, eight miles away.

ALL OF OUR STABLES REQUIRE CURRENT NEG. COGGINS, CURRENT HEALTH PAPERS, & OWNERSHIP PAPERS.

WHITE PIGEON
Circle J Farm **Phone: 616-483-7898**
William & Laura Jungjohan **1-888-225-9872**
9620 Barker Road [49099] Directions: I-80/90 (Indiana Toll Road), Exit 101, head north on Ind. 15, turns into Michigan 103 approx. 4 miles. Look for Barker Road sign, turn right, fourth place on left; look for fencing. **Facilities:** 9 - 12' x 12' stalls, two 12-acre pasture lots, 100' x 200' riding arena, two 50' x 100' pens, feed/hay and trailer parking available. Vet/farrier on call. Reservations preferred. **Rates:** $6 per night; $10 per night with feed. **Accommodations:** Plaza Motel, Tower Motel, Maplecrest Motel in White Pigeon, about 6 miles away.

WILLIAMSTON
Silohetti Manor **Phone: 517-655-3561**
Gene Schneider
3725 Norris Road [48895] Directions: Located 8 miles from I-96. Call for directions. **Facilities:** 33 indoor 12' x 12' box stalls, 200' x 72' indoor arena, 250' x 100' outdoor arena, feed/hay & trailer parking. Training of hunt seat, Western pleasure, dressage, & jumping done at stable. Call for reservation. **Rates:** $20 per night. **Accommodations:** The Williamston Inn 1/4 mile from stable.

WEST BRANCH
Log Haven Bed, Breakfast and Barn **Phone: 989-685-3527**
Gail & Paul Gotter **E-mail: gotter@m33access.com**
1550 McGregor Rd (48661) Directions: I-75 exit 212 into West Branch. Second traffic light, right onto Fairview Rd. North 14 miles, then 1 3/4 miles east on McGregor Rd. Also accessible from M-33 Rose City. **Facilities:** 3 12x12 indoor box stalls, 1 12x12 outdoor stall, 48'x36' turnout, 1 and 1/2 acre pasture, feed/hay available, trailer parking, adjacent to State and Federal Forest with miles of riding trails. **Rates:** $15 per night. **Accommodations:** B&B on site-Call for rates. Motels in West Branch(15 miles) and Rose City(8miles)

YPSILANTI
Sandy Hills Farm **Phone: 313-485-3939**
Tom Rudnicki
9101 Cherry Hill [48198] Directions: Take Geddes Exit off of US 23. Call for easy directions. **Facilities:** 50 indoor stalls, large indoor & outdoor arenas, pastures, monthly boarding & training available, feed/hay included & trailer parking available. Breeding program & horses for sale. Standing-at-stud: "Cabin Bar Command," AQHA. Tack shop on premises. **Rates:** $20 per night; weekly rate negotiable. **Accommodations:** Knights Inn & Fairfield Inn 6 miles from stable.

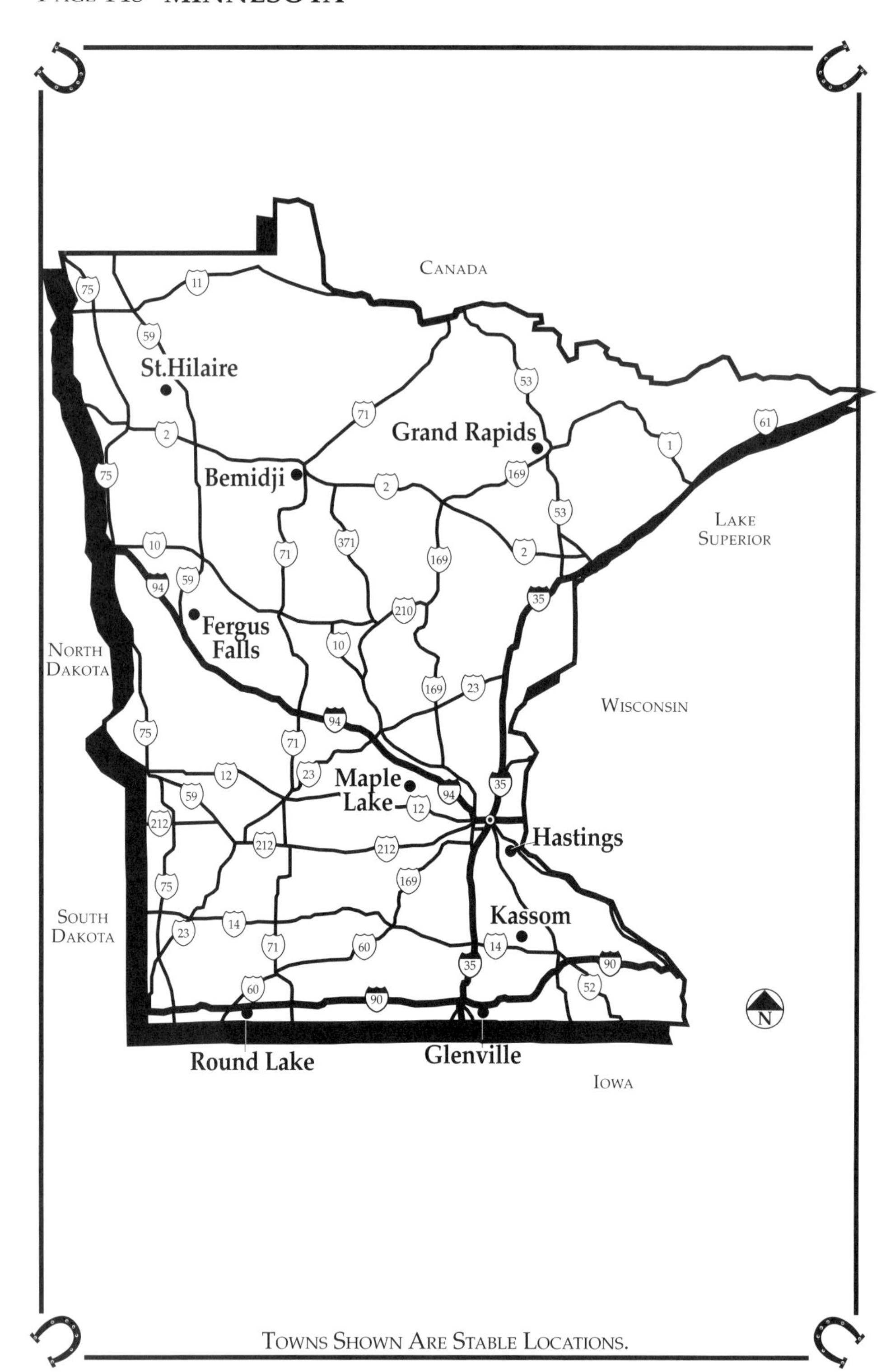

TOWNS SHOWN ARE STABLE LOCATIONS.

ALL OF OUR STABLES REQUIRE CURRENT NEG. COGGINS, CURRENT HEALTH PAPERS, & OWNERSHIP PAPERS.

BEMIDJI
Maple Ridge Farm & Stable **Phone: 218-751-8621**
Daniel & Peggy Nickerson
3000 Adelia Drive SE [56601] Directions: 6 miles off of Hwy 2. Call for directions. **Facilities:** Farm is 80 acres with 12 indoor box stalls, 3 run-in sheds with separate paddocks, 100' x 200' outdoor arena, 6 separate paddocks, large pasture, locked tack room, full lounge, feed/hay & trailer parking available. Arabian stallion standing-at-stud: "Khalim Pasha." Saddlebred stallion standing-at-stud: "Carnival's Loaded Dice." Training of horses & riders at all levels. Call for reservation. **Rates:** $20 per night. **Accommodations:** Edgewater Motel.

FERGUS FALLS
Corriente Ranch **Phone: 218-736-5134**
Brad & Cheri Brause
Rt. 4, Box 109 [56537] Directions: Call for directions. **Facilities:** 9 indoor stalls, two 12' x 14' paddocks, three 30' x 80' outdoor paddocks. Will also take cattle. Camper hook-up. Please call for reservation. **Rates:** $15 per night; $75 per week. **Accommodations:** Super 8 is 11 miles from ranch.

GLENVILLE
Margaret & Gaylen Schewe **Phone: 507-448-3786**
Rt. 2, Box 103 [56036] Directions: 10 miles south of Albert Lea, MN on 35N, Exit 2. Take right off ramp, go 1/8 mile. Take left on County 18 & go 1.5 miles south. Take right on County 83 & go 1 mile. Take right into driveway with red gate. **Facilities:** 3 indoor stalls, 30' x 60' pasture with 8' gates, feed/hay available. Will also take cattle. Call for reservation. Water & electric hook-up for campers. **Rates:** $15 per night. **Accommodations:** Albert Lea on I-90.

GRAND RAPIDS
K & K Stable, Inc. **Phone: 218-245-3814**
Kathy Horn
1901 Scenic Drive [55744] Directions: From US Hwy 2 in Grand Rapids: Go north on State Hwy 38 for 6 miles; go to City Rd 49 & take right; road turns into City 59. Stable is 10 miles down on left. Look for sign. **Facilities:** 6 - 8 indoor box stalls, indoor riding arena, 2 pens with lean-tos which could accommodate 3-6 horses, full-size outdoor arena, pasture/turnout area, lots of riding trails, feed/hay & trailer parking. Camping nearby and on premises. Horses must be wormed within a month. **Rates:** $15 box stall, $10 pen, per night. **Accommodations:** Rustic sleeping lodge available on grounds. 20 minutes from Grand Rapids where there are many motels.

ALL OF OUR STABLES REQUIRE CURRENT NEG. COGGINS, CURRENT HEALTH PAPERS, & OWNERSHIP PAPERS.

HASTINGS
Schoen's Chimney Rock Farm **Phone: 612-438-3121**
Harry P. Schoen **Evenings & Weekends: 437-4698**
21637 Joan Avenue [55033] Directions: 8 miles south of Hastings and 2 miles west of US Hwy 61. Call for further directions. **Facilities:** 11 indoor box stalls, 2 indoor tie stalls, indoor heated barn, pasture/turnout area, feed/hay and trailer parking available. Water & electric camper hook-up. **Rates:** $15 per night. **Accommodations:** Bed & Breakfast on premises.

KASSON
Turn Crest Stables **Phone: 507-634-4474**
Gene & Vicki Holst
Rt. 1, Box 47 [55944] Directions: 10 miles west of Rochester off Hwy 14. Call for directions. **Facilities:** 24 indoor stalls, 14-acre pasture, 4 acres of paddocks, 56' x 152' indoor arena, 90' x 150' outdoor arena. Riding lessons for English hunt seat, jumping, and Western pleasure for all levels. Also horses for sale. Call for reservation. **Rates:** $20 per night. **Accommodations:** Howard Johnson's & Holiday Inn in Rochester, 10 miles from stable.

MAPLE LAKE
Freedom Stables Inc. **Phone: 320-963-3351**
Kevin & Laura Holen **Fax: 320-963-6616**
2868 90th Street NW [55358] Directions: Exit 183 off of I-94: go south on Co. Rd. 8 for 7.5 miles. Turn east on County Rd. 106 (90th St.) for 2.5 miles. Stable is on left side. **Facilities:** 72 indoor 12' x 11' stalls, 13 paddocks & pastures, 70' x 200' heated indoor arena, 100' x 200' outdoor arena. Hay, grain and bulk bedding available. Trailer friendly parking including semi-truck pull through and loading. 24 hour emergency vet and farriers service available locally. Emergency pick-up can be arranged. Hot and cold water wash stalls, full lounge w/vending. On site camping. Reservations requested, short notice and late night/emergency stopovers OK. Owners reside on premises. Pets welcome. **Rates:** $15 per night; $90 per week. **Accommodations:** Many nearby motels and restaurants. Owners will assist in finding an appropriate facility. Our facility is a favorite stop for national equine transport companies.

ROUND LAKE
Painted Prairie Farm **Phone: 507-945-8934**
Ralph & Virginia Schenck
RR 1, Box 105 [56167] Directions: From I-90: Exit 50 (Hwy 264), Round Lake Exit; go south 4.5 miles. Driveway to farm on left - arch with wagon wheels & brick pillars. **Facilities:** 25 indoor stalls ranging in size from 10' x 10' up to 10' x 20'. Some are wood, some pipe, some with turnout paddocks attached. Paddocks, 30' x 120' runs, & grass pasture available for large groups. Trailer parking & large rigs OK. Alfalfa/grass and good water. Electric camper hook-up available at $5 per night. This farm breeds Paint Horses. Standing-at-stud: "Barlink Tuff Spade" and "Sonny Jet Storm." **Rates:** $15 for stall, $7.50 for paddock per night including feed. **Accommodations:** Prairie House Bed & Breakfast on premises plus motels in Worthington, 10 miles from farm.

ALL OF OUR STABLES REQUIRE CURRENT NEG. COGGINS, CURRENT HEALTH PAPERS, & OWNERSHIP PAPERS.

ST. HILAIRE

Roger Johnsrud **Phone: 218-681-8069**

Rt. 1, Box 28A [56754] Directions: 5 miles from St. Hilaire on Hwy 59. Call for directions. **Facilities:** 4 indoor stalls on a 240-acre farm. 5 & 40 acres of pasture/turnout, 25' x 25' outside corral, working arena. Breeding, training & breaking of horses done at stable. Thoroughbred stallion standing-at-stud: "Wild Charamda." Quarter horse stallion standing-at-stud: "Black Horse." Call for reservation. **Rates:** $15 per night. **Accommodations:** Best Western located 12 miles from stable.

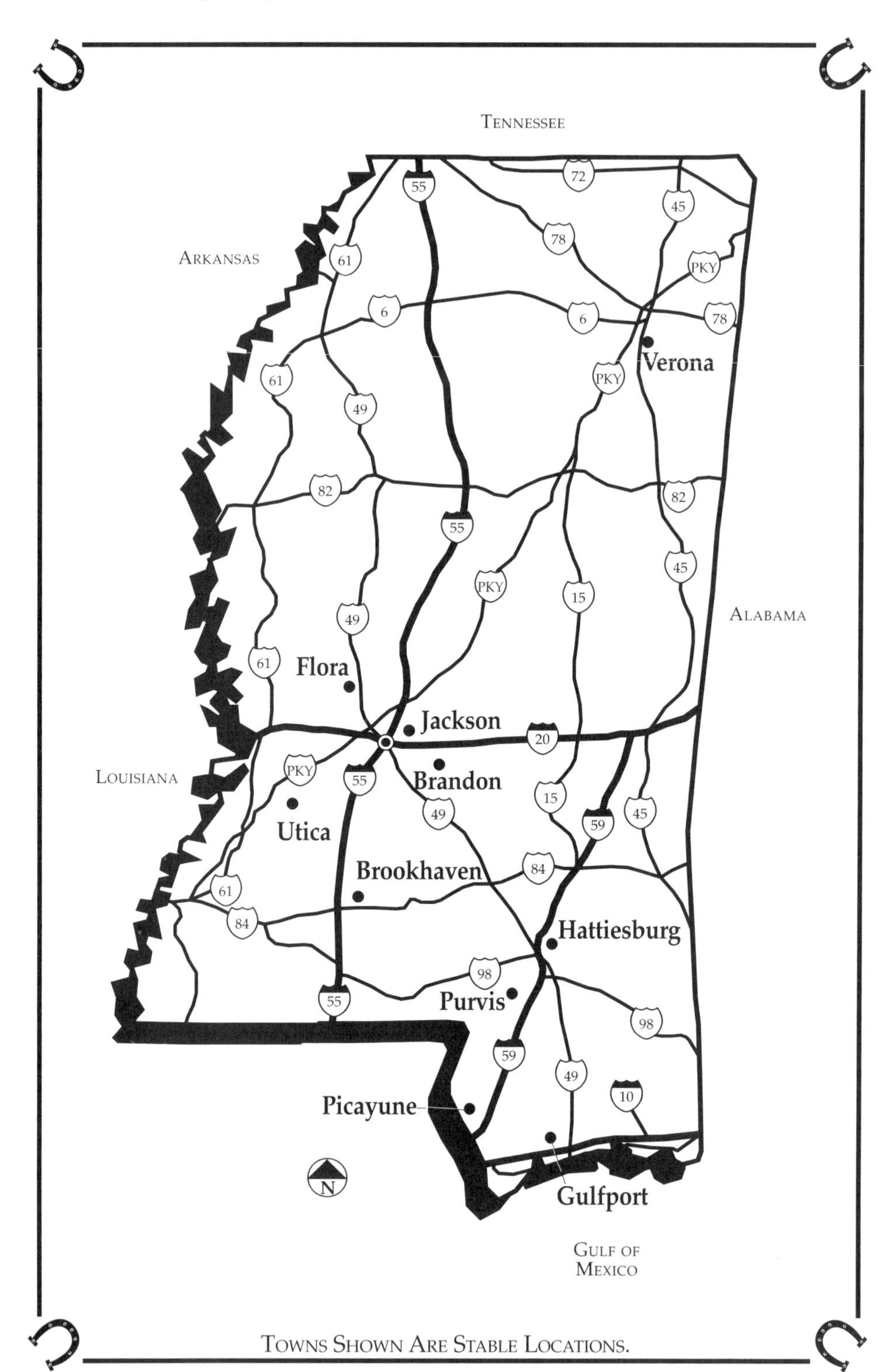

TOWNS SHOWN ARE STABLE LOCATIONS.

ALL OF OUR STABLES REQUIRE CURRENT NEG. COGGINS, CURRENT HEALTH PAPERS, & OWNERSHIP PAPERS.

BRANDON

Hilltop Painted Acres **Phone: 601-825-2094**
John & Barbara Blough **601-940-0246**
607 North Street [39042] Directions: 12 miles east of Jackson right off of I-20 Exit 59 from East go downtown Brandon to monument turn North at courthose on North St. Follow across RR to farm. **Facilities:** 20 indoor stalls, 25 acres of boarded pasture, 75 acres rental pastures, walker, pipe arena, 200+ acres for riding trails. Breeding, training, & farrier available; veterinarian available. Pens available for rodeo cattle and parking for large rigs. Negative coggins & health papers. Camper hook-ups. Call for reservations. **Rates:** $15 including feeding per night; $60 per week. **Accommodations:** Rustic cabin with barn, $45 per night. Days Inn in Brandon, 3 miles from stable; Comfort Inn in Pearl, 10 miles away.

BROOKHAVEN

Herndon Quarterhorses **Phone: 601-833-9247**
Dr. Natalie Herndon **Barn: 833-9214**
201 West Meadowbrook Dr. [39601] **Office: 833-1977**
Directions: Exit 42 off of I-55. Call for directions. **Facilities:** 32 indoor stalls in insulated barns, 30 outdoor paddocks with sheds, 72 acres of pasture, 100' x 200' indoor arena, round pen, hot walker. 110-acre facility with access for big rigs. Breeding & training center for champion horses for cutting, halter, & all around performance. Two stallions standing-at-stud with World Champion AQHA bloodlines: "Top Impressive" and "Shadows Impressed." Home of Billy Herndon & Joleen Struts, AQHA Performance Champions, World Champions & also "Chief Investor," Natalie Herndon. Please call 24 hours ahead. **Rates:** $20 per night. **Accommodations:** Best Western, Ramada Inn, Hampton Inn, & Days Inn 1 mile from stable.

FLORA

Winterview Farms **Phone: 601-879-3468**
Paty Nail
140 Spring Road [39071] Directions: I-55 to Exit 108, west on Hwy 463; go 3 miles & take left on Robinson Spring Road; go 4.5 miles & turn right on Robinson Spring Road Extension; go 1 mile & turn left on Spring Road. **Facilities:** 10 indoor stalls, 3 1-acre paddocks, round pen, cross-country course, feed/hay & trailer parking available. Fox hunting and polo ponies available. **Rates:** $20 per night. **Accommodations:** Comfort Inn at Ridgeland, 10 miles from stable.

GULFPORT

Harrison County Fairgrounds **Phone: 601-832-0080**
Mike McMillan
15321 Countfarm Road [39503] Directions: I-10 west to Exit 28, north 7.5 miles. **Facilities:** 96 - 9' x 9' wood stalls, trailer parking available. **Rates:** $10 per night. **Accommodations:** Holiday Inn and Best Western at Gulfport, 12 miles from fairgrounds on Hwy 49.

ALL OF OUR STABLES REQUIRE CURRENT NEG. COGGINS, CURRENT HEALTH PAPERS, & OWNERSHIP PAPERS.

GULFPORT

Shady Oaks Stables **Phone: 601-832-0435**

Ronnie Bourgeois

12726 Wolf River Road [39503] Directions: From I-10: Take Exit 28N and go 2 miles to 4-way stop; take left; go 1/4 mile and take right; stable is 2.25 miles on right. **Facilities:** 8 indoor stalls, 20 1-acre fields, 60' x 125' working pen, feed/hay & trailer parking available. 2 full camper hook-ups @ $13.95 per night. **Rates:** $15 per night. **Accommodations:** Many motels 5 miles from stable.

HATTIESBURG

Boots & Saddles Stables **Phone: 601-583-6726**

Larry Mills

2782 Old Richton Road [39465] Directions: Call for directions. **Facilities:** 12 indoor stalls, lighted arena, show equipment, pasture/turnout, trailer parking and feed/hay available. Call for reservations. **Rates:** $7 per night; weekly rate available. **Accommodations:** Motel 6 & Holiday Inn in Hattiesburg, 8 miles from stable.

JACKSON

High Point Farm **Phone: 601-362-5345**

Marcie Lockett

1235 Stigger Road [39209] Directions: 9 miles from I-220; 15 miles from I-20 & 10 miles from I-55. Call for directions. **Facilities:** 26 indoor stalls, 4 outside paddocks, 2 outside rings, jump course, 24-hour security. Trains hunter/jumpers at all levels and sells horses. Call for reservations. **Rates:** $20 per night. **Accommodations:** Motels 10 miles from stable.

Mississippi State Fairgrounds **Phone: 601-961-4000**

Tommy Strickland, assistant director

1207 Mississippi Street [39202] Directions: High Street exit off I-55, 2 miles north of I-20 and I-55. **Facilities:** 750 indoor 10' x 10' stalls, feed/hay & trailer parking available. Fairground hosts numerous horse activities and events, over 40 per year. **Rates:** $5 per night. **Accommodations:** Ramada Inn, Wilson Inn, Red Roof Inn within 30 minutes.

PICAYUNE

Circle M Riding Stables **Phone: 601-798-7677**

David Megehee

24 Circle M Lane (off Liberty) [39466] Directions: Exit 6 off of I-59. In Picayune take Hwy 43 North for 3 miles. Sign on right. **Facilities:** 2 indoor stalls, feed/hay & trailer parking. Overnight boarding in emergency situations only. Please call first. **Rates:** $18 per night. **Accommodations:** Heritage Inn & Majestic Inn 3 miles away on I-59.

ALL OF OUR STABLES REQUIRE CURRENT NEG. COGGINS, CURRENT HEALTH PAPERS, & OWNERSHIP PAPERS.

PURVIS

South Mississippi Veterinarian Complex **Phone: 601-794-8884**
William Ewell
5396 US Hwy 11 [39475] Directions: I-59 to Purvis exit; go toward Purvis on 589 to Hwy 11N; go 2 miles, complex on right. Sign out front, well-lit. **Facilities:** 10 to 20 stalls, 35 paddocks, 3-4 acres of pasture/turnout. Veterinarian on site 24 hours a day. Please call ahead. **Rates:** $15 per night; $70 per week. **Accommodations:** Hampton Inn & Days Inn in Hattiesburg, 7 minutes away.

UTICA

Big Sand Campground, Inc. **Phone: 601-885-8068**
David & Sherril Strong, Darrell & Loretta Baker
3412 Reedtown Road (39175) Directions: From Vicksburg: I-20 take exit 1-c, go straightfor 19.2 miles to 4 way stop.Turn rightand, go two miles and turn right onto Ross Road. Go 1/2 m to the second drive on the left. From Jackson: Natchez Trace Parkway, take second Utica exit (right hand turn). Go to stop sign, and turn left. Go one mile to 4 way stop, and turn left. Go 2 miles, and turn left onto Ross Road. Go 1/2 mile to the second drive on the left. From Crystal Springs: I-55, take exit 72 (Hwy 27) to Utica. At stop sign, go straight to downtown Utica and 3 way stop. Turn left, go 9.2 miles to 4 way stop, then turn left onto Ross Road. From Port Gibson: Natchez Trace Parkway to Utica exit (first exit past Rock Springs, approx. 5-7 miles) Take exit to first stop sign, and turn left. Go to 4 way stop and turn left onto Ross Road. Go 1/2 mile to the second road on left. **Facilities:** 6 12x12, 12x14 indoor stables, 150x75 turnout, trailer parking available, RV hookup, Pipe Arena. Borders Natchez Trace National Scenic Trail (Horse-back & Hiking Only). Negative Coggins **Rates:** $10 per night. **Accommodations:** Bunkhouse in Barn with hot showers. Vicksburg, MS (19.2 miles) Fairfield Inn-Days Inn-EconoLodge.

VERONA/TUPELO

Lee County Agri-Center **Phone: 601-566-5600**
Frank Swanger, General Manager
Larry Schulz, Operations Manager
5395 Hwy 145 South [38879] Directions: Located 15 miles from Hwy 78. Follow signs on Hwy. 45 to the Agri-Center or call. **Facilities:** 110 permanent indoor 10' x 10' box stalls, 53 temporary show stalls, corrals available if notified in advance, 138' x 240' heated indoor arena, lockable stalls. Facility is on 100 acres and has 17 acres of parking. Hours: 7:30 A.M. - 4:30 P.M. Call in advance for stall reservation. 35 RV camper hook-up with water & electricity. RV dump station on grounds. $10 per day. $20 per hour for 200' x 95' outdoor arena. **Rates:** $10 per night. **Accommodations:** Town House Motel in Tupelo, 6 miles from stable.

Towns Shown Are Stable Locations.

ALL OF OUR STABLES REQUIRE CURRENT NEG. COGGINS, CURRENT HEALTH PAPERS, & OWNERSHIP PAPERS.

BOONVILLE

Littles Four Oaks Farm **Phone: 660-882-8048**
John & Kathy Little
22045 Boonville Rd [65233] Directions: I-70 to SR 87 (Exit 106), south on SR87 for 1.8 mi., left on Boonville Rd. (gravel). Stable 2.4 mi. on right. **Facilities:** 10 indoor 12' x 12' stalls/c paddocks, 10 acres of pasture/turnout, round pen, trailer parking and campers welcome. Pasture riding on 25 acres. Call for reservations. **Rates:** $20/per horse; $125/week. **Accommodations:** B&B on premises; Motels at Boonville, 10 miles from stable.

BOURBON

Meramec Farm B&B **Phone: 573-732-4765**
Carol Springer, David Curtis **or: 573-732-3080**
E-mail: mfarmbnb@fidnet.com Web : www.wine-mo.com/meramec.html
208 Thickety Ford Road [65441] Directions: 1 hour west of St. Louis off I-44. Call for further directions. **Facilities:** 5 large indoor stalls, pasture/turnout, feed/hay & trailer parking available. Guided rides on farm trails and Forest Service trails. "Farm to farm" rides available with meals included. 5 different all-day tours available. **Rates:** $10 per night; $50 per week. **Accommodations:** B&B for 14 available on premises. Motel in Bourbon, 9 miles from farm.

BRANSON

South Branson Stables **Phone: 501-426-2473**
David Bird
Rt. 1, Box 226, Omaha, Arkansas [72662] Directions: Located just 20 minutes south of Branson on US Hwy 65. Call for directions. **Facilities:** 10 indoor 10' x 14' stalls, 4 acres of fenced pasture, feed/hay & trailer parking for any size trailer available. Preferably no stallions. Call for reservations. Many tourist activities and sites in Branson. **Rates:** $20 per night; discounted weekly rate. **Accommodations:** Big Oaks Motel in Omaha, 3 miles from stable.

CARTHAGE

Royalty Arena **Phone: 417-548-7722**
Mike Pickard **or 417-358-7711**
9895 Cork Lane [64836] Directions: I-44 & Country Rd. #10, Exit 22: Go 1/2 mile north. **Facilities:** 200 inside stalls, 10 outside pens, 2 small grass lots, large parking lot & 20 camper hook-ups. Shavings, feed/hay available. **Rates:** $10 per night. **Accommodations:** Days Inn in Carthage, 8 miles from arena; Tara Motel in Joplin, 6 miles from arena, Holiday Inn, Joplin Mo. $43.00/night.

CHESTERFIELD

J. M. Pierce Stables **Phone: 314-394-4733**
J. M. Pierce
2315 Baxter Road [63017] Directions: I-64 (Hwy 40) to Clarkson Road south to Baxter Road. Stable is 2 miles on right. **Facilities:** 3 indoor stalls, pasture/turnout, feed/hay & trailer parking available. **Rates:** $20 per night. **Accommodations:** Doubletree & Residence Inn both in Chesterfield, 2 miles from stable.

ALL OF OUR STABLES REQUIRE CURRENT NEG. COGGINS, CURRENT HEALTH PAPERS, & OWNERSHIP PAPERS.

CHILLICOTHE

Indian Creek Equine Center **Phone: 660-646-6227**
William "Bill" Hinkebein
Route 4 [64601] Directions: Located 12 miles from town, north on Hwy 65 to Rte. 190 W, west 9 miles, north 3.5 miles of gravel road. **Facilities:** 6 indoor 10' x 10' stalls, 24 stalls at nearby fairgrounds, fenced pastures, feed/hay, trailer parking available. Fox trotters, stallion standing-at-stud: "Hickory's Country Gold," top in nation. North American Trail Ride Conference Competitive Trail Ride. Horses for sale. **Rates:** $20 per night; weekly rates available. **Accommodations:** Motels 10 miles away in Chillicothe.

COLUMBIA

Midway Expo Center **Phone: 573-445-8338**
C.W. Adams
I 70 Hwy 40 [65202] Directions: Exit 121 off of I-70. 2 hours east of Kansas City & 2 hours west of St. Louis. **Facilities:** 420 indoor 10' x 10' stalls, 4 indoor arenas (100' x 200' and 50' x 80'), outdoor arena 300' x 150', outdoor riding area. Horse shows & sales held here. Travel plaza with gas & diesel, tire & auto repair, 24-hr restaurant & convenience store. **Rates:** $20 per night. **Accommodations:** Budget Inn on site as well as electric & water hook-ups for RVs - no sewer.

CUBA

Blue Moon RV Park & Horse Motel **Phone: 573-885-3622**
Liz Barton **Toll Free: 877-440-CAMP**
355 Highway F [65453] **Fax: 573-885-3752**
Directions: I-44 to exit 203 N on Hwy F, 1500 ft. Paved access. **Facilities:** 6 stalls in covered barn, 1 stall open. 3- 12′ x 10′ stalls, 3- 12′ x 12′ stalls. Feed and hay available. Round pen 50′, arena 60′ x 84′. Blue Moon Equestrian School with lessons and training. All types of livestock accepted. **Rates:** $15 per night, $75 per week. **Accommodations:** Motel 8 in Cuba (5miles), Comfort Inn in St. James (8miles)
.

KANSAS CITY

Benjamin's Ranch **Phone: 800-43-RANCH**
Bob Faulkner, operator **800-437-2624**
6401 E. 87th Street, [64138] Directions: Located in S.E. Kansas City, one block east of I-435 on 8th Street. Call for directions. **Facilities:** 27-12′ x 12′ indoor stalls with 20′ x 20′ turnout, pens, arena, isolation area. Vet. available & security 24 hours a day. Complete horse care available. **Rates:** $20 per night. **Accommodations:** Day's Inn across from stable (816-765-4331)

ALL OF OUR STABLES REQUIRE CURRENT NEG. COGGINS, CURRENT HEALTH PAPERS, & OWNERSHIP PAPERS.

KANSAS CITY / OAK GROVE AREA
Thunder Ridge Farm **Phone: 816-373-1777**
Deborah Lascuola, Gary Roe, manager
9224 South Campbell [64064] Directions: 8 mi. south of I-70. Take I-70. 20 minutes east of Kansas City MO. **Facilities:** 15 indoor 12′ x 12′ wood & panel (wire) stalls with 12′ x 15′ outdoor turnout, 4 stud stalls, 10-acre pasture with 2 ponds, round pen, outdoor arena, hay & trailer parking available. Two miles north of 1,800-acre trail park. Minimum 2-hour notice required. **Rates:** $15 per night. **Accommodations:** Scottish Inn at I-70 & BB Highway, 8 miles from farm.

KEARNEY
Over the Hill Ranch, Inc. **Phone: 816-628-5686**
Bill & Connie Green
P.O. Box 743 [64060] Directions: 5 miles from I-35. Call for directions. **Facilities:** Eight 10' x 20' indoor stalls with bedding, 60' x 200' indoor arena, feed/hay, camper parking with electrical hook-up, 24-hr security. Owner is professional farrier; vet and feed store close by. Jesse James hometown, historical James Farm 7 miles away. Watkins Mill & Smithville Lake nearby. Call for reservations. **Rates:** $15 per night. **Accommodations:** Bed & breakfast on premises with rooms in private guest house adjacent to barn.

MARSHFIELD
Cliff Hartman Farm **Phone: 417-859-2200**
Mary Chris Hartman
Route 2, Box 260F [65706] Directions: Northview Exit off of I-44. Call for directions. **Facilities:** 35 indoor pipe box stalls, indoor arena, round pen, 50 acres of pasture on a total of 450 acres. Farm has 4 superior quarter horse stallions. **Rates:** $15 per night. **Accommodations:** Motels 7 miles away.

MARSHFIELD
Stevens Farm Inn Horse & Rider **Phone: 417-859-6525**
Walter and Sharon Stevens **E-mail: pony1459@aol.com**
5484 State Hwy 00 [65706] **www.usipp.com/stevensfarm**
Directions: Less than .5 mile off I-44. From Exit 96, go south on B, east on 00, first place on left. **Facilities:** 8 indoor 8' x 10' box stalls, indoor arena, feed/hay and trailer parking available. Farm is in the Paint and Quarter horse business. Farm faces old historic Route 66. Call for reservations, especially for Saturday. **Rates:** $15 per night per house. **Accommodations:** Suite in barn overlooking arena accommodates 4; house accommodates 4. Small motel within 5 miles. Breakfast by request.

ALL OF OUR STABLES REQUIRE CURRENT NEG. COGGINS, CURRENT HEALTH PAPERS, & OWNERSHIP PAPERS.

MARTHASVILLE
Gramma's House Bed & Breakfast **Phone: 314-433-2675**
Judy & Jim Jones
1105 Hwy D [63357] Directions: From I-70 East, take Foristel Exit. From I-70 West, take Warrenton Exit. Also accessible from I-44. Call for directions. **Facilities:** Barn area with wood-fenced space outside barn with trough, 5 acres of fenced pasture/turnout area, feed/hay available with prior notice. **Rates:** $10 per night. **Accommodations:** Bed & Breakfast on premises: 150-year-old farmhouse with private & shared baths and full complimentary breakfast. Also a private cottage. Located in a beautiful and historic area. Overnight boarding only for B&B guests: $75/dbl. in house; $90/dbl. for private cottage.

MOSCOW MILLS
Shenandoah Stables **Phone: 314-356-9205**
David & Debra Young
116 Majestic Lane [63362] Directions: 10 miles north from I-70 or 60 miles south of Hwy 54 & 61 from Hanibal, MO. Right on Hwy 61. **Facilities:** 30 - 10' X 12' indoor wood stalls. Hay, feed and trailer parking available. 2 acre turnout lot. 100' X 200' indoor riding arena, 150' X 350' lighted outdoor arena. Restrooms, snack bar and phone. **Rates:** $15.00 per night $70.00 per week. **Accommodations:** Oak Grove Inn (Troy) 3 miles North of stable. 1-800-435-7144 (special rate if horse boarded with us).

PHILLIPSBURG
Oak Grove Farm **Phone: 417-589-2186**
Gary & Doris Anglehart
18296 Dallas County Line Rd. (65722) Directions: 6 1/2 miles northwest of !-44. Call for Directions **Facilities:** 4 10x12 indoor and 2 10x12 outdoor stables, 3 acre turnout, feed/hay, trailer parking available. Electric and water available for trailers with living quarters. A nice quiet spot to rest and relax. 24-48 hour notice requested. **Rates:** $10 per night. **Accommodations:** Lebanon, MO-12 miles

RAYMONDVILLE
Golden Hills Trail Rides & Resort **Phone: 800-874-1157**
Charles Golden, owner **or: 417-457-6222**
Charlotte Golden-Gray, contact
19546 Golden Drive [65555] Directions: Off Hwy 63 in Houston, MO take Hwy B to Raymondville. Go thru town. 3/4 mile away are signs for High Point Dr. & "Golden Hills Trail Rides." Follow signs. 2 miles off blacktop. **Facilities:** 600 covered stalls with central lighting and water nearby, round pen, arena. Riding trails. Tack store on premises. Minimum of 12-hr notice required. **Rates:** $10 per night. **Accommodations:** 2 large bunkhouses on premises; restroom/shower facilities, 21-room motel: $35 per person per night, $5 each add'l person. Campgrounds on site: $10/night including electrical hook-up.

ALL OF OUR STABLES REQUIRE CURRENT NEG. COGGINS, CURRENT HEALTH PAPERS, & OWNERSHIP PAPERS.

ST. LOUIS

Towne & Country Stables, Inc. **Phone: 314-391-7896**

Louise Shapleigh

527 Weidman Road, Ballwin [63011] Directions: From I-70, take Exit 210A, take Hwy 40-61 towards Kirkwood, go approx. 25 miles to I-270 South, go 1 mile to first exit, west on Manchester Rd. for 2 miles, north on Weidman Rd., 1/2 mile to stable on west side. **Facilities:** Box stalls (limited number available), holding pens, indoor & outdoor arenas, camper hook-up. Reservation required; do not attempt to go thru electric gate — get assistance. **Rates:** $20 per night; discount for 3 or more; $85 per week. **Accommodations:** Several hotels within 15 minutes of stable.

WATSON

H-Bar Ranches **Phone: 816-993-2200**

Jeff Holmes

R 4, Box 143 [64496] Directions: Exit 16 off of I-29 to Rt. B. 1 mile. **Facilities:** 6 indoor stalls, 6 outdoor stalls, pasture/turnout, feed/hay, & trailer parking available. Ranch borders Brickyard Hill Wildlife Area, which is 3,500 acres of public hunting, fishing, etc. 24-hr restaurant at Exit 111 on I-29 nearby. **Rates:** $15 for inside stall, $10 for outside stall. **Accommodations:** Rock Port Inn in Rock Port, 5 miles from stable.

WAYNESVILLE

Johnson Stables **Phone: 573-774-6512**

22985 Reporter Road [65583] Directions: Halfway across Missouri. 3 miles off I-44 at Exit #156. **Facilities:** 23 indoor stalls ranging from 10 X 12 to 14 X 14 with bedding @ $15.00 per night. Outdoor shelter $10.00 per night. 3 holding pens for turn out/exercise. Lighted indoor and outdoor arenas. **Accommodations:** Electric/water hook-ups for campers. Many motels, gas stations and restaurants close by. Questions/reservations, please call.

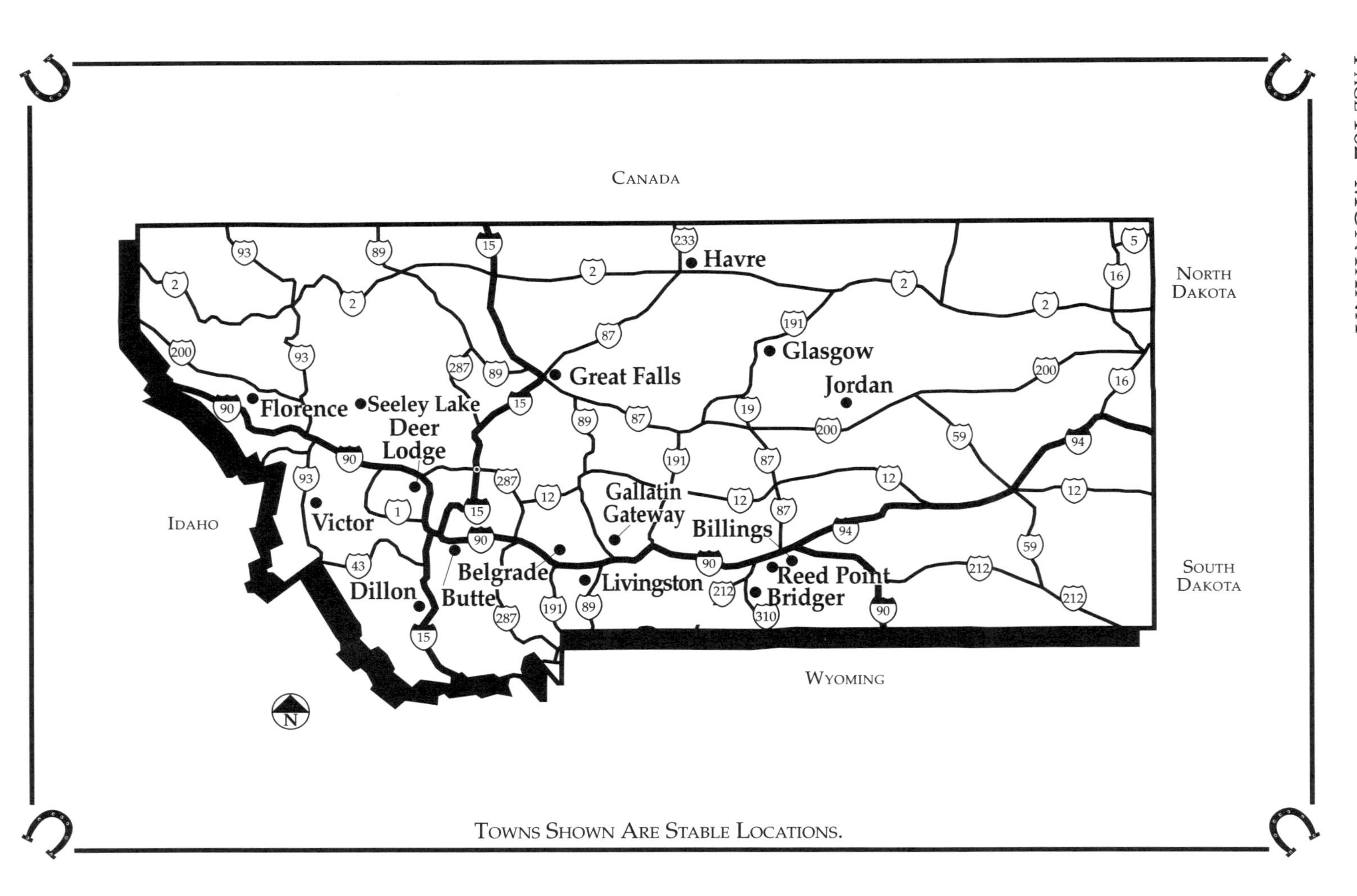

Towns Shown Are Stable Locations.

ALL OF OUR STABLES REQUIRE CURRENT NEG. COGGINS, CURRENT HEALTH PAPERS, & OWNERSHIP PAPERS.

Strip in DJ Bar Ranch 1/2 Page Ad Here

BELGRADE

DJ Bar Ranch **Phone: 406-388-7463**
Jehnet **Fax: 406-388-7443**
5155 Round Mountain Road [59714] Directions: 13 miles from Belgrade. Call for directions. **Facilities:** 320-acre farm/ranch, 3 indoor stalls, 1 to 160 acres of pasture available, feed/hay & trailer parking on premises. Reservations required. **Rates:** $15 per night. **Accommodations:** Luxury guest house available on premises: 3 bedrooms, 2 baths, complete kitchen. 1 week minimum stay required from 6/1 until after Labor Day. Other times are subject to discussion. Also, Super 8 in Belgrade 13 miles away.

BILLINGS

Chuck Larsen **Phone: 406-248-7652**
3306 Becraft Lane [59101] Directions: From I-90: Turn south to flashing light; turn left to Becraft Lane; turn right & follow Becraft Lane 1.2 mile to 3306; turn right & follow road 1/2 mile to barn. **Facilities:** 14 indoor 12' x 20' stalls, eighteen 12' x 60' runs with shed row, five 16' x 48' pens with shed, large outdoor arena, hay available, & trailer parking on premises. All stalls & pens have fresh running water. Reservation preferred. Owner of Chuck Larsen, Livestock Transporter, which is licensed in 44 states (excluding only northern New England). Can transport up to 10 head at a time. Call for further information. **Rates:** $12 per night. **Accommodations:** Many motels in Billings.

ALL OF OUR STABLES REQUIRE CURRENT NEG. COGGINS, CURRENT HEALTH PAPERS, & OWNERSHIP PAPERS.

BILLINGS
June's Horse Motel — **Phone: 406-252-9563**
June Nagel — **or: 406-248-4944**
406 Johnson Lane [59101] Directions: Exit 455 off I-90, .5 mile south on Johnson Lane. **Facilities:** 2 indoor stalls, 3 stalls with 18' x 30' outside runs, corral at barn and working round pen, hay grown on premises, sawdust for bedding, trailer parking based on size and weather, security camera. Call for reservations. **Rates:** $15 per night. **Accommodations:** Motels within 4 miles .

BRIDGER
Circle of Friends Bed & Breakfast — **Phone: 406-662-3264**
Dorothy Sue Phillips
Rt. 1, Box 1250 [59014] Directions: 40 miles SW of Billings via I-90, MT state highway 212 and 310. Call for specific directions. **Facilities:** 2 large corrals, loafing sheds & auto waterer, 2 acres of pasture/turnout, feed/hay & trailer parking available. **Rates:** $10 per night; $50 per week including feed & full care. **Accommodations:** Bed & Breakfast on premises with 3 bedrooms and full breakfast included. Bridger Motel within 3 miles.

BUTTE
No Excuses Arena — **Phone: 406-782-4540**
Paula Scott & Audrey Chamberlin — **Arena: 406-494-6547**
21 Elgin Drive- Elk Park (59701) — **Cell: 406-490-6000**
Directions: 2 miles N of junction I-15 and I-90, 5 miles N of Butte, visible from I-15 (call for specific directions). **Facilities:** 19 indoor, matted box stalls, outdoor runs, pasture. Feed/hay available and 100 x 200 arena. **Rates:** $20 per night indoor stalls. **Accommodations:** B & B, Motels 5 miles.

DEER LODGE
Mountain View Arena — **Phone: 406-846-1989**
Alex & Kayo Fraser, owners — **e-mail: horsebk@imine.net**
255 Boulder Road [59722] Directions: You can see the arena from I -90. Take exit#184 go east to first right, Boulder Rd. Take second drive on left, next to Vet clinic (Westbound traffic). Turn left at the stop sign and take the first right. The left next to vet clinic (Eastbound traffic) **Facilities:** 30+ indoor stalls 12'x12', no pasture or runs. 6 outdoor pens for full time boarders. 70'x230' indoor arena. Hay and trailer parking available on premises. Driving lessons, training, team roping, horse shows and clinics. We will bring in other clinicians depending on interest. Equine book and art for sale on premises and some feed and horse supplies available. We request visiting horses be current on all vaccinations. **Rates:** $15 per night, extra shavings for sale. Reservations appreciated but not required. **Accommodations:** Motels, restaurants, campgrounds less than 1 mile from stable.

ALL OF OUR STABLES REQUIRE CURRENT NEG. COGGINS, CURRENT HEALTH PAPERS, & OWNERSHIP PAPERS.

DILLON

Five Bar J — **Phone (Day): 406-683-9715**

Lanie or Cecil Jones — **Phone (Eve): 406-683-2153**

4225 Anderson Lane [59725] Directions: Take Exit 63 off of I-15 to the intersection with Hwy 41, then go 7 1/2 miles North on Hwy 41; turn left, proceed 1/2 mile on Anderson Lane. Call for further directions. **Facilities:** 6 indoor barn stalls, 6 outdoor paddocks with wood rail construction (smallest paddock is 30' x 30'). Outdoor riding arena, round pen, hay and trailer parking available. **Rates:** $15 per night. **Accommodations:** Motels available in Dillon, 7 miles from stable. Reservations required.

FLORENCE

Parsons' Pony Farm — **Phone: 406-273-3363**

Suzi Kimzey Parsons

5710 Yarrow Drive [59833] Directions: West of Florence on Hwy 93. Call for directions. **Facilities:** 3 enclosed stalls, 3 open stalls with access to paddocks, 3 pastures from 1 to 3 acres each, feed/hay and trailer parking available. Camper hookup. Beautiful extensive trails nearby in Bitterroot Mts., dog kennels available. Cart & saddle ponies, lessons, training, trail rides just for kids. B&B on premises, tennis court. **Rates:** $10-$12 per night. **Accommodations:** B&B on premises; 2 guest bedrooms with shared bath. Days Inn in Lolo, 10 miles from farm.

GALLATIN GATEWAY

Wild Rose Bed & Breakfast — **Phone: 406-763-4692**

Dennis & Diane Bauer

1285 Upper Tom Burke Road [59730] Directions: Call for directions. Located 2 miles south of Gallatin Gateway. **Facilities:** 2 corrals (one large, one small), both with water, 3-5 acres of pasture/turnout, feed/hay and trailer parking available. **Rates:** $15 per night. **Accommodations:** B&B on premises. Other accommodations 2-15 miles from stable.

GLASGOW

Jack & Andie Billingsley — **Phone: 406-367-5577**

Box 768 Billingsley Road [59230] Directions: 10 miles west of Glasgow. Call for directions and reservations. **Facilities:** 3 inside box stalls, 3 inside stalls with runs, 4 large pens one with shed, indoor and outdoor arenas, feed/hay and trailer parking available. Plenty of open range for riding; roping clinics. **Rates:** $10 per night for outside pen; $15 per night for inside stall. **Accommodations:** Modern cabins with private shower and bathroom on premises.

GREAT FALLS

B – C Stables — **Phone: 406-761-7426**

Keith Lewis

31 - 60th Street South [59405] Directions: 5 miles from I-15 and 4 blocks from by-pass. Call for directions. **Facilities:** 33 indoor stalls, heated auto waterer, 120' x 180' outdoor arena, paddocks, breaking arena, feed/hay & trailer parking on premises. Call for reservation. **Rates:** $10 per night. **Accommodations:** Highwood Motel 1 mile from stable.

ALL OF OUR STABLES REQUIRE CURRENT NEG. COGGINS, CURRENT HEALTH PAPERS, & OWNERSHIP PAPERS.

GREAT FALLS

Rocky Mountain Animal Medical Center **Phone: 406-727-8387**
Dr. Bruce MacDonald
1401 N.W. Bypass [59404] Directions: 1/2 mile off I-15. Call for directions. **Facilities:** 8 indoor stalls with outdoor runs, 14' x 150' turnout, feed/hay & trailer parking. "SCOTCH AN' SODA" Appaloosa World Champion All Around Performance Horse at Center. Electric and water hook-up for campers available. Call for reservation. **Rates:** $12 per night. **Accommodations:** Days Inn & Heritage Inn less than 1 mile from Center.

Skyline Horse Hotel & Vet Clinic **Phone: 406-761-8282**
Nora or John Seekins **Fax: 406-761-6900**
Junction of Bootlegger Trail & Haver Highway [59403] Directions: 2 miles off I-15. Call for directions. **Facilities:** 7 indoor stalls, 1 outdoor arena, 3 outdoor paddocks, feed/hay & trailer parking on premises. Clean facility and accommodating staff. Call for reservation. Vet services available. **Rates:** $20 per night. **Accommodations:** Days Inn 1 mile from clinic (406-727-6565) Best Western Heritage Inn (406-727-7200) Ask for special Skyline Horse Hotel rates.

HAVRE

The Great Northern Fairgrounds **Phone: 406-265-7121**
Mike Spencer
1676 Highway 2 West [59501-6104] Directions: Right on Hwy 2 west of town. **Facilities:** 35 metal 10' x 10' indoor stalls, 200' x 140' indoor arena, 180' x 300' outdoor arena, 20 camper/RV hook-ups, trailer parking available. **Rates:** $5 per night; $20 deposit. Must clean stall. **Accommodations:** Campers $5 per night no utilities, $12.40 with utilities.

JORDAN

Sand Creek Clydesdales Ranch Vacations **Phone: 406-557-2865**
Wade & Bev Harbaugh
Hwy 200 East [59337] Directions: Hwy 200 runs eastto west through Montana, ranch is east of Jordan, near mile marker 220 (7 miles out of Jordan) then 3 miles south of hwy. **Facilities:** 3 indoor stalls and 4 outdoor pens, pasture/turnout area. Feed/hay available, trailer parking. Ranch vacations, overnight lodging and wagon trains. **Rates:** $10 per night, $70 per week. **Accommodations:** Lodging available at Ranch

LIVINGSTON

Park County Fairgrounds **Phone: 406-222-4185**
Donna Goldner
46 Vista View Drive [59047] Directions: Livingston Exit off I-90. Park Street to Main Street to Fairgrounds. **Facilities:** 30 indoor 8' x 10' stalls, 10 outdoor 10' x 10' and 12' x 12' metal stalls, 180' x 300' outside rodeo area, trailer parking available, motor homes okay but no sewer or electricity. Open all year. Gates open 24 hrs with caretaker on property. Fair last week in July. **Rates:** $5 per night. **Accommodations:** Motels within 1/2 mile.

ALL OF OUR STABLES REQUIRE CURRENT NEG. COGGINS, CURRENT HEALTH PAPERS, & OWNERSHIP PAPERS.

REED POINT

S Bar K Ranch — **Phone: 406-326-2280**

Blanche Davis, Joe Davis, Mary Berry

#1 Dead End Road [59069] Directions: 2 miles from I-90. Exit 392 off of I-90. Go north through town and across railroad tracks and Yellowstone River Bridge. Continue north until you see a cattle guard, turn right onto Dead End Road, first place on left. **Facilities:** Working cattle ranch. 4 indoor 8' x 10' box stalls, 1 15' x 23' covered barn pen, 3 - 38' x 10' runs with cover, 3 solid-board 16' x 22' box pens, 18' x 46' barn-connected box pen, 2 corrals approx. 500 sq. yds, no turnout/pasture, feed/hay & trailer parking available. 640-acre area for riding. Proof of Influenza, Rhinopneumonitus shots & worming. **Rates:** $10-$20 per night. with hay; $75 per week **Accommodations:** On premises: 2 bedrooms with bath, double beds, & breakfast, $75 per night. Super 8 motels in Columbus (20 miles east) and Big Timber (26 miles west).

SEELEY LAKE

Horseshoe Hills Guest Ranch — **Phone: 406-677-2276**

Wayne & Rena Heaton

6190 Woodworth Road [59868] Directions: Scenic Hwy 83, N. off Hwy 200. At MM 7 take Woodworth Road, 5 miles to ranch entrance. Call for more details. **Facilities:** 17 covered, 10' X 20' outdoor, 18 indoor, 10' X 12' heated stalls. More with temporary panels. Corrals, 40' and up. Pastures 1/2 acre and up. Breeding, boarding & training. Paint and quarterhorses for sale at all times, all levels. Unlimited trails on state & national forest and Bob Marshall Wilderness. **Rates:** $6 to $15 per horse. **Accommodations:** Ranch rooms and camp sites on premises, (no pets in rooms - kennels available), B&B 1 mi., Seeley Lake Motels 20 minutes to town.

VICTOR

Bear Creek Lodge — **Phone: 406-642-3750**

Roland & Elizabeth Turney

1184 Bear Creek Trail [59875] Directions: Call for directions. **Facilities:** 3 large indoor stalls, 3 pens, 5 acres pasture adjacent to the Selway-Bitterroot Wilderness (1.6 million acres) with numerous riding trails starting from the property or within a short drive from the lodge. **Rates:** Free to our guests. **Accommodations:** On premises: small secluded, exquisite lodge featuring fine dining; $21 single and $300 double per night with meals.

Five Brooks Ranches — **Phone: 406-642-6335**

Fred C. Vaughn

2291 Meridian Drive [59825] Directions: 1/2 mile south of Victor on Hwy 93 on N. Fork Bear Creek. **Facilities:** 6 paddocks, 20-acre pasture/turnout, trailer parking available on premises. Will feed in AM & PM. Please make reservations. **Rates:** $10 for paddock, $15 for stall, per night. **Accommodations:** Many motels from Hamilton to Missoula on Hwy 93.

NOTES AND REMINDERS

SAY STRANGER....
YOU AIN'T FROM
AROUND THESE
PARTS ARE YOU?
FREDLEARY

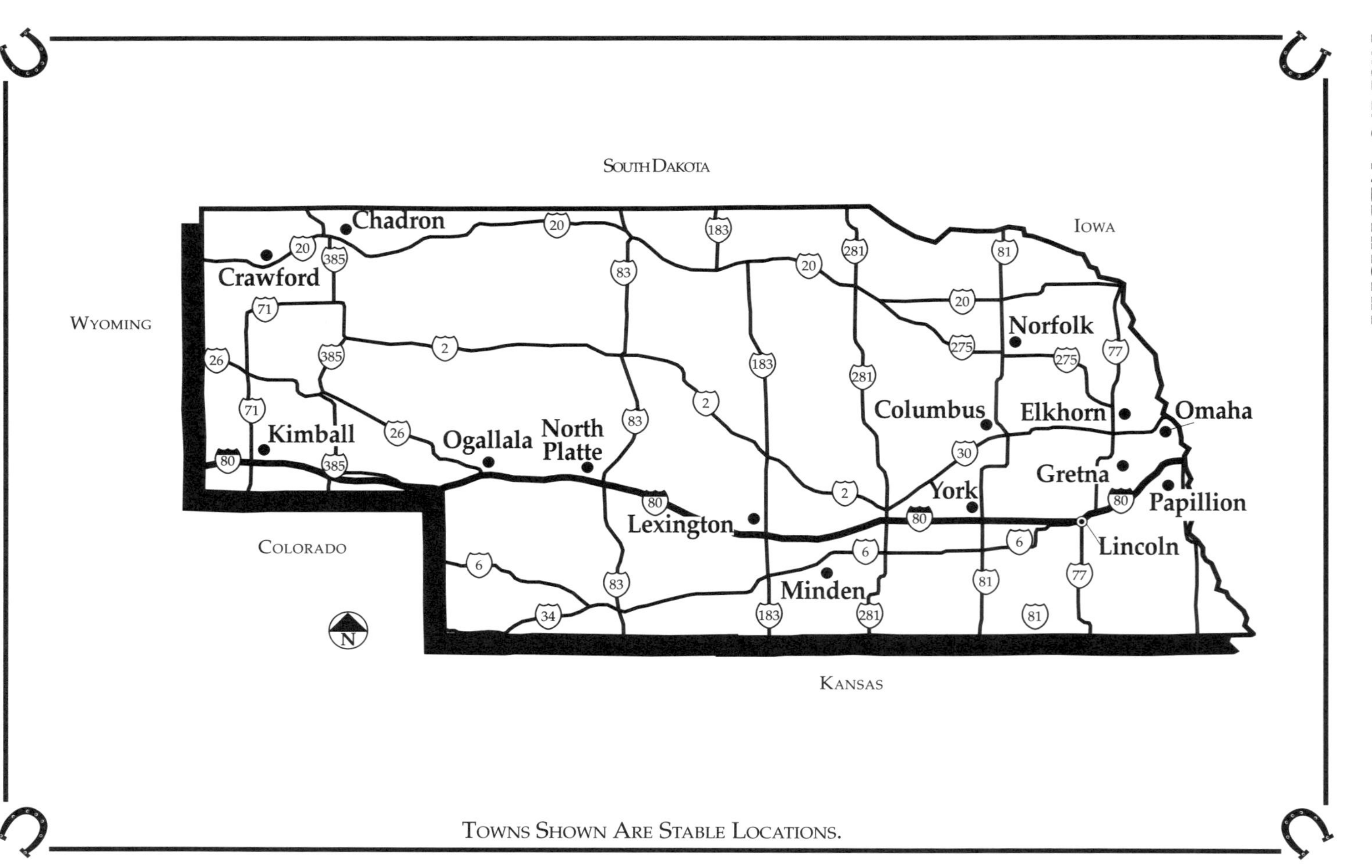
South Dakota
Iowa
Wyoming
Colorado
Kansas
Chadron
Crawford
Kimball
Ogallala
North Platte
Lexington
Minden
Columbus
Norfolk
Elkhorn
Omaha
Gretna
York
Papillion
Lincoln
N
Towns Shown Are Stable Locations.

ALL OF OUR STABLES REQUIRE CURRENT NEG. COGGINS, CURRENT HEALTH PAPERS, & OWNERSHIP PAPERS.

CHADRON
Panhandle Veterinary Clinic **Phone: 308-432-2020**
John E. Gamby, DVM
985 Hyway 385 South(69337)
Directions: 1/2 mile south of Hwy 20 on Hwy 385
Facilities: 2 indoor stalls (10′ x 12′) 2 outdoor stalls 12′ x 12′. **Rates:** $10 per night, $60 per week. **Accommodations:** Best Western (across from Hwy), Days Inn (1/4 mile), Super 8 (1 mile). General veterinary service.

COLUMBUS
Columbus Racetrack/Fairground **Phone: 402-564-0133**
Frank Zuroski **402-564-6746**
15th Street & 10th Avenue [68601] Directions: From US 30 or US 81, follow signs to racetrack. **Facilities:** 800 indoor stalls, arena, feed/hay & trailer parking available. This is a large thoroughbred racetrack with over 800 stalls. **Rates:** $6 per night. **Accommodations:** Sleep Inn 1/2 mile from track.

Strip in Ash Creek Ad Here

CRAWFORD
Ash Creek Ranch Vacations
Phone: 308-665-1580
Gary & Nancy Fisher
Ash Creek, Inc.,
617 West Ash Creek Road [69339]
Directions: Call or write for brochure. **Facilities:** 4 indoor stalls, 20 acres of pasture/turnout, feed/hay & trailer parking available. 1,800-acre working ranch, thousands of acres of adjoining National Forest land, beautiful area for trail riding. **Rates:** $10 per night, $60 per week. **Accommodations:** 100-yr.-old ranch house with two bedrooms, kitchen, living room, indoor bathroom. Sleeps up to ten. $75 per night for two; $6 each additional person. No housekeeping provided.

ALL OF OUR STABLES REQUIRE CURRENT NEG. COGGINS, CURRENT HEALTH PAPERS, & OWNERSHIP PAPERS.

CRAWFORD

Fort Robinson State Park **Phone: 308-665-2900**
Jim Lemmon, contact
Box 392 [69339] Directions: Located 3 miles west of Crawford on Hwy 20. **Facilities:** 120 indoor stalls, no pasture/turnout area, trailer parking but no feed available. **Rates:** $6 per night. **Accommodations:** Cabins to rent on site from April to November: 2 to 9 bedrooms starting at $54 per night with lodge rooms at $28 to $33 per night.

ELKHORN

Quail Run Horse Center **Phone: 402-289-2159**
22021 West Maple Road [68022] Directions: Take Maple St. Exit off of I-680. Call for directions. **Facilities:** 45 indoor stalls, arena, feed/hay included, trailer parking available. A teaching, training, & show facility specializing in English & hunter/jumper. Call for reservation & availability. **Rates:** $15 per night. **Accommodations:** Motels in Omaha 10 minutes from stable.

GRETNA

T S Arabians, Inc. **Phone: 402-332-4328**
Ted Smalley
22603 Fairview Road [68028] Directions: Exit 432 off of I-80. Call for directions. **Facilities:** 47 indoor stalls, round pen, 70' x 176' indoor arena, 4 outdoor runs with pasture, heated & air-conditioned lounge. Breeds, sells, & trains Arabians, quarter horses & thoroughbreds. Riding lessons for all levels. Call for reservation. **Rates:** $15 per night. **Accommodations:** Motels 7 miles from stable.

KIMBALL

Mary Webb **Phone: 308-235-2164**
Rt. 28, S. Hwy 71 [69145] Directions: 3 miles south of I-80. Call for directions. **Facilities:** Corrals & 14 acres of pasture. Trailer parking available. Call for reservation. **Rates:** $15 per night. **Accommodations:** Super 8 & Interstate Inns in Kimball, 3 miles from stable

LEXINGTON

D C Stables **Phone: 308-324-6303**
Dennis & Cyndi Ocken
204 W. River Road, [68850] Directions: 1 mile west of I-80. Call for directions. **Facilities:** 8-12 x 12' indoor stalls, round pen, outdoor arena, trailer parking available. Riding trails to Platte River. Health papers. **Rates:** $10 per pen. $20 per stall. **Accommodations:** Super 8, Comfort Inn, & Days Inn.

ALL OF OUR STABLES REQUIRE CURRENT NEG. COGGINS, CURRENT HEALTH PAPERS, & OWNERSHIP PAPERS.

LINCOLN
K/B Stables **Phone: 402-465-5855**
Kenneth & Berna Stading
6100 N. 98th [68507] Directions: Please call ahead for reservations. Take Waverly Ext. going west. Turn left on 98th & go about 1 mile. **Facilities:** 1 stall plus paddocks and inside arena. Feed & hay available. Trailer parking available. Pasture available. **Rates:** $15.00. **Accommodations:** Quality Inn in Lincoln, 2 miles from stable.

MINDEN
South Wind Ranch **Phone: 308-832-0431**
Dudley & Peggy Benson
575 32nd Road [68959] Directions: 6 indoor & outdoor stalls, pens on 40 acres, feed/hay & trailer parking. Ranch is easy to find and is in a tourist area. Call for reservation. **Rates:** $15 per night. **Accommodations:** Pioneer Inn 5 miles from ranch.

NORFOLK
Norfolk Livestock Market **Phone: 402-371-0500**
Ask for Junior
1601 South 1st Street [68701] Directions: Take Hwy 81 to Omaha Ave.; go east on Omaha 1 mile & turn right and go 2 blocks past tracks and you will see sign. Also easily accessible from Hwy 275 from Omaha. **Facilities:** 30 indoor stalls, pasture/turnout pens, feed/hay & trailer parking on site. **Rates:** $2 per night. **Accommodations:** Holiday Inn & Super 8 in downtown Norfolk 1 mile from stable.

NORTH PLATTE
Remuda Stables **Phone: 308-532-4359**
Ron & Dawn Andersen
Rt 3, Box 257 [69101] Directions: Easy access -- Please Call. **Facilities:** Total: 24 stalls. 10 available for overnight use. 14 open front shed, 12' X 13'/w 40' runs. 3 - 22' X 30', 7 - 12' X 12' outside with good protection. Feed, hay and trailer parking available. No pasture available. 150' X 250' arena and 60' round pen. **Rates:** $11 per night per horse. **Accommodations:** 5 miles to restaurants and motels.

OGALLALA
Baltzell Veterinary Hospital **Phone: 308-284-4313**
C.W. or David Baltzell
1710 West 4th Street [69153] Directions: Located off of I-80. 1/4 mile north of Hwy 30. Call for directions. **Facilities:** 3 indoor stalls, 1 acre of fenced pasture/turnout, feed/hay & trailer parking on site. Medical services available at clinic. Call for reservation. **Rates:** $15 per night; group rates & feed available. **Accommodations:** Many motels 1 mile from hospital.

ALL OF OUR STABLES REQUIRE CURRENT NEG. COGGINS, CURRENT HEALTH PAPERS, & OWNERSHIP PAPERS.

OGALLALA
Peterson Stables **Phone: 308-284-8235**
K.C. Peterson
851 Rd. West D North, Ogallala [69153] Directions: Located off I-80, 4 miles west on Hwy 30, 1/2 miles north . **Facilities:** 20 Box stalls with turnouts, indoor arena, feed, hay and trailer parking available. **Rates:** $10 a night per horse.

OMAHA
Prairie Gem Stables **Phone: 402-426-2882**
Robert & Linda Brown
Hwy 133, Rt. 32, Box 279 [68142] Directions: Call for directions. Easy access off I-680 north of Omaha; 5.5 miles north on Hwy 133. **Facilities:** 6 box stalls 12' x 8', 3-5 open stalls 12' x 16', 5 dry lots, 4 acres pasture, large outdoor arena, round pen, large corrals, feed/hay available, trailer parking. 90+ acres of excellent rolling hills trails, hills overlook surrounding valleys, creeks, Cunningham Lake with trails 3.5 miles from stable. Riding instruction (Western, English, jumping, dressage). Christian Fellowship riding group. Vet/farrier on call. **Rates:** $10-$20 per night depending on stall. **Accommodations:** Ramada, LaQuinta Inn in Omaha, 7-8 miles from stable; Rath Inn in Blair, 10 miles from stable.

PAPILLION
Oakleaf Stables **Phone: 402-339-2519**
13309 S. 72 [68046] Directions: Call for directions. **Facilities:** Indoor & outdoor stalls, turnout with stalls, feed/hay included, trailer parking on site. No arrivals after 5 P.M. Call for availability & they request advance notice. **Rates:** $10 per night. **Accommodations:** HiWay House, Ben Franklin within 6 miles of stable.

SPRINGFIELD
Highland Stables **Phone: 402-253-2550**
Jonnie & Howard Surland
12211 Fairview Road [68059] Directions: Exit 440 off of I-80. Call for directions. **Facilities:** 4 indoor stalls, some outdoor stalls, pasture/turnout area, feed/hay included, trailer parking. 1-2 days notice if possible. **Rates:** $10 for indoor stall, $7 for outdoor stall per night. **Accommodations:** Park Inn 3 miles from stable.

YORK
Diamond B, Inc. **Phone: 402-362-5439**
Bryan & Diane Buss
Rt. 1, Box 154-A [68467-9781] Directions: Call for directions. **Facilities:** 3+ box stalls, turnout pens, 70' x 230' indoor arena, 100' x 300' outdoor arena, feed/hay available, trailer parking on site. Camping allowed. Please call evenings or weekends for reservation. **Rates:** $15 per night; $75 per week. **Accommodations:** Staehr Motel in York 3 miles from stable.

NOTES AND REMINDERS

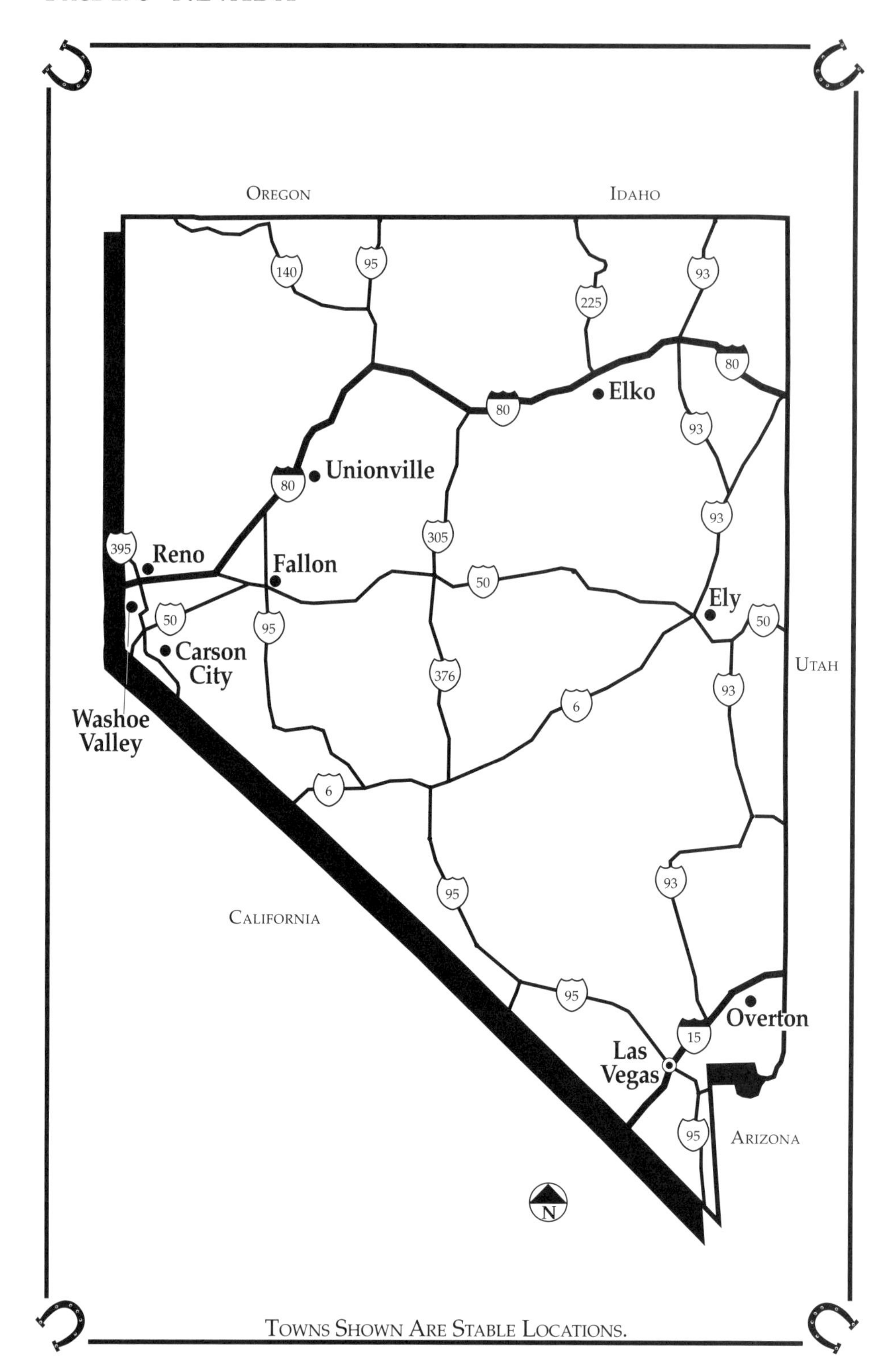
OREGON
IDAHO
140
95
225
93
80
Elko
80
93
Unionville
80
93
305
395
Reno
Fallon
50
Ely
50
50
95
Carson City
UTAH
376
93
6
Washoe Valley
6
93
95
CALIFORNIA
95
Overton
15
Las Vegas
95
ARIZONA
N
TOWNS SHOWN ARE STABLE LOCATIONS.

ALL OF OUR STABLES REQUIRE CURRENT NEG. COGGINS, CURRENT HEALTH PAPERS, & OWNERSHIP PAPERS.

CARSON CITY

Equest Training Center **Phone: 702-849-0105**

Vicki Sherwood

805 Washoe Drive [89704] Directions: Located on Hwy 395. Call for directions. **Facilities:** 15 indoor stalls, 14 outdoor paddocks, dressage arena, jumper field, 60' x 100' covered arena, round pen, wash rack. 24-hr care. Training for dressage, jumping & 3-day events at all levels. Overnight boarding in emergencies only. **Rates:** $20 per night; call for discounted weekly rate. **Accommodations:** Motels 1 mile from stable.

Old Washoe Stables **Phone: 702-849-1020**

Michael Stockwell

1201 Hwy 395 [89704] Directions: Call for directions. **Facilities:** At least one indoor stall, 6 - 12' x 12' holding pens, 10' x 20' indoor corral. Guided horseback rides at $20/hr. and children's riding. Advance reservations required. **Rates:** Will be competitive. **Accommodations:** Round Hill Station in Carson City, 7 miles from stable.

ELKO

Elko County Fairgrounds **Phone: 702-738-7925**

Ask for Jeanie or Angelo

P.O. Box 2067 [89803] Directions: Located off of I-80. Call for directions. **Facilities:** 365 indoor stalls, arena, 5/8 mile race track. Hay & trailer parking available. Open 24 hours. 1 horse per stall. Check in at double-wide trailer at entrance. **Rates:** $10 per night. **Accommodations:** Motels 1/4 mile away.

ELY

White Pine County Fairgrounds **Phone: 702-289-4691**

Sterling Wines

McGill Highway, Rt. 93 [89315] Directions: Call for directions. **Facilities:** 200 inside stalls, 6 outside stalls, no feed/hay. Trailer parking available. Open 24 hours. **Rates:** $10 per night, inside stall; $5 per night, outside stall. **Accommodations:** Motels 2 miles from fairgrounds.

FALLON

Challenger Horse Stables **Phone: 702-423-2888**

Murray Morin

2245 Coleman Road [89406] Directions: Less than 1 mile from I-80. Take Trinity Junction Exit then call for further directions. **Facilities:** 6 covered holding pens on 8 acres, turnout. Call for reservations. **Rates:** $16 per night.

ALL OF OUR STABLES REQUIRE CURRENT NEG. COGGINS, CURRENT HEALTH PAPERS, & OWNERSHIP PAPERS.

LAS VEGAS

Bamberry Stables **Phone: 702-361-6620**

Don Bamberry

7475 Rogers [89139] Directions: Call for directions and reservations. Let phone ring and leave message. Quiet neighborhood 1 mile from I-15. **Facilities:** 12 outdoor pipe corrals 20' x 20' with shade, 100' x 130' arena, alfalfa extra, trailer parking. **Rates:** $15 per night per horse. **Accommodations:** Silverton Hotel & Casino with RV parking 1 mile away; Vacation Village, Hotel & Casino 2 miles away.

Beckridge Ranch **Phone: 800-704-8127**

Richard and Gina Beck

8375 Gilespie Street [89123] Directions: I-15 to Blue Diamond Exit. East to Las Vegas Blvd. South to Windmill. East to Gilespie St. South 1-1/2 blocks to Beckridge Ranch. **Facilities:** 10 - 16′ X 16′ outside stalls. 10′ X 10′ shedrow w/runs. Alfalfa & grass hay, Trailer parking available. 1 acre turn out arena. 4 horse barn with wash rack. Quiet neighborhood for riding. **Rates:** $12.00 per night or $75.00 per week. **Accommodations:** Silverton Hotel, Casino and RV campground 2 miles away. "Strip" casinos 5 miles away.

OVERTON

Tanglewood Ranch Horse Motel & Boarding **Phone: 888-387-8656**

Kim Fox

585 N. Moupa Valley Blvd. [89040] Directions: Half way between Las Vegas, NV & St. George UT. Take exit 93 off I-15. Go 10 miles to ranch. Call anytime. **Facilities:** Outdoor 10 - 20′ X 20′ & 4 - 16′ X 16′ pipe corrals. Lighted arena, round pen, wash rack, 2 acre turnout, Open range land nearby. Trailer parking available. **Rates:** $10 per night, $12 per night with feed. Long term boarding rates available. **Accommodations:** Best Western right across field.

RENO

Long Ears Long Walk Ranch **Phone: 702-677-7046**

Nancy Jackson

3205 Indian Lane [89506] Directions: From I-80 and 395: Go north on 395 seven miles to Golden Valley Road Exit. Go east to first left (Estates), right on Indian Lane. **Facilities:** 40 - 12' x 24' stalls with shade or box stalls, 4 arenas, tack rooms, feed/hay available, trailer parking. Open space to trail ride. Tennessee Walker stallion standing at stud. **Rates:** $10-$20 per night. **Accommodations:** Miners Inn 15 minutes from stable.

RENO

Reno Livestock Events Center **Phone: 702-688-5751**

1350 N. Wells Avenue [89512] Directions: From I-80, take Wells Avenue North Exit. **Facilities:** 660 indoor stalls with rubber mats, holding pens, indoor/outdoor arenas, camper hookup and parking. **Rates:** $19 per night for box stalls, $10 per night for 5 nights; $11 per night for corrals. Shavings available for $6 per bag. **Accommodations:** Motel 6, Days Inn, Holiday Inn within 2 blocks.

ALL OF OUR STABLES REQUIRE CURRENT NEG. COGGINS, CURRENT HEALTH PAPERS, & OWNERSHIP PAPERS.

UNIONVILLE
Old Pioneer Garden **Phone: 702-538-7585**
Mitzi & Lew Jones
2805 Unionville Road [89418] Directions: 20 minutes from Exit 149, Unionville Exit off I-80. **Facilities:** 3 indoor 12' x 10' stalls connecting with paddocks, 1/2-acre fenced pasture, feed/hay, traller parking available. Trail riding, gold mines nearby, nearby Winnemucca has gambling (1 hr away). Only 18 people in this ghost town. **Rates:** Horses free when you stay at inn. **Accommodations:** Inn on premises, $75-$85 per night including breakfast, other meals available. Separate house available for rent.

WASHOE VALLEY
Franktown Meadows **Phone: 702-849-1600**
Janice
4200 Old Hwy 395 North [89704] Directions: 20 miles south of I-80. Call for directions. **Facilities:** On 41 acres. 63 indoor stalls, 12 pasture/turnout areas, indoor arena, paddock, outside arena, jump course, dressage court. Feed/hay & trailer parking available. Training of hunter/jumper & dressage for horses & riders. Call for reservations. **Rates:** $15 per night. **Accommodations:** Motels 6 miles from stable.

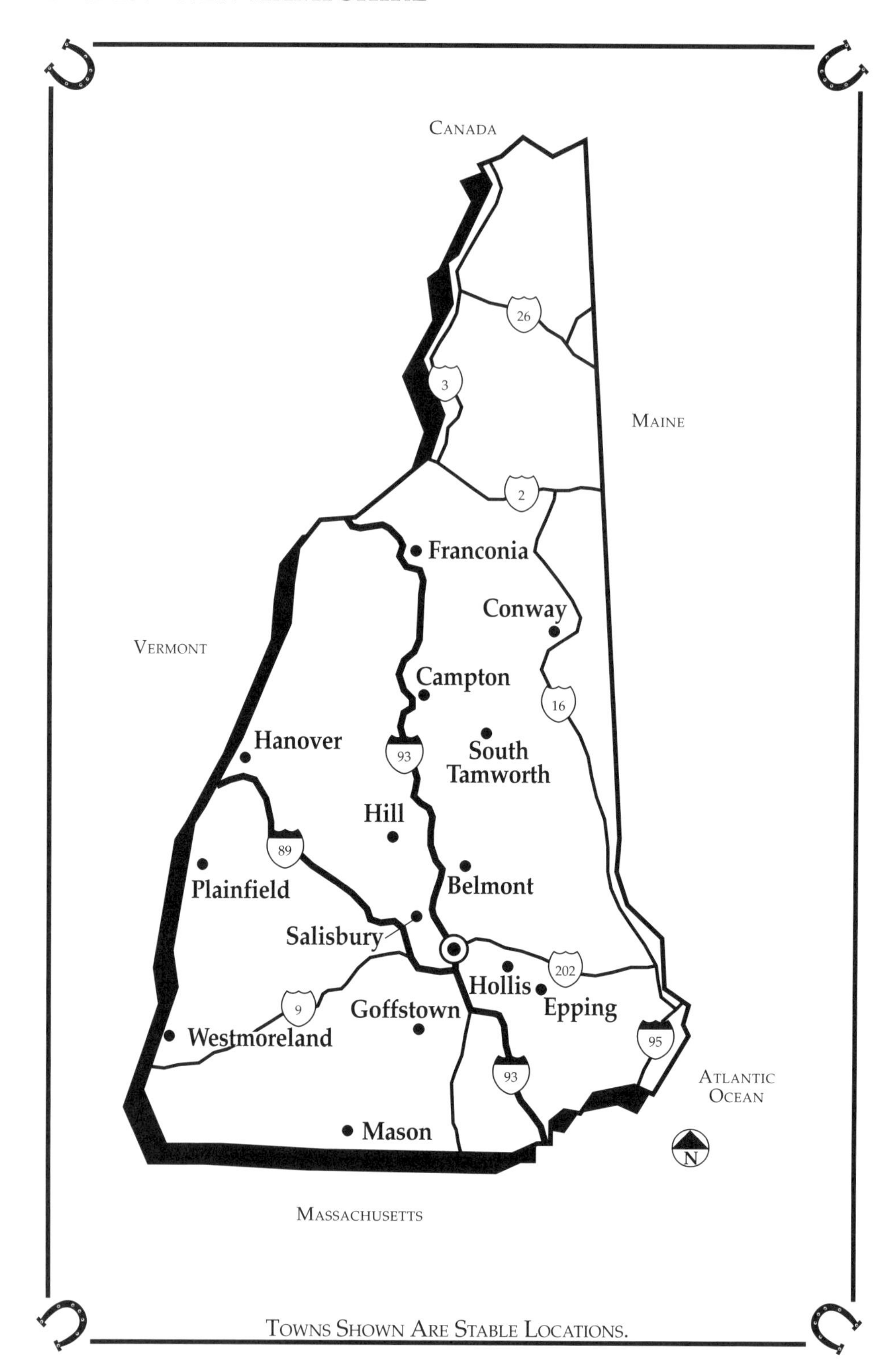

TOWNS SHOWN ARE STABLE LOCATIONS.

ALL OF OUR STABLES REQUIRE CURRENT NEG. COGGINS, CURRENT HEALTH PAPERS, & OWNERSHIP PAPERS.

BELMONT
Hardscrabble Stable **Phone: 603-267-7065**
,Terry Coyman
403 Hurricane Road [03220] Directions: Exit 20 off of I-93. Call for directions. **Facilities:** 10 indoor stalls, 8 acres of pasture, round pen, 1-acre paddock on 32.5 acres. 450 acres of trails nearby. Hay & sleigh rides and carriage-drawn weddings. Call for reservations. **Rates:** $15 per night; $90 per week. **Accommodations:** Motels 5 miles from stable.

CAMPTON
White Winds Farm **Phone: 603-726-4664**
Lena Johnson
Ellsworth Road [03223] Directions: Call for directions. **Facilities:** 21 indoor stalls, 70' x 170' heated indoor arena, heated wash & tack room, large outdoor arena, 30 acres of pasture, 4 paddocks with run-in sheds, trails. Arabian stallion standing-at-stud: "Special Agent." Foals for sale. Borders national forest and near Waterville Valley ski resort area. Call for reservations. **Rates:** $20 per night; ask for weekly rate. **Accommodations:** Days Inn 2 miles away.

CONWAY
The Foothills Farm Bed & Breakfast **Phone: 207-935-3799**
Kevin Early
P.O. Box 1368 [03818] Directions: Call for directions. **Facilities:** 3 indoor 10' x 10' stalls, 2 outdoor 12' x 12' stalls, three 1-acre pastures, one turnout area, feed/hay & ample trailer parking available. Quiet riding on marked trails that range from 8 to 22 miles in length - no busy roads to cross. Guided trail rides available. **Rates:** $18 per night. **Accommodations:** 4-bedroom B&B on premises: $48/dbl including full breakfast.

EPPING
Rum Brook Farm **Phone: 603-679-5982**
Meg Preston
44 Hedding Road (Rte. 87) [03042] Directions: From I-495, take Rte 125 north, east on 87. From I-95, take Rte 101 west to Rte 125 north, right on 87. Farm is 1/3 mile on 87. **Facilities:** 30 indoor 10' x 12' box stalls, large and small paddocks, indoor arena, feed/hay, trailer parking. Breeding facility; standing-at-stud: "Immortal Command" and "Serenity March Time." Morgan horses for sale. Equitation program. **Rates:** $15 per night, $94 per week. **Accommodations:** Best Western in Exeter, 15 miles away; Epping Motel in Epping, 1 mile away; motels in Hampton Beach, 20 miles away.

ALL OF OUR STABLES REQUIRE CURRENT NEG. COGGINS, CURRENT HEALTH PAPERS, & OWNERSHIP PAPERS.

FRANCONIA
Bungay Jar Bed and Breakfast **Phone: 603-823-7775**
Kate Kerivan **Toll Free: 1-800-421-0701**
P.O. Box 15, Easton Valley Road [03580] **Fax: 603-444-0100**
Directions: From I-93 North, take Exit 38. South on Hwy 116 for 5.5 miles. Bungay Jar on left just past Sugar Hill Road. From I-91 North, take Exit 17. Take Hwy 302 east through Woodsville for about 12.5 miles. Turn right onto Hwy 117. From Franconia, go south on 116. Bungay Jar is 5.5 miles on left. **Facilities:** Turnout only, no stalls; hay & pasture/turnout available, bring own grain; trailer parking available. 15 acres in White Mountains with National Forest and AMC trails out the back pasture. Dressage, hunter/jumper clinics nearby. **Rates:** $15 per night. **Accommodations:** B&B on site, $75-$150 per night (dbl), higher during foliage season. Country breakfast included.

GOFFSTOWN
Welch Farms **Phone: 603-497-2004**
Paul Welch
11 Welch Lane [03045] **Directions:** Located just north of Manchester. Call for directions. **Facilities:** 12 indoor stalls, 40' x 80' indoor arena, 3 pastures & 2 paddocks on over 200 acres. Breeds & sells Holsteiner warmbloods. Call for reservations. **Rates:** $15 per night. **Accommodations:** Motels 6 miles away.

HANOVER
Velvet Rocks Farm **Phone: 603-643-2025**
Marilyn Blodgett
24 Trescott Road [03750] **Directions:** Call for directions. **Facilities:** Up to 6 indoor box stalls with rubber mats, riding ring, no pasture/turnout, feed/hay & trailer parking available. Call for reservations. **Rates:** $15 per night. **Accommodations:** Radisson & Sheraton in White River Junction, about 5 miles from stable.

HILL
Gloria King's Stables & Trail Rides and Boarding **Phone: 603-934-5740**
Leon & Gloria King
RR 1, Box 1191 [03243] **Directions:** Exit 23 off of I-93. Call for directions. **Facilities:** 31 indoor stalls, pasture, feed/hay & trailer parking available. Tack store on premises. Acres to ride on plus river crossing. **Rates:** Rides, $20 per hr; $35, two hrs. Boarding: $15 per night; $75 per week; $125 per month. **Accommodations:** B&B 5 miles from stable.

HOLLIS
Glory Hole Ranch **Phone: 603-465-2672**
Rich Lasal
Wheeler Road [03049] **Directions:** Only 45 minutes from Boston, call for directions. **Facilities:** 10 quality 12'x12' indoor stalls, 10 acre pasture, feed/hay available, trailer parking. Brand new barn w/indoor & outdoor arenas. 5 minutes from Nashue and near major highways. **Rates:** $20 per day. **Accommodations:** Many hotels/motels in the area.

ALL OF OUR STABLES REQUIRE CURRENT NEG. COGGINS, CURRENT HEALTH PAPERS, & OWNERSHIP PAPERS.

MASON
Hearthstone Farm **Phone: 603-878-3046**
Barbara Baker
610 Greenville Road [03048] Directions: Call for directions. **Facilities:** 6 indoor stalls, 60' x 180' indoor ring, 3 paddocks on 55 acres. Buys & sells horses and offers full training & lessons. Parking for big rigs available. Call for reservations. **Rates:** $15 per night. **Accommodations:** B&Bs within 10 miles of stable.

PLAINFIELD
MNMS Stables **Phone: 603-675-2915**
Sue Zayatz **Home: 675-5662**
Westgate Road [03781] **NH Residents: 800-871-2915**
Directions: Call for directions. **Facilities:** 13 indoor 12' x 12' stalls, 10 outside paddocks, 10-acre pasture, 64' x 120' indoor arena, outside ring, hunter/jumper course. Horse transportation throughout New England. Call for reservations. **Rates:** $15 per night; ask for weekly rate. **Accommodations:** Motels 7 miles from stable.

SALISBURY
Horse Haven B&B **Phone: 603-648-2101**
Velma Emery
462 Raccoon Hill Road [03268] Directions: I-93 N from Concord. Exit 17 onto Rt 4. 9 miles to Salisbury. Call for details. **Facilities:** Beautiful scenic New Hampshire country side. Thoroughbred foals frolicking on the hillside. 35 acre farm. Horse stalls for B&B guests only. Outdoor 12' X 12' outdoor pipe frame corrals Free. **Rates:** Large indoor bedded stall $15.00 per night. **Accommodations:** Double Occupancy/w shared bath $65.00. Single occupancy/w shared bath, $35.00 to $50.00, Extra person, $15.00. Room rates include complete continental breakfast.

SOUTH TAMWORTH
Red Horse Hill Farm **Phone: 603-323-7275**
Diana Louis, Dominic Bergen
Bunker Hill Road [03883] Directions: Call for directions. 40 minutes from I-93 or 20 minutes from Route 16. **Facilities:** 13 indoor 11' x 11' stalls with rubber matting, 1/4 to 1-1/2 acre turnout paddocks, hay, bedding, trailer parking. Instruction and training. Dressage ring, cross-country course, access to miles of trails. **Rates:** $15 per night, $70 per week. **Accommodations:** B&B on premises.

WESTMORELAND
Singin' Saddles Ranch **Phone: 603-399-7003**
Carole Fletcher
462 Glebe Road [03467] Directions: 3 miles off Rt. 9 and 12 miles off I-91. Call for directions. **Facilities:** 32 indoor stalls, 6 paddocks, 60' round pen, 80' x 140' outside arena, 60' x 120' indoor arena, auto waterers with heaters & heated observation room. Sales of Quarter horses & Paints. Paint stallion standing-at-stud: "Heza Night Train." Trick horse training and horses available for shows. Call for reservations. **Rates:** $15 per night; ask for weekly rate. **Accommodations:** Motels 7 miles from stable.

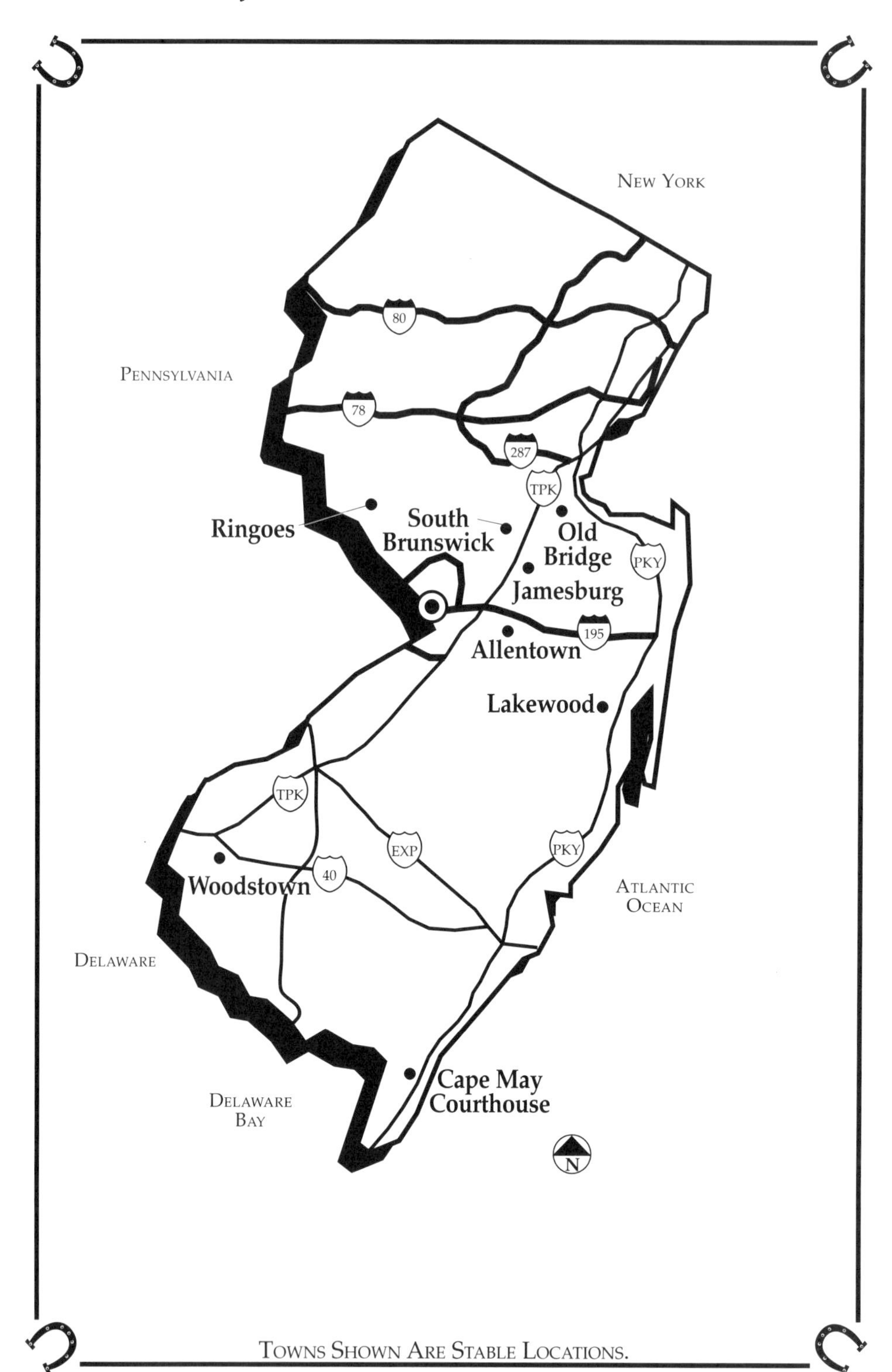

Towns Shown Are Stable Locations.

ALL OF OUR STABLES REQUIRE CURRENT NEG. COGGINS, CURRENT HEALTH PAPERS, & OWNERSHIP PAPERS.

ALLENTOWN
O-NO Acres **Phone: 609-259-2334**
Sandra Baggitt
45 Imlaystown Road [08501] Directions: Exit 11 off of I-95. Call for directions. **Facilities:** 10 indoor stalls, 6 paddocks, feed/hay & trailer parking available. Call for reservations. **Rates:** $15 per night. **Accommodations:** Motels 10 miles from stable.

CAPE MAY COURTHOUSE
Triple R Ranch **Phone: 609-465-4673**
Mary Ruffing
210 Stagecoach Road [08210] Directions: Exit 9 off of Garden State Parkway. Call for further directions. **Facilities:** 20 indoor box stalls, 80' x 200' indoor riding arena, turnout areas, tack store on premises. English & Western lessons and trail rides on ranch. Surrounded by campgrounds. **Rates:** $15 per night. **Accommodations:** Motels 2 miles from ranch.

JAMESBURG
Superior Horse Farm **Phone: 908-521-4969**
Sharon Farmer
Old Forge Road [08831] Directions: Exit 8A off of N.J. Turnpike. Call for directions. **Facilities:** 14 box stalls, 6 turnout paddocks, 60' x 100' outdoor arena, split rail fencing, wash rack, trail riding. Western pleasure riding, complete boarding, Western lessons available and horses for sale. Call for reservations. **Rates:** $20 per night. **Accommodations:** Motels 5 miles away.

LAKEWOOD
Lakewood Boarding Stables **Phone: 908-367-6222**
Bill M. Eak
436 Cross Street [08701] Directions: Exit 21 off of 195. Take 527 south to 528 east. Take right on Cross St. Stable is approx. 1 mile on right. **Facilities:** 85 indoor 10' x 12' stalls, 2-1/4 acre paddocks, 2 acres of pasture/turnout, giant indoor arena, quarantine area, feed/hay & trailer parking available. Stallions can be accommodated. 600 acres of riding trails. Riding lessons given. Reservations required. Tack shop on premises. **Rates:** $20 per night. **Accommodations:** Ramada Inn in Lakewood, 5 minutes from stable.

OLD BRIDGE
Royal Farms Horse Center **Phone: 732-251-9810**
Jerry & Rosemary Jacks
1564 Englishtown Road [08857] Directions: Convenient access from Rt. 18, Rt. 9, & NJ Turnpike. Call for directions. **Facilities:** 20 indoor stalls, 2 lighted rings for rodeo & dressage, riding trails. Training of standardbreds, trotters, & pacers done at Center plus English & Western lessons. Call for reservations. **Rates:** $15 per night. **Accommodations:** Motels 2 miles from stable.

ALL OF OUR STABLES REQUIRE CURRENT NEG. COGGINS, CURRENT HEALTH PAPERS, & OWNERSHIP PAPERS.

RINGOES
Cross Creek Farm **Phone: 908-806-3248**
Loretta McCay
45 Dutch Lane [08551] Directions: 5 minutes off 202; easy access to Rts. 95, 78, and 287. Call for exact directions. **Facilities:** 10 indoor 10' x 10' stalls, 120' x 250' ring, 3 acres of pasture, wash stall, tack room, feed/hay available, trailer parking. Trail riding. Camping with bathroom accommodations, 3 campsites for trailers, water hookups. Petting zoo. **Rates:** $20 per night, $75 per week. **Accommodations:** Cottage on premises. Ramada Inn in Flemington, 3 miles from stable; other motels and B&B within 5 miles.

NORTH BRUNSWICK
Farrington Farms **Phone: 732-821-9844**
Gary Ippoliti
28 Davidson Mill Road [08902] Directions: Exit 8A off of NJ Turnpike. Call for directions. **Facilities:** 60 indoor stalls, round pen, 2 outdoor arenas, 100' x 250' & 100' x 100', 7 outdoor paddocks, indoor ring, jump course. Training of horses & riders for hunter/jumper, hunt seat, & English & Western pleasure. Breeds thoroughbreds. Call for reservations. **Rates:** $25 per night. **Accommodations:** Motels 2 miles from stable.

WOODSTOWN
Victorian Rose Farm Bed & Breakfast **Phone: 609-769-4600**
947 Rt 40 [08098] Directions: At mile marker 8 1/2 on Rt 40 just east of Delaware Memorial Bridge. Or, Exit #1 off of the New Jersey Turnpike. **Facilities:** Turnout pasturage only, no stalls, feed/hay available, trailer parking. **Rates:** For B&B guests: $5 per night, $10 per night with feed. Overnight stabling only: $20 per night. **Accommodations:** 4-bedroom Victorian B&B on site, 2 rooms w/private bath, 2 rooms share a bath.

NOTES AND REMINDERS

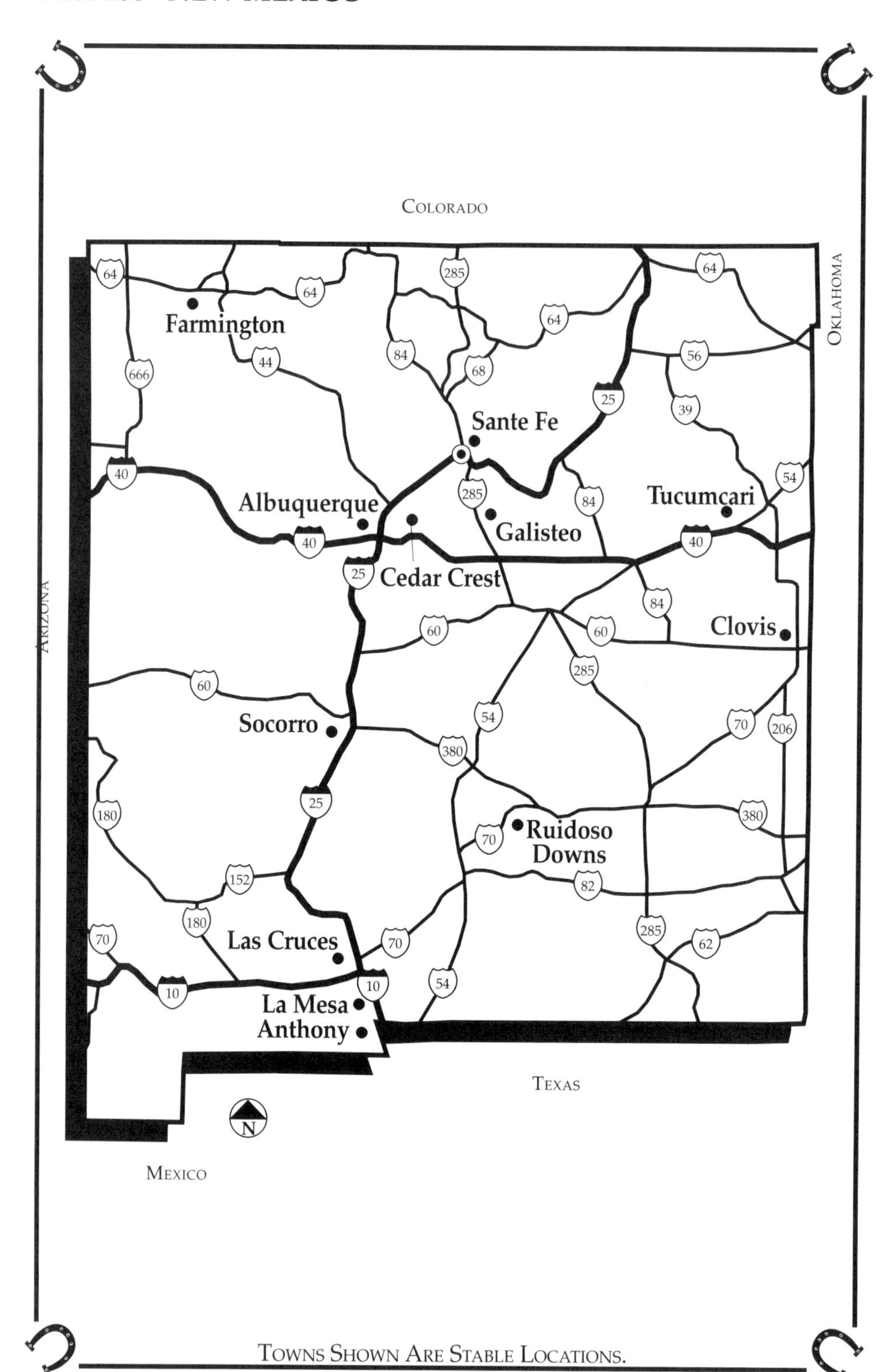
Colorado
Oklahoma
Arizona
Texas
Mexico
Farmington
Sante Fe
Albuquerque
Galisteo
Tucumcari
Cedar Crest
Clovis
Socorro
Ruidoso Downs
Las Cruces
La Mesa
Anthony
N
Towns Shown Are Stable Locations.

ALL OF OUR STABLES REQUIRE CURRENT NEG. COGGINS, CURRENT HEALTH PAPERS, & OWNERSHIP PAPERS.

ALBUQUERQUE

Blakley's Spur Stable **Phone: 800-305-1851 or 505-877-1851**
Clarinda Blakley **E-mail: spurstable@aol.com**
2029 Lakeview SW [87105-6103] Directions: South of I-40 on I-25, Exit 220; go west 2.5 miles to Isleta Blvd. (3rd light); go south 1 mile to Lakeview; go west 1/4 mile. Stable is on right. **Facilities:** 7 indoor stalls, 6 outdoor stalls, & 15 stalls with runs, horse walker, riding trails, feed/hay & trailer parking available. Vet on call. Camper hook-up. Owners on premises. Reservations required! **Rates:** $15 per night; $70 per week. **Accommodations:** Guest house on premises. Motel 6 in Albuquerque, 5 miles from stable.

Heartlane Farms, Inc. **Phone: 505-345-7072**
Julie & Bob Luzicka
6730 Rio Grande Blvd. NW [87107] Directions: From I-40 West: Take Rio Grande Exit; go north (right) for 4 miles. Barn is on east side, past fire station with sign out front. **Facilities:** 12 indoor stalls with shavings & outdoor runs,8 indoor stalls, 6 outdoor stalls, 24' x 48' turnouts, large shaded arena, wash rack, crosstie and grooming area, feed/hay & trailer parking available. **Rates:** $25 per night for indoor stalls; $15 per night for outdoor stalls. $100 per week. Owner on premises. **Accommodations:** Sheraton Old Town & Best Western Inn at Rio Rancho, 5 miles away.

Town-n-Country Feed & Stables **Phone: 505-296-6711**
Judy Davis
15600 Central SE [87123] Directions: Exit 170 off of I-40E. Call for directions. **Facilities:** 10 open pens. Complete feed & tack store on premises. Open 7 days a week. **Rates:** $10 per night.

ANTHONY

Spur-C Ranch **Phone: 505-874-3603**
Boyd & Nancy Carson
Rt. 1, Box 617-L [88021] Directions: Exit 8 off of I-10. Go west about 3/4 mile to stop light; turn right on Doniphan; go 1/2 mile to stop light, turn left on Borderland Rd.; go 3.2 miles; at the third stop sign the road T's; sign for ranch is to the right & across the road; follow the signs on the dirt road to ranch gate that is 1.1 mile from pavement. **Facilities:** Call ahead for availability. Stalls are outdoor 50' x 50' with shades, 130' x 330' turnout, feed/hay & trailer parking. **Rates:** $10 per night. **Accommodations:** Motels in El Paso, 10 miles away.

ALL OF OUR STABLES REQUIRE CURRENT NEG. COGGINS, CURRENT HEALTH PAPERS, & OWNERSHIP PAPERS.

CEDAR CREST/ALBUQUERQUE

Cedar Crest Country Cottage & Stables — **Phone: 505-281-5197**
Donald & Annette Romeros — **Web: www.cedarcrestcottage.com**
47 Snowline Road, P.O. Box 621 [87008] — **E-mail: draeccc@aol.com**
Directions: Just 15 minutes from Albuquerque. East I-40 to Exit 175. Hwy 14 to Cedar Crest. Go north 4 miles (4 lane hwy). Turn left just past mile marker 4. Go 1/2 mile past stop sign. White pipe fence. Check in - 1st driveway on left. #39, The Romeros. **Facilities:** 4 outdoor and 2 indoor covered stalls, 200′ x 200′ turnout paddock. Hay & trailer parking available. Adjacent to Cibola National Forest & trails. **Rates:** $10/$15 per horse. From $75 Double occupancy in cottage - Call for reservations. **Accommodations:** 3 bedroom 2 bath cottage. Fully self sufficient kitchen.

CLOVIS

Amigo Del Caballo Horse Motel — **Phone: 505-742-1033**
Bob Meisenheimer
945 Curry Road-E [88101] Directions: 3 miles west of the Texas state line on Hwy 60-70 & 84; 4 miles east of Curry County Fairgrounds. Turn north at sign and go north 1/3 mile. **Facilities:** All pipe and concrete stalls with a "Shoofly" fly control system in all 7 stalls. Each 13′ x 13′ stall opens into an additional 13′ x 27′ run. There are two 50′ x 50′ six foot high pens, two 50′ x 120′ pens 4′ high. More pens 50′ x 300′ and larger. A 335′ x 205′ lighted roping arena as well. Feed/hay and trailer parking available with 2 RV water and electrical hookups. Owner lives on site. **Rates:** $15 per night. $10 per day weekly rate. **Accommodations:** Bishops Inn, Clovis Inn, Comfort Inn, Days Inn, Holiday Inn, Kings Inn, LaVista Inn, Motel 6, Sands Motel on Hwy 60-70-84, 3-4 miles from stable.

Curry County Fairgrounds — **Phone: 505-763-6502**
Gary Hillis, caretaker — **or: 762-8827**
600 South Norris [88101] Directions: At the junction of US 60, US 70, & US 84. Call for directions. **Facilities:** 12 outdoor 20' x 20' pens with no shelter, 18 indoor stalls, rodeo arena. 24-hr security. Rodeos, bull-riding, barrel racing, roping, circus, & Special Olympics are held at the fairgrounds. **Rates:** Donation requested for outside pens; $5 for inside stalls per night. **Accommodations:** Days Inn 1/2 mile from fairgrounds.

FARMINGTON

LAS BRISAS — **Phone: 505-564-8948**
Donnie & Sherry Pigford — **505-327-1855**
2446 LaPlata Hwy (87401) Directions: Hwy 64 to Shiprock. Right on Hwy 170 (La Plata Hwy) 1 mile N on right side of Hwy. **Facilities:** 7 indoor stalls with 16′ runs and 9 outdoor stalls with covers and runs. Feed available. Trailer parking. Hot walker, wash rack, arena, round pen. 1 camper hook-up with water and electric $15 per night. Rates: $10 per night for outdoor stall and $15 for indoor. **Accommodations:** Several motels 1-5 miles away.

ALL OF OUR STABLES REQUIRE CURRENT NEG. COGGINS, CURRENT HEALTH PAPERS, & OWNERSHIP PAPERS.

FARMINGTON

McGee Park Fairgrounds **Phone: 505-325-5415**

Jim Parnell, Director

41 Road 5568 [87401] Directions: Fairgrounds is visible from Hwy 64 behind race track. Call for directions. **Facilities:** 100 outside covered 12' x 12' stalls, 100' x 290' indoor coliseum, 100' x 200' outdoor arena. Concerts, rodeos, roping, horse shows, etc. held at fairgrounds. Open 24 hours. Office hours: 8-4:30 M thru F. **Rates:** $5 per night. **Accommodations:** Motel 6.5 miles away.

LA MESA

Armstrong Equine Services **Phone: 505-233-2208**

Joe B. Armstrong

Rt. 1, Box 303B [88044] Directions: Exit 155, Vado, off of I-10. Call for further directions. **Facilities:** 53 indoor stalls, wash racks, arenas, walker, paddock, electric hook-up for self-contained campers. Feed/hay & trailer parking available. Horse training and stud service available. **Rates:** $20 per night; $100 per week. **Accommodations:** Motels in Las Cruces & in Anthony, Texas.

RUIDOSO DOWNS

Tull Stansell Estate **Phone: 505-378-4503**

Cheryl & June McCutcheon **Barn Manager: 505-378-8188**

P.O. Box 57 [88346] Directions: Located off of Hwy 70. Call for directions. **Facilities:** 9 indoor stalls, 5 stalls with runs, 10 paddocks, hot walker. Boards & sells llamas. Reservations requested. **Rates:** $5 per night; $21 per week. **Accommodations:** RV campground nearby; Inn at Pine Springs, 2 miles away.

SANTA FE

Nix Farm **Phone: 505-471-8630**

Gene Nix

Rt. 10, Box 85E [87501] Directions: From I-25, take Exit 282N. Call for further directions. **Facilities:** 25-30 stalls, corrals, arena, and holding pens. Room for camper with hook-up. Advance reservations required. **Rates:** $15 per night; $75 per week. **Accommodations:** Motels within 1 mile.

SOCORRO

Haley Hacienda & Hay Burners **Phone: 505-835-0711**

Scotty & Pat Haley

Box 154 [87801] Directions: Call for directions. **Facilities:** 5 indoor stalls, 6 paddocks, pasture ring, feed/hay & trailer parking available. English & Western pleasure lessons at all levels for horses & riders. Call for reservation. **Rates:** $15 per night. **Accommodations:** Motels 1 mile from stable.

ALL OF OUR STABLES REQUIRE CURRENT NEG. COGGINS, CURRENT HEALTH PAPERS, & OWNERSHIP PAPERS.

TUCUMCARI
Western Drive Stables **Phone: 505-461-0274**
Jim and Marlene Haller
4194 Quay Rd 63 [88401] Directions: North of I-40 exit 331/Camino del Coronado. Close to Route 66. Call for directions. **Facilities: 20-** 12' x 18' covered outdoor and boxed stalls, stallion stalls available. Holding pens, paddocks, exercise area, space for trailers, electric hook-up only. All weather driveway & feed & bedding available. Vet close by. Owners on premises. Reservations welcome but not necessary. **Rates:** Please call. **Accommodations:** Close to motels and restaurants.

NOTES AND REMINDERS

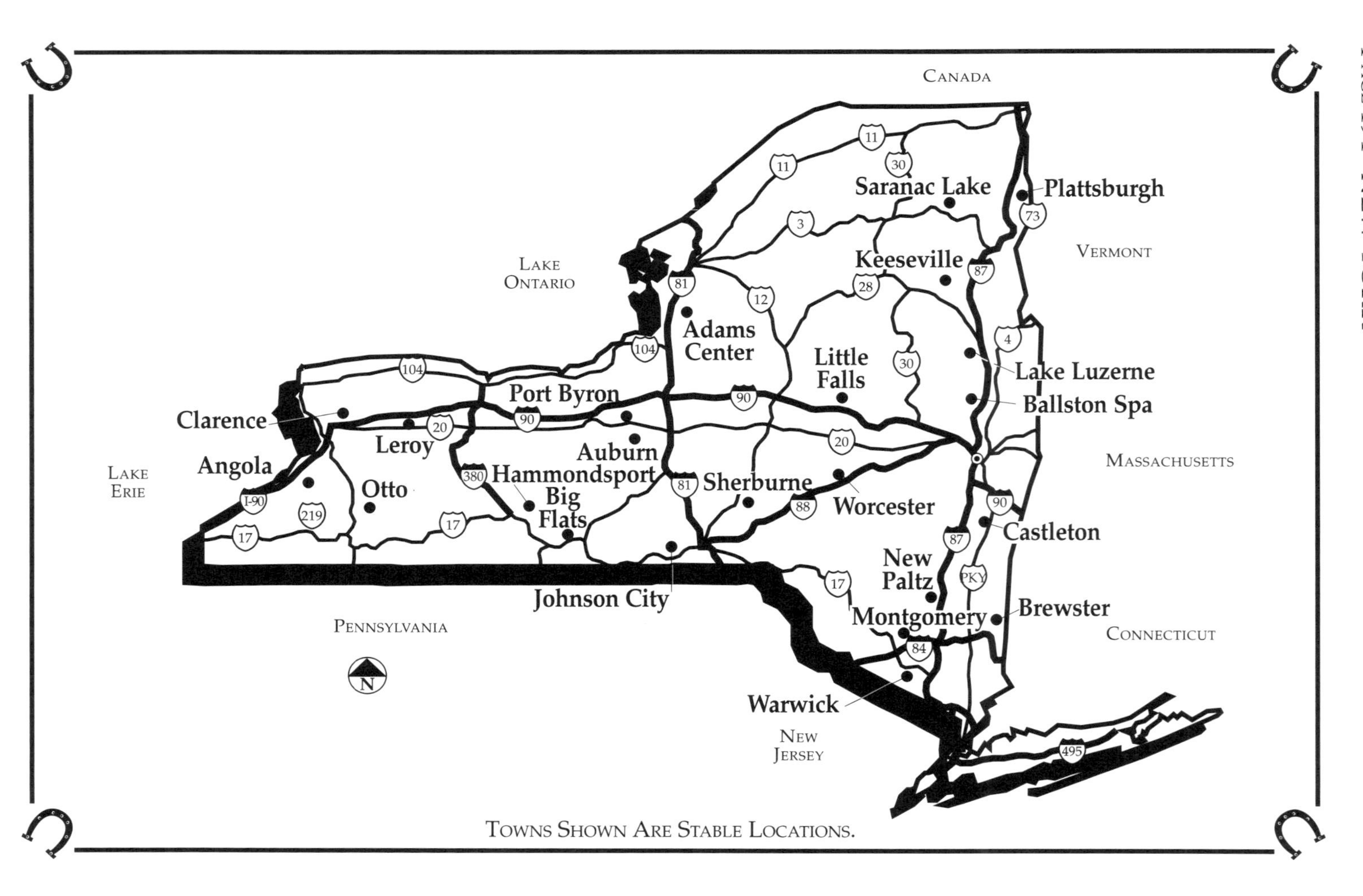

Towns Shown Are Stable Locations.

ALL OF OUR STABLES REQUIRE CURRENT NEG. COGGINS, CURRENT HEALTH PAPERS, & OWNERSHIP PAPERS.

ADAMS CENTER
Royal Stables **Phone: 315-583-6429**
Mary Ramsey
Green Settlement Road [13606] Directions: 1/2 mile off of I-81 at Exit 42. Call for directions. **Facilities:** 27 indoor stalls, 2 acres of pasture/turnout, 60' x 110' indoor arena, 100' x 200' outdoor arena, round pen, 4 paddocks, feed/hay & trailer parking on premises. Beautiful, scenic riding trails. Full care boarding facility that also offers training of horses & riders in English, Western, jumping, etc. at all levels. Must have rabies certificate. Call for reservation. **Rates:** $15 per night. **Accommodations:** Days Inn & Ramada Inn 6 miles from stable.

ANGOLA
Rosie Acres Boarding Stables **Phone: 716-337-3490**
Mare Russo
9315 Versailles Plank Road [14006] Directions: Exit 57A off of I-90. Call for directions. **Facilities:** 40 indoor box stalls, 60' x 140' indoor arena, 100' x 120' outdoor arena, wash rack, 6 paddocks, 5 separate fenced pasture areas, 26 acres of cut trails. Lounge, kitchen facilities, & bathrooms. Training of horses & riders offered at stable. Horses for lease. Call for reservation. **Rates:** $25 per night; weekly rate available. **Accommodations:** Motels 8 miles away.

AUBURN
Christmas Tree Stables **Phone: 315-255-6138**
Sherry Lombardo **or: 315-776-8952**
RD #5 Northrup Road [13021] Directions: Located 10 miles off NY State Thruway (I-90). Call for directions. **Facilities:** 34 indoor stalls, 5 acres of fenced pasture with run-in sheds, 60' x 150' indoor arena, 120' x 240' outdoor arena, riding trails, feed/hay & trailer parking on premises. Heated tack room with bathrooms. Training of horses & riders in English & Western at all levels. 6,000 Christmas trees grown on 82 acres at stables. Call for reservation. **Rates:** $15 per night. **Accommodations:** Dilaj Inn on Rt. 34 & Holiday Inn both 5 miles from stable.

BALLSTON SPA
Rolling Meadows Farm **Phone: 518-885-3248**
Susan McGrath
161 White Road [12020] Directions: Call for directions. **Facilities:** 3 indoor 10' x 12' stalls, outdoor paddock with run-in shed, indoor arena, feed/hay & trailer parking. Well-behaved dogs only on farm. Please call for reservation. **Rates:** $15 per night; $75 per week. **Accommodations:** Bed & Breakfast on premises with 2 guest rooms, $55-$95 per night depending on season. No smoking.

ALL OF OUR STABLES REQUIRE CURRENT NEG. COGGINS, CURRENT HEALTH PAPERS, & OWNERSHIP PAPERS.

BIG FLATS
Gale's Equine Facility **Phone: 607-796-9821**
Gale Wolfe
219 Sing Sing Road [14845] Directions: Call for directions. **Facilities:** 5 indoor stalls, two 100' x 200' turnout areas, indoor arena, feed/hay & trailer parking available 24-hr advance notice preferred. **Rates:** $15 per night. **Accommodations:** Howard Johnson's & Holiday Inn in Horseheads, 15 minutes from stable.

BREWSTER
Big Elm Farm **Phone: 914-279-6736**
113 Big Elm Road [10509] Directions: Call for directions. **Facilities:** 51 indoor 10' x 12' and 12' x 14' stalls, all size paddocks, 3 outdoor rings, insulated indoor ring 150' x 100', feed/hay, trailer parking available. Full-service facility on 85 acres, trails on property, call for reservations. **Rates:** $30 per night, $180 per week. **Accommodations:** Motels within 5 minutes.

CASTLETON
Brookside Stables **Phone: 518-479-4363**
Donald Bushnell
2475 Kraft Road [12033] Directions: I-90 E to Exit 10: Take right off exit; at light, take left & go .7 mile; take left on Kraft Rd.; Drum Veterinary on corner; follow road to end; stable is at the dead end. **Facilities:** 5 indoor 12' x 12' box stalls, large & small pasture/turnout areas, large indoor & outdoor riding arenas, feed/hay & trailer parking available. Stable offers pony parties for birthdays. **Rates:** $15 per night. **Accommodations:** 3 or 4 motels in Schodack, 5-8 minutes from stable.

CLARENCE
N.C. Bechtel Training Stables, Inc. **Phone: 716-741-3688**
Nancy Bechtel **Home: 716-741-9228**
5505 Old Goodrich Road [14031] Directions: 1.5 miles off of Rt. 5 & 3 miles east of I-78. Call for directions. **Facilities:** 35 indoor box stalls, 80' x 120' indoor ring, 150' x 280' outdoor sand ring, 2 paddocks, 3 pastures, jump course, feed/hay & trailer parking available. Training of horses & riders in hunter/jumper at all levels. Horses for sale. Call for reservation & availability. **Rates:** $15 per night. **Accommodations:** Motels 3 miles from stable.

ALL OF OUR STABLES REQUIRE CURRENT NEG. COGGINS, CURRENT HEALTH PAPERS, & OWNERSHIP PAPERS.

HAMMONDSPORT

Donameer Farm **607-569-2115**

Cynthia Harrison, Neal Esposito **E-mail: donameer@empacc.ney**

7417 Smallige Road (14840) Directions: I-86 to exit 40 (Savona) 415N to Robie Rd. Follow 2 miles, take left at fork onto Velie Rd. Velie Rd. Turns into Smallige Rd. Farm on left. **Facilities:** 10- 12x12 covered stalls. Hay, trailer parking available. 10 fenced in acres, 8- 100′ X 100′ foot paddocks. 12-30 amp/water hookups, dump station, trails and dirt roads. 10 miles from Sugar Hill State Park; 50 miles of trails. Well-behaved dogs are welcome. Negative Coggins. **Rates:** Nightly- $20/night/horse, $25 hookup. Weekly-$10/night/horse, $20 hookup. **Accommodations:** 1 furnished bedroom cottage, $55/night. Days Inn/Bath and Vinehurst Motel each within 5 miles.

JOHNSON CITY

Kit-Mar Farm **Phone: 607-798-1465**

John Cummings

Ask for Mary or Kitty Cummings, 810 East Maine Road [13790] Directions: 5 minutes off I-81, I-88, & Rt. 17. Call for directions. **Facilities:** 30 indoor stalls in 2 barns, 60' x 120' indoor arena, 150' x 150' outdoor arena, fenced paddocks, several acres of pasture, wash racks, feed/hay & trailer parking. Breeding of Appaloosas at farm. 2 registered Appaloosa stallions standing-at-stud. Also, training of horses & riders in hunt & stock seat. Call for reservation. **Rates:** $15 per night; $75 per week. **Accommodations:** Motels 5 miles from stable.

KEESEVILLE

AuSable Meadows Farm **Phone: 518-834-7660**

Denise Lussier

468 Dugway Road [12944] Directions: 10 minutes from Exit 33 off of I-87. Call for directions. **Facilities:** One 12' x 12' indoor box stall plus 1/8 acre of isolated pasture/turnout, feed/hay and trailer parking available. Call in advance (24 hours) for reservation. **Rates:** $20 per night. **Accommodations:** Bed & breakfast in 1840 farm house on premises. 2 rooms with private baths. No smoking and no pets please. Also motels in Keeseville, 4 miles from farm.

LAKE LUZERNE

Bennett's Riding Stable **Phone: 518-696-4444**

Lawrence & Bonnie Bennett

RR 2, Box 208 Gage Hill Road [12846] Directions: From I-87: Take Rt. 9N south towards Lake Luzerne for 5 miles. Stable is on the left-hand side. **Facilities:** 4 indoor stalls, 5 acres of pasture/turnout, sweet feed & hay available, trailer parking on premises, easy access to beautiful state riding trails plus many other trails within short driving distance. Portable round pen also available. Reservation required. Guided horseback riding on their horses from 1 hour to all day. **Rates:** $15 per night; $100 per week. **Accommodations:** Kastner's Motel & Pine Point Motel & Cottages, & Nancy Lee Motel all within walking distance in Lake Luzerne.

ALL OF OUR STABLES REQUIRE CURRENT NEG. COGGINS, CURRENT HEALTH PAPERS, & OWNERSHIP PAPERS.

LEROY

Jensen Stable **Phone: 716-768-8452**

Abigail Jensen

7077 West Main Street [14482] Directions: I-90 to LeRoy Exit, south on Rte. 19 into LeRoy, turn right at light onto Rte. 5, stable 1 mile on right. **Facilities:** Box stalls, pastures, paddocks, indoor arena available. Feed, hay and trailer parking available. Certified instructor. Late arrivals welcome. **Rates:** $25 per night. **Accommodations:** Motels and Hotels colose by.

LITTLE FALLS

Diamond Hill **Phone: 315-429-3527**

Adam (Mike) Miller **Barn Manager: 315-429-3514**

RD 1, Dairy Hill Road [13365]

Directions: Exit 29A on NY State Thruway (Little Falls) to Rt. 169; make right at flashing light; make left at Rt. 167; travel thru Dolgeville; bear left on State St.; go 3.4 mi. to stop sign; cross intersection; go 1.9 mi. Diamond Hill on left. **Facilities:** 15 indoor 12' x 12' box stalls, 7 paddocks of various sizes that can accommodate 20 horses, auto waterers, wash bay, jumping ring, sand warm-up ring, indoor arena, beautiful riding trails on property, tack shop on premises. To meet your needs, reservations are necessary. "We appreciate early arrivals." In the case of delays, please call ahead. This equestrian center has been host to dressage clinics, charitable events, shows and tack & horse auctions. **Rates:** $20 per night; $18 per day weekly rate. **Accommodations:** Best Western in Little Falls, 8 miles away, Adriana's B & B in Dolgeville, 5 miles awau, Herkimer Campgrounds 14 miles from stable.

MONTGOMERY

Highland Farm **Phone: 845-361-2204**

Yvonne Turchiarelli

2101 Rt. 17K [12549] Directions: From I-84 W: Take Exit 5, right turn onto Rt. 208 N; go 1 mile to light at intersection of 208 & 17K; turn left onto 17K & go 5.6 miles to farm on left. **Facilities:** 24 indoor stalls, 8 outdoor stalls, several varied sizes of pasture/turnout, paddocks, feed/hay & trailer parking available. This farm has a clean barn & experienced help. Vet nearby. **Rates:** $25 per night. **Accommodations:** Comfort Inn in Newburgh, 12 miles; Harvest Inn in Pine Bush, 8 miles from stable.

NEW PALTZ

Million Dollar Farm Ltd. **Phone: 914-255-TROT**

Frank Heyer

300 Springtown Road [12561] Directions: NY Thruway (I-87) Exit 18: Left on 299; go thru New Paltz for 1.8 miles over bridge; take first right on Springtown Rd.; go 3.2 miles; farm is on right. **Facilities:** 6 indoor box stalls, 23 indoor straight 5' x 10' stalls, 125' x 220' pasture/turnout, feed/hay & trailer parking on premises. "We have the best trails in the county." (NYC aqueduct, Mohonk Trails, Railtrail, Minnewaska State Park, plus trails on the farm). **Rates:** $10 for straight stall; $15 for box stall. **Accommodations:** Super 8, Day Stop Inn, Motel 87, & EconoLodge all in New Paltz, 5 miles from stable.

ALL OF OUR STABLES REQUIRE CURRENT NEG. COGGINS, CURRENT HEALTH PAPERS, & OWNERSHIP PAPERS.

OTTO

R & R Dude Ranch **Phone: 716-257-5140**

Alice Ferguson

8940 Lange Road [14766] Directions: 8 miles west off Hwy 219; 45 miles south of Buffalo. Call for directions. **Facilities:** 27 box stalls (12' x 15', 12' x 30', 10' x 12'), paddocks (30' x 30', 1/2 acre, 2 acre), 100 X 200 outdoor arena, feed/hay, trailer parking, pasture/turnout from 1/2 acre to 5 acres. Horse camp in summer, 9 miles of endurance trails, creek for exercising horses or swimming. Horseback riding offered (12 miles of trails). 12,000 acres of state land trails available. Arab stud available. **Rates:** $10 per night. **Accommodations:** Bed & Breakfast on premises. Jaccuzi.

PLATTSBURGH

Cedar Knoll Farm **Phone: 518-561-6003**

Susan Castine **or: 518-561-8391**

246 Spellman Road [12901] Directions: Exit 40 off I-87, 1/4 mile west; 30 miles south of Canadian border. **Facilities:** 4 indoor 12' x 12' box stalls, 40-acre pasture, 200' x 300' paddock, feed/hay, trailer parking available, call for reservations. **Rates:** $25 per night, $100 per week. **Accommodations:** Restaurant and motel within 1/4 mile.

PORT BYRON

Snug Horse Haven **Phone: 315-776-8243**

Jeri Marshall

RD #1 Maiden Lane Road [13140] Directions: Off of I-90 at Exit 40. Call for directions. **Facilities:** 4 indoor stalls, two 75' x 100' paddocks, large round pen, 120' x 200' outdoor ring, full jump course, feed/hay & trailer parking on premises, horse training available at this small comfortable facility. Call for reservation. **Rates:** $15 per night. **Accommodations:** Port Forty Motel & Best Western 5 miles away.

SARANAC LAKE

Sentinel View Stables **Phone: 518-891-3008**

Carol Shante

Harriettstown Road [12983] Directions: Located on Rt. 86. Call for directions. **Facilities:** 5 indoor stalls, 3 holding paddocks, riding trails on top of the Adirondack Mountains, trailer parking available. English and Western riding lessons offered. Call for reservation. **Rates:** $10 per night; ask for weekly rate. **Accommodations:** Motels 3 miles from stable.

SHERBURNE

Brookfield Trail System **Phone: 607-674-4036**

Robert Patrick, Supervising Forester

P.O. Box 594 [13460] Directions: Off Rt. 12. North of Sherburne between Sherburne & Bridgewater. Call for directions. **Facilities:** Covered & uncovered tie stalls in camping area. 130-mile trail system within 13,000 acres of state forest with 25 miles of maintained dirt access roads throughout forest. Camping free without permit for up to 3 nights. Large group activities must obtain permit. Handicapped mounting platform available. General store next to camping area. Food & hot showers available. **Rates:** No charge unless it is a large group.

ALL OF OUR STABLES REQUIRE CURRENT NEG. COGGINS, CURRENT HEALTH PAPERS, & OWNERSHIP PAPERS.

WARWICK

Meadowood Farm, Inc. **Phone: 914-986-7387**
Cindy Van der Plaat
82 Belcher Road [10990] Directions: 20 minutes off of Rt. 17. Call for directions. **Facilities:** 14 indoor box stalls, two 2-acre pastures, two 1/2-acre pastures, 60' x 160' indoor arena, dressage arena, jumping field, round pen, 5 outdoor paddocks, heated wash stall, trailer parking on premises. Training of horses & riders in event riding and hunter/jumper. Call for reservation. **Rates:** $15 per night; ask for weekly rate. **Accommodations:** Motels within 7 miles of stable.

WORCESTER

Robinson Stables **Phone: 607-397-8275**
Paul Robinson
Brady Road [12197] Directions: Exit 19 off of I-88. Call for directions. **Facilities:** 4 indoor stalls, many acres of pasture, 200' x 200' outdoor ring, riding trails, feed/hay & trailer parking available. Buying & selling of horses offered at stable. Call for reservation. **Rates:** $15 per night; $75 per week. **Accommodations:** Motel 8 & Holiday Inn in Oneonta, within 20 minutes.

DON'T LOOK NOW SHERIFF BUT THAT OZ GANG IS BACK IN TOWN AN' LOOKS LIKE THEIR SPOILIN' FER A FIGHT!
MERCHANTILE
SAL
NAILS
FRED LEARY

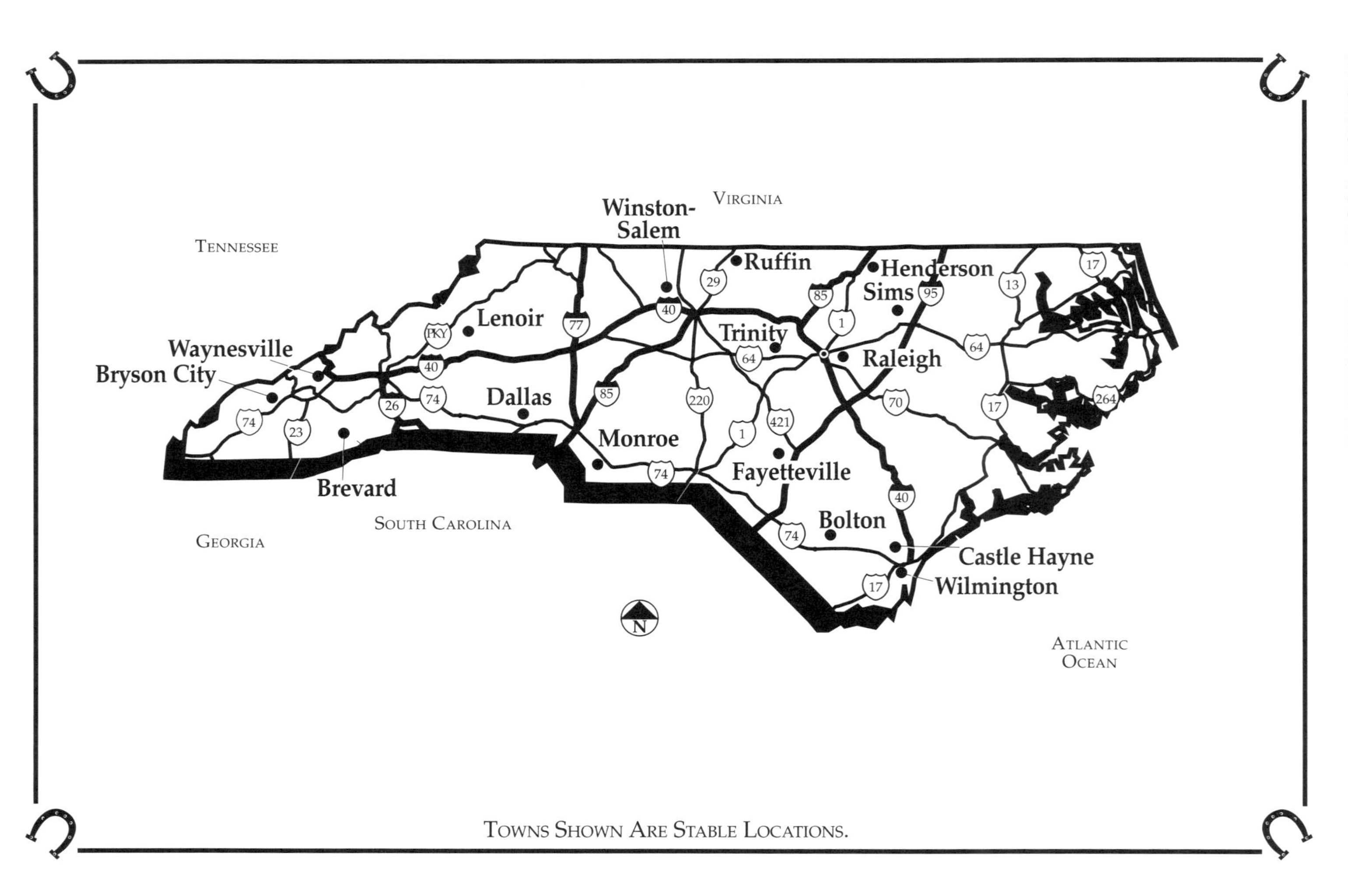

TOWNS SHOWN ARE STABLE LOCATIONS.

ALL OF OUR STABLES REQUIRE CURRENT NEG. COGGINS, CURRENT HEALTH PAPERS, & OWNERSHIP PAPERS.

BOLTON
Deerfield Stables **Phone: 910-655-2814**
Diana Smith
Rt. 1, Box 235 Hwy 74-76 [28423] Directions: 30 minutes from I-40 and located on Hwy 74-76. Call for directions. **Facilities:** 5 indoor stalls, 8 acres of pasture, feed/hay & trailer parking, near Myrtle Beach. Reservation required. **Rates:** $10 per night; $50 per week. **Accommodations:** Bed & Breakfast on premises with 3 rooms.

BREVARD
Las Praderas Stables and Cottages **Phone: 704-883-3375**
Nancy Searles
Rt. 1, Box 12-A [28712] Directions: From I-26: Take Asheville Exit 9 (Airport exit) then Rt. 280W to center of Brevard (Broad & Main approx. 18.5 miles from airport); take 276 South for 7.1 miles then left onto See Off Mountain Rd.; go 1.7 miles to Las Praderas sign. **Facilities:** 33 indoor box stalls, feed/hay & trailer parking available, no tractor-trailers, please, at least 24-hr notice. **Rates:** $20 per night; $100 per week. **Accommodations:** Beautiful lodging available for a 2-night minimum stay. Also Imperial Motor Lodge in Brevard, 10 miles from stable. Call for free brochure.

BRYSON CITY
Double Eagle Farm **Phone: 704-488-9787**
Greg & Karen Crisp
50 Sawmill Creek Road [28713] Directions: 1 mile off Hwy 74 (4-lane), easy access. Call for directions. **Facilities:** 14 indoor 12′ x 12′ stalls each with window, skylight, ceiling fan; several 1/2-acre turnouts; feed/hay & trailer parking available. Close to many trails in Great Smoky Mountains National Park, Nantahala National Forest, Pisgah National Forest. Trail information & maps available. Call for reservations. **Rates:** $20 per night; weekly rate negotiable. **Accommodations:** Many motels nearby. Will mail list.

CASTLE HAYNE
Castle Stables, Inc. **Phone: 910-675-1113**
Debi Mastrangelo
1513 Sidbury Road [28429] Directions: 2-3 miles from Exit 414 off of I-40. Call for directions. **Facilities:** 10 indoor 12' x 12' & 12' x 14' stalls, 11 paddocks & pastures, 2 indoor wash racks with hot water, 2 riding rings, 1 with jumps and lighting. 13 miles from beach. No stallions without prior discussion. 24-hr advance notice. **Rates:** $15 including feed; weekly rate negotiable. **Accommodations:** Fairfield Inn & Days Inn 5 miles from stable.

ALL OF OUR STABLES REQUIRE CURRENT NEG. COGGINS, CURRENT HEALTH PAPERS, & OWNERSHIP PAPERS.

CLAYTON

The Grove at Rock Ridge **Phone: 919-553-3255**
Linda Sewall **Stable: 919-291-5617**
124 Canyon Road [27520] Directions: From I-95: take Exit 116 (Hwy 42); go west to Rock Ridge-School Road & turn right; go 1/2 mile to Rock Ridge-Sims Road & turn right; go .9 mile to 1st intersection (Boykin Road) & turn left. Farm is 1/2 mi. on right. **Facilities:** 10 indoor stalls on center aisle of barn, six 1/2-acre paddocks, riding ring, feed/hay if arranged in advance, and trailer parking, call first number for reservations - if you get machine, please leave a message and your call will be returned. **Rates:** $15 per night; $50 per week. **Accommodations:** Hampton Inn and Comfort Inn in Wilson, 10 miles & campground nearby.

DALLAS

Rose Hill Farm & Equestrian Training Center **Phone: 704-922-0866**
Sandra Digby
702 Ike Lynch Road [28034] Directions: From I-85, take 321N to Hardin Rd. exit; turn left & go straight 1.5-2 miles. Hardin Rd. turns into Ike Lynch Rd. **Facilities:** 5 indoor stalls, pasture/turnout areas, feed/hay & trailer parking available. Historic areas and Dallas Horse Park 5 minutes away, as much notice as possible. **Rates:** $15 including feed. **Accommodations:** Hampton Inn & Holiday Inn within 10 minutes.

FAYETTEVILLE

Arrowhead Farms **Phone: 910-425-3631**
Phyllis L. Jonke
1803 Strickland Bridge Road [28304] Directions: From I-95 S, take Exit 56. From I-95 N, take Exit 40. Take Rt 301 to Owen Drive; take Owen Dr. to Raeford Rd; left on Raeford to Strickland Bridge; take left & stable is 1.5 miles on left. **Facilities:** 15 indoor 12' x 12' stalls, pasture/turnout available, trailer parking & feed/hay if needed, please call for reservations and call to cancel if you cannot make it. **Rates:** $15 per night; $75 per week. **Accommodations:** Comfort Inn, Innkeeper, & Fairfield Inn all about 5 miles from stable.

HENDERSON

Burnside Plantation **Phone: 252-438-7688**
Agnes & George Harvin **Web: burnsideplantation.com**
960 Burnside Road [27537] Directions: 10 miles from I-85. Call for directions. **Facilities:** 3 indoor 12' x 14' & 1 indoor 14' x 14' stalls, pasture/turnout. Many miles of trails with maps/guide provided. Within minutes of Kerr Lake - a premium fishing & boating facility. Owners are breeders of registered Morgan horses for sport & pleasure. **Rates:** $15 per stall per night; ask for weekly rate. **Accommodations:** Restored 19th-century guest house on property.

ALL OF OUR STABLES REQUIRE CURRENT NEG. COGGINS, CURRENT HEALTH PAPERS, & OWNERSHIP PAPERS.

LENOIR

Moore's Horseplay Ranch **Phone: 704-757-9114**

Everette & Kathy Moore

Rt. 2, Box 727, 1154 Woodrow Place [28645] Directions: I-40 to 64/90. Less than 1 mile off Hwy 64/90. In the foothills of Blue Ridge Parkway. Call for directions. **Facilities:** New stall barn, 4 indoor 10' x 10' stalls, 6 outdoor 3' x 8' tie-up stalls, corrals, feed/hay & trailer parking available. Room for oversized trailers. Ranch offers riding lessons & miles of riding trails in a beautiful vacation area offering camping, fishing, hiking, etc., unlimited area for horse camping for portable stalls, cross ties, & picket lines, 9 indoor stalls in camping area. Western Town & tack shops nearby. Beautiful setting for horse vacations. Home of black & white national spotted saddle horse stallion: "Sepper Go Devil" (Go Boy). Must have reservations. **Rates:** $10-$20 per night. **Accommodations:** Log cabins for rent on premises plus large campsites. Also Holiday Inn & Days Inn in Lenoir, 7 miles from ranch.

MONROE

Kathy's Stables **Phone: 704-753-4397**

Kathy Hardy

7605 Carriker Williams Road [28110] Directions: Hwy 74 to Hwy 601. Call for further directions. **Facilities:** 6 indoor stalls, two 3-acre fenced pastures on a total of 10 acres. Feed/hay & trailer parking available. Reservation required. **Rates:** $20 per night; ask for weekly rate. **Accommodations:** Motels 10 miles from stable.

RALEIGH

Triton Stables **Phone: 919-847-4123**

Ellen Welles **Evenings: 919-847-5446**

9901 Macon Road [27613]

Directions: From I-85 S, take Exit 86 for Creedmore/Buttoner; take Hwy 50 towards Raleigh for 12 miles; at stop light turn left on Norwood; go 1.75 miles and take left on Macon, stable is 1 mile on left. **Facilities:** 5 indoor stalls, no pasture/turnout. Call for reservation. **Rates:** $15 per night w/o feed.

RUFFIN

Chestnut Hill Stables & Riding School **Phone: 910-939-7126**

Carole Moore

630 Mayfield Road [27326] Directions: From I-29 N, take Mayfield Road exit; turn left at end of ramp. From I-29 S take right at end of ramp. Stable is 3/4 mile on left. **Facilities:** 10 indoor stalls, pasture/turnout, feed/hay at additional charge, and trailer parking available. Reservation preferred but will help in an emergency with no notice. **Rates:** $10 per night. **Accommodations:** Holiday Inn and Comfort Inn 8 miles from stable.

ALL OF OUR STABLES REQUIRE CURRENT NEG. COGGINS, CURRENT HEALTH PAPERS, & OWNERSHIP PAPERS.

TRINITY

Never Done Acres **Phone: 910-475-1914**

Michael O'Neill & Pamela Merritt

3124 Finch Farm Road [27370] Directions: 4 miles south of Exit 106 (Finch Farm Road) off of I-85. Call for directions. **Facilities:** 3 indoor stalls, three 2-acre pastures, lighted riding ring, feed/hay, & trailer parking available. Vet & farrier on call. Camper hook-ups. Dog pen available. Call in advance. **Rates:** $15 for stall per night; $5 for pasture. **Accommodations:** Bed & Breakfast on premises, $20 per room & no smoking in house. Days Inn & Ramada Express in Thomasville, 6 miles away.

WAYNESVILLE

Dye-Na-Mite Show Barn **Phone: 704-627-2666**

Martha Dye, owner, Nancy Campbell, manager

300 Witch Way [28786] Directions: 2.5 miles from I-40 at Exit 24. Call for directions. **Facilities:** 10 indoor 12' x 12' stalls, 84' x 100' indoor arena, small pasture, feed/hay & trailer parking available. Reservation required. **Rates:** $15 per night; $75 per week. **Accommodations:** Several motels 6 miles away.

WILMINGTON

Wolffdune Farms **Phone: 910-763-9999**

Julia May Gannon

2528 Castle Hayne Road [28401] Directions: Located on Hwy 117 off of I-40. Call for directions. **Facilities:** 10 indoor stalls, 4 board-fenced pasture/turnout areas, feed/hay & trailer parking. 10 miles from beach. 24-hr advance notice required. **Rates:** $12 per night; weekly rate negotiable. **Accommodations:** Howard Johnson's, Hilton, & Holiday Inn within 5 miles.

NOTES AND REMINDERS

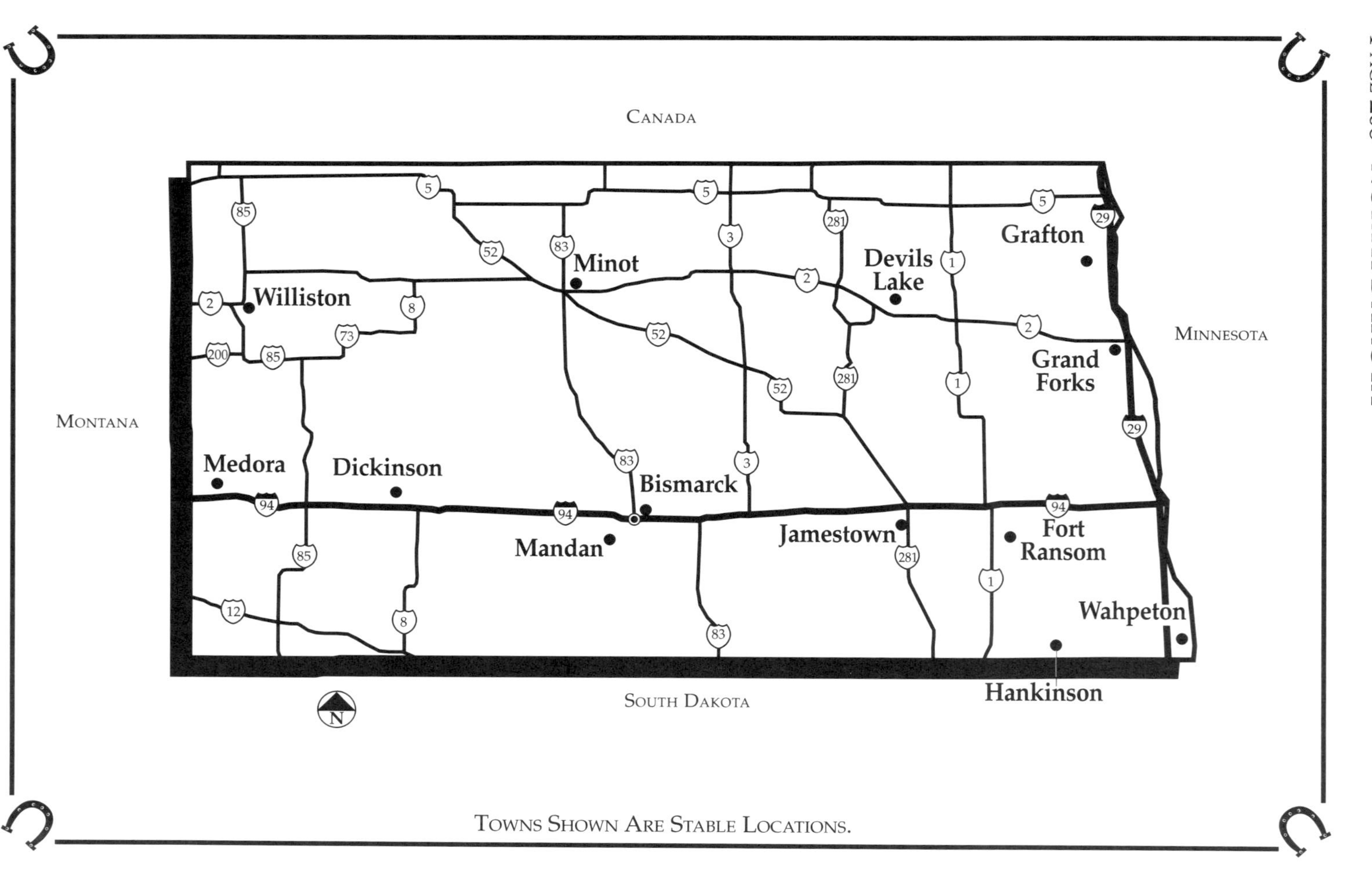
Canada
Montana
Minnesota
South Dakota
Williston
Minot
Devils Lake
Grafton
Grand Forks
Medora
Dickinson
Bismarck
Mandan
Jamestown
Fort Ransom
Wahpeton
Hankinson
N
Towns Shown Are Stable Locations.

ALL OF OUR STABLES REQUIRE CURRENT NEG. COGGINS, CURRENT HEALTH PAPERS, & OWNERSHIP PAPERS.

BISMARCK
Selland Indoor Arena **Phone: 701-255-1420**
Lee Selland
6600 N.E. 26th Street [58501] Directions: Located off of I-94. Call for directions. **Facilities:** 68 indoor stalls, outside pens, indoor arena, 2 outdoor arenas, feed/hay & trailer parking available. Open 24 hours daily. Call ahead for reservation. **Rates:** $10 per night; weekly rate available. **Accommodations:** Embassy Suites & Ramada Inn 5 minutes from stable.

DEVILS LAKE
Rush Valley Farms **Phone: 701-662-4386**
Kevin Frith
Rt. 3, Box 340 [58301] Directions: Located off of Old Hwy 2, two miles outside of town. Call for directions. **Facilities:** 12 indoor stalls, pasture/turnout area, feed/hay & trailer parking available. Call for reservation. **Rates:** $10 per night. **Accommodations:** Many chain motels 2 miles from stable.

DICKINSON
Dakota Stables **Phone: 701-225-0240**
Dr. Kim Brummond
Radar Base Road [58601] Directions: Take second Dickinson exit off of I-94. Call for directions. **Facilities:** Up to 10 indoor stalls, outdoor pens, feed/hay at extra charge, trailer parking available. Stalls are large & heated. This stable is also veterinary clinic. Same-day notice needed. **Rates:** $6 per night. **Accommodations:** Super 8 & Hospitality Inn within 3 miles of stable.

FORT RANSOM
Fort Ransom State Park **Phone: 701-973-4331**
John Kwapinski
5981 Walt Hjelle Parkway [58033-9712] Directions: Follow signs. 12 miles off Hwy 46, 34 miles from I-94 (Exit 292), 34 miles south of Valley City. **Facilities:** 16 - 16' x 16' board-fenced pens, room to set up portable corrals, water and trailer parking available, electric hook-ups nearby. 900-acre park in Sheyenne River Valley; wooded, hilly park with 5 miles of riding trails. **Rates:** $4 per night per horse. Vehicle entrance fee $4 per vehicle. **Accommodations:** Primitive camping near corral area $11 per night, $16 per night with electricity. Campsites with electricity are 1 mile from horse corral. Island Park Motel and Super 8 Motel in Lisbon, 21 miles away. Viking View Resort (cabins) 2 miles south of park.

GRAFTON
Thompson Stables **Phone: 701-352-2732**
Charles Thompson **701-352-2410**
527 West 15th Street [58237] Directions: Call for directions. **Facilities:** 2 indoor stalls, pasture/turnout area, feed/hay & trailer parking available. Call for reservation. **Rates:** $15 per night; weekly rate available. **Accommodations:** Super 8 two minutes from stable.

ALL OF OUR STABLES REQUIRE CURRENT NEG. COGGINS, CURRENT HEALTH PAPERS, & OWNERSHIP PAPERS.

GRAND FORKS

Kuster's Wagon Wheel Stables **Phone: 701-772-6526**

Myron & Joyce Kuster

Rt. 1, Box 226 [58201] Directions: Located 1 3/4 mile southwest of Grand Forks city limits (I-29 & 32nd Ave. S). Call for directions. **Facilities:** 24 indoor stalls, 55' x 100' indoor riding arena, trailer parking. Call for reservation. **Rates:** $10 per night; negotiate longer stays, includes feed/hay. **Accommodations:** Many motels approx. 2 miles into town.

HANKINSON

Country Acres Veterinary Clinic **Phone: 701-252-7133**

Dr. Barb Looysen

8279 37R Street S.E. [58401] Directions: Less than 1 mile from the Jamestown Exit off of I-94. Call for directions. **Facilities:** 4 indoor box stalls, 4 acres of fenced pasture/turnout, holding pen, feed/hay & trailer parking. Complete veterinary care available on premises. Call for reservation. **Rates:** $15 per night. **Accommodations:** Comfort Inn, Dakota Inn, & Super 8 all 1 mile from clinic.

MEDORA

Little Missouri Horse Co. **Phone: 701-623-4496**

Wally Owen

P.O. Box 8 [58645] Directions: Off of I-94, 1 mile south of Medora, Theodore Roosevelt National Park and Sally Creek State Park. **Facilities:** 6 - 8 X 10 outdoor stalls with primitive camping next to Little Missouri River. Well-marked riding trails. **Rates:** $5 per horse. **Accommodations:** $8 per night camping, also 2 bedroom fully equipped cabin (sleeps 6) with private corrals. **Rates:** $95.75 per night plus $5.00 per horse. Motels within 6 miles.

MINOT

Sonny's Stables **Phone: 701-839-5351**

Sonny Ehr

Rt. 5, Box 410 [58701] Directions: Located 1/2 mile from major junction in Minot. Call for directions. **Facilities:** At least 6 indoor stalls, 100 acres of pasture/turnout, feed/hay & trailer parking available. Excellent location. Farrier on premises. This is a training facility for barrel racing & roping. Call for reservation. **Rates:** $10 per night; weekly rate available. **Accommodations:** Most major motels 1/2 mile from stable.

WAHPETON

Sherven's Stables **Phone: 701-642-2544**

Ron Sherven

17880 Hwy 13 [58075] Directions: 8 miles east of I-29. Call for directions. **Facilities:** 10 indoor stalls, pasture/turnout, trailer parking, & feed/hay available. Overnight boarding in emergency situations only. Call ahead for inquiries. **Rates:** $10 per night; weekly rate negotiable. **Accommodations:** 3 to 4 motels within 3 miles.

ALL OF OUR STABLES REQUIRE CURRENT NEG. COGGINS, CURRENT HEALTH PAPERS, & OWNERSHIP PAPERS.

WILLISTON

Don Horob Stables & Feed Grounds **Phone: 701-572-3047**

Don Horob

Rt. 4, Box 60 [58801] Directions: Call for directions. **Facilities:** 3 outdoor & indoor pens able to hold 20 horses, feed/hay & trailer parking. Has cattle on lot but can make room when someone needs a layover. Call ahead for reservation. **Rates:** $15 per night. **Accommodations:** Travel Host Motel & International Inn within 1/2 mile of stable.

TOWNS SHOWN ARE STABLE LOCATIONS.

ALL OF OUR STABLES REQUIRE CURRENT NEG. COGGINS, CURRENT HEALTH PAPERS, & OWNERSHIP PAPERS.

ASHTABULA

Koch Show Horses **Phone: 440-224-2097**

Jim & Anne Koch

2560 Plymouth-Gageville [44004] Directions: I-90 to Kingsville exit (#235, Rt. 193 South) to second crossroad (Plymouth - Gageville) turn right. Farm is 1.8 miles on left. Large illuminated farm sign. **Facilities:** 35 stalls, all indoor. Size ranges from 10' X 10' to 12' X 12'. Outdoor turnout w/large run-in shed available. Feed, Hay and trailer parking available. Large circular drive and parking area. Seven individual turnouts w/split rail fencing, 1 acre each. Large indoor riding arena, Indoor Round Pen. Owner operated. Onsite supervision of horses at all times. The horse business is our only business. Forty years professional equine experience. **Rates:** $15 per day, $80 per week. **Accommodations:** Three motels 3 to 7 miles away.

BATAVIA

East Fork Stables & Lodge **Phone: 513-797-7433**

George Wisbey

2215 Snyder Road [45103] Directions: Call for directions. **Facilities:** 32 indoor stalls, indoor arena, several fenced paddocks with run-in shelters, 50 miles of trails, feed/hay & trailer parking available. Horse rentals and trail rides. Reservation required and at least 24-hr notice. **Rates:** $20 per night. **Accommodations:** 1 cabin on premises available for rent. Holiday Inn & Red Roof Inn within 15 minutes.

BLUE ROCK

McNutt Farm II/Outdoorsman Lodge **Phone: 740-674-4555**

Don R. & Patty L. McNutt

6120 Cutler Lake Road [43720] BED BREAKFAST & BARN, Overnight for humans and their critters. Advance reservation required with 50% deposit, balance in cash on arrival. No alcoholic beverages. **Directions:** 11 miles from I-70, 35 miles from I-77, & 55 miles from I-71. **Facilities:** Secure stalls and trailer parking. Check-in 5-7 P.M.; check-out by 9 A.M. Overnight boarding is for lodge guests only. **Rates:** $40 per night per person, $10 per night per horse, pets allowed at $5.00. For weekenders or by the week, ask about our vacationers rates. **Accommodations:** Rooms at the Lodge type farm house, Log Cabin, Carriage House, the Cellar building at a vacation destination offering trail rides, hiking, fishing, etc. We pride ourselves on serving your needs and the needs of your traveling equine.

CENTERVILLE

Menker's Circle 6 Farm **Phone: 513-885-3911**

Robert Menker

11090 Yankee Street [45458] Directions: Call for directions. South suburban area of Dayton; 2.1 miles south of I-675, less than 1 mile from I-75; 35 miles north of Cincinnati. **Facilities:** 78 indoor 10' x 12' stalls, various sized turnouts, 60' x 130' indoor arena, outdoor arena, round pen, feed/hay, trailer parking, and self-contained campers welcome. Specializing in quarter horses. Call for reservations. **Rates:** $25 per night. **Accommodations:** Motels and shopping within 2 miles of stable.

ALL OF OUR STABLES REQUIRE CURRENT NEG. COGGINS, CURRENT HEALTH PAPERS, & OWNERSHIP PAPERS.

CIRCLEVILLE
Synergy Farm **Phone: 614-474-1129**
Pat Whalen-Shaw
7041 Zane Trail Road [43113] Directions: US 23 S from Columbus to Tarlton Rd. light (past Circleville exits); turn east to Kingston Pike; turn south & go to Hayesville Rd and take left; 1 mile go straight onto Zane Trail. Farm is 1.2 mile east of split. **Facilities:** 6 stalls, 2 with attached paddocks, 100' x 200' paddock, 50' x 100' pasture, grain/hay available, entry to farm too narrow for big rigs but can take 4-horse or possible 6-horse gooseneck. Please call ahead. Farm is headquarters of Optissage, Inc. specializing in equine/canine/feline massage. **Rates:** $15 per night. **Accommodations:** B&B across street; TraveLodge & Comfort Inn within 6 miles.

CLEVELAND
Pinehaven Stables **Phone: 216-235-3200**
Malcolm Cole
7611 Lewis Road [44138] Directions: 2 miles from I-480 & 4 miles from I-71. Call for directions. **Facilities:** 2 indoor stalls, 1/2 acre turnout, 55' x 120' turnout, feed/hay & trailer parking. Call for reservation. **Rates:** $10 per night; $70 per week. **Accommodations:** Red Roof Inn & Holiday Inn nearby.

COLUMBUS
Liberty Farm **Phone: 614-279-0346**
Kathryn Osborn
2620 Fisher Road [43204] Directions: 1 mile off of I-70 on west side of Columbus. Call for directions. **Facilities:** 30 indoor 12' x 12' box stalls, grooming stall, indoor arena, outdoor ring, 8 acres of pasture, full jump course on total of 17 acres. Training of horses & riders in hunt seat. Buying & selling of hunter/jumpers. Call for reservation. **Rates:** $20 per night; $77 per week. **Accommodations:** Motels 5 miles from stable.

DUBLIN
Dublin Stables **Phone: 614-764-4643**
Ginette Feasel
3910 Summit View Rd. W. [43016] Directions: 1.5 miles north of 270, off exit 20, Sawmill exit. Turn left at Summit View Rd.W. light, 4th drive on right. **Facilities:** Fortynine 8x10 to 10x16 indoor stalls. 60'x120' indoor arena, lighted 110'x250' outdoor arena, six pastures, 4+ acres and 2+ acres. Timothy/grass, 10% sweet feed, trailer parking available. **Rates:** $15 per day $75 per week. **Accommodations:** Please call.

FINDLAY
Lazy Creek Horse Farm **Phone: 419-427-2400**
Chris Lyon
11355 Hancock County Road [45840] Directions: Exit 23 off of I-75. Call for directions. **Facilities:** 17 indoor stalls, 4 holding pens, 3 pasture areas, indoor arena, feed/hay & trailer parking available. Training of horses & riders in English & Western. **Rates:** $10 per night. **Accommodations:** Motels 3 miles from stable.

ALL OF OUR STABLES REQUIRE CURRENT NEG. COGGINS, CURRENT HEALTH PAPERS, & OWNERSHIP PAPERS.

FREDERICKTOWN
Heartland Country Resort & Stables **Phone: 419-768-9100**
Dorene Henschen (owner), Ben Daniel (mgr.) **or: 419-768-9300**
2994 Township Road 190 [43019] Directions: Exit 151 off I-71, east on 95 for 2 miles; in Chesterville, south on 314 for 2 miles; left (east) on County Road 179 to intersection of Township Road 190. **Facilities:** 10 wood indoor 12' x 12' stalls some with outdoor turnouts; 20 acres of pasture, indoor and outdoor arenas, feed/hay, trailer parking. Trail riding, lessons in Western, English, hunt seat, barrel racing, or calf roping; horses available to rent with certified guides and instructors. Numerous activities available at award-winning resort. Call for reservations. **Rates:** $12 per night. **Accommodations:** B&B on premises, $85-$125 per night.

HILLIARD
Country Squire Farms **Phone: 614-529-0055**
Kathy Morgan
3687 Alton & Darby Creek Road [43026] Directions: 3 miles from I-70. Call for directions. **Facilities:** 9 indoor stalls, 3-4 acre pasture, seven 1/2- to 1-acre pastures, indoor arena, indoor exercise track, 2 outdoor arenas with jumps, 1/2-mile race track, plus 140 acres for riding. Call for reservation. **Rates:** $15 per night; $75 per week. **Accommodations:** Red Roof Inn in Hilliard, 3 miles from stable; Comfort Inn in Columbus, 3 miles from stable.

JACKSON
Henderson's Arena Complex **Phone 614-988-4700**
Jerry L. Henderson
800 Van Fossan Road, County Road 23 [45640] Directions: On County Rd. 23, service-type road parallel to east/west 4-lane Hwy SR 32, almost midway to north/south 4-lane Hwy SRs 35 & 23. **Facilities:** 60 indoor 10' x 12' stalls aluminum fronts concrete aisle, 2 faucets in barns, two 130' x 250' riding rings, feed/hay available, trailer parking, camper hook-ups. Western wear store on grounds; saddle shop less than 3 miles away. **Rates:** $10 per night. **Accommodations:** Days Inn, Comfort Inn in Jackson, 10 miles from stable.

JACKSONTOWN
Beechwood Forest **Phone: 800-317-5157**
Jeff Edwards
10050 Jacksontown Road [43030] Directions: 300 feet north of I-70, on SR 13. **Facilities:** 8 indoor box stalls, pasture/turnout area, trailer parking. Stable along with a 60-site RV park with full hook-ups. **Rates:** $15 per night. **Accommodations:** In addition to RV park, 2 motels within 4 miles.

ALL OF OUR STABLES REQUIRE CURRENT NEG. COGGINS, CURRENT HEALTH PAPERS, & OWNERSHIP PAPERS.

KIRTLAND

Dorchester Farms **Phone: 216-256-9254**

Amy Battersby

8560 Billings Road [44094] Directions: Exit 306 (Mentor/Kirtland) off I-90, south 3.5 miles to Billings Road on left, facility 1/4 mile on right. **Facilities:** 50 indoor 10' x 11' box stalls with bedding, turnout, 80' x 200' indoor arena, 380' x 196' outdoor arena, trailer parking available. Hunter/jumper facility, lessons, training, selling. **Rates:** $20 per night, $25 per night with turnout. **Accommodations:** Motels within 3 miles of stable.

LANCASTER

Colt's Place Stables **Phone:614-687-6364**

Amy L. Allen

2855 Wheeling Road N.E. [43130] Directions: Call for directions. **Facilities:** 4 -10' X 12' indoor stalls (can be 2 - 12' X 20') opening on 70' X 120' indoor arena. Alfalfa or Timothy hay available, no grain. 105' X 150' with 4' high, high tension wire electric fenced turnout with minimum grass. 60' round pen. Carrie Cradock Resident Equine Massage Therapist. Facility available for clinics. **Rates:** $15 per night, $90 per week. **Accommodations:** Motels in Lancaster, 15 min away, Amerihost, Best Western & Knights Inn. 25 min drive to Butterfly Inn, Bed & Breakfast.

LONDON

Old Horsefeathers Stables **Phone: 614-852-1011**

Tami L. Martin

420 Deck Road [43140] Directions: Access from I-70 and I-71. 25 minutes from Ohio State Fairgrounds, Columbus. Call for directions. **Facilities:** 15 large indoor box stalls, varying sizes from small grassless paddocks to 2-acre grass pasture, feed/hay available, trailer and camper parking. **Rates:** $12 per night; $70 per week. **Accommodations:** Winchester House Bed & Breakfast 5 miles away. London Motel (5 miles) and Trails Inn and Holiday Inn Express (10 miles off I-70) in London. Royal Inn in Mt. Sterling, 10 miles away.

LUCASVILLE

Silverstone Farm **Phone: 740-259-5919**

Tom Bombolis

735 Cook Road [45648] Directions: Located off of Rt. 23 N. Call for directions. **Facilities:** 25 indoor box stalls, 3 turnout paddocks, indoor arena, outdoor ring, wash racks, 2 miles of trails. Trains & shows horses & riders. Call for availability & reservations. **Rates:** $15 per night. **Accommodations:** Motels within 5 miles of stable.

ALL OF OUR STABLES REQUIRE CURRENT NEG. COGGINS, CURRENT HEALTH PAPERS, & OWNERSHIP PAPERS.

MASSILLON
Lanewood Acres Boarding Stables **Phone: 216-837-2136**
William & Carol Kuhlins
14100 Kimmens Road SW [44647] Directions: 5 minutes from I-77 & Rt. 30; 45 minutes from I-71. Call for directions. **Facilities:** 3 run-in stalls leading into 30' x 40' paddocks, board-fenced 50' x 100' arena, feed/hay & trailer parking available. No stallions. Advance reservation required. This is a boarding stable near Amish country. **Rates:** $15 per night; $80 per week. **Accommodations:** Motel 6 & Super 8 in Masillon, 10 minutes away plus nearby Amish-cooking restaurants.

MILLERSBURG
Kricket Hill Farm **Phone: 330-674-2430**
Kay E. Earney
6825 TR 346 [44654] Directions: 4 miles north of Millersburg on SR 83. Call for directions. **Facilities:** 5 indoor 12' x 12' stalls, one 12' x 14' stallion stall, 2 separate pastures, large outside paddock, feed/hay & trailer parking available. Hook-up for self-contained camper. Please call for reservation. **Rates:** $15 per night; $75 per week. **Accommodations:** B&B on premises, $35 per room.

NEW LEBANON
Nickel & Dime Stable **Phone: 513-687-3632**
Carolyn Palmer
10582 Mile Road [45345] Directions: From I-70W take Rt. 49 Exit; south on Hoke Rd.; go to dead end, right on Westbrooke to STOP sign; left on Diamond Mill Rd.; cross Rt. 35 & take right on next street which is Mile Rd. **Facilities:** 6 indoor stalls, three 1/2-3 acres pasture/turnout, 1/2 acre pasture with separate barn, riding trails nearby. Dealer for UltraGuard vinyl fencing. All notices of arrivals made by 5 P.M. **Rates:** $15 per night. **Accommodations:** Camper possibly available at $15 per night; camping on premises for $5 per night; Days Inn in Brookville, 10 miles from stable.

PATASKALA
Hawks Nest Farm **Phone: 330-336-8104**
Bill Thompson **Phone: 740-964-2404**
8056 Outville Road [43062] Directions: 2 miles north of I-70, Exit 122, Kirkersville - Baltimore. **Facilities:** 9 indoor 12′ x 12′ stalls, 100′ x 100′ and 75′ x 75′ pasture/turnout areas, feed/hay and trailer parking available. Riding trails. **Rates:** $15 per night. **Accommodations:** Bunk house on the property with complete sleeping/bath facilities. **Rates:** $55.00 per night. Also numerous motels in Columbus, 15 minutes from farm.

PICKERINGTON
New Beginning Stables **Phone: 614-862-6206**
Al Brodfor
5785 Stemen Road [43147] Directions: Exit 256 off of I-70. Call for directions. **Facilities:** 26 indoor stalls, 8 paddocks, 70' x 120' indoor arena, 170' x 200' outdoor arena. Trains horses & riders at all levels. Breeds quarter horses & thoroughbreds. Call for reservation. **Rates:** $15 per night. **Accommodations:** Motels 6 miles.

ALL OF OUR STABLES REQUIRE CURRENT NEG. COGGINS, CURRENT HEALTH PAPERS, & OWNERSHIP PAPERS.

SEVILLE

Flowers' Fabled Stables **Phone: 330-887-5482**
Denny and Kay Flowers
6320 Greenwich Road [44273] Directions: 2 miles from I-71 and I-76 interchange. Call for directions. **Facilities:** 4-6 indoor 10′ x 10′ box stalls, 3 half-acre pastures, round pen, equine massage therapy (certified through Equissage), trails, outdoor arena, riding lessons, hay available, trailer parking with electrical hook-up available. **Rates:** $15 per night. **Accommodations:** Comfort Inn in Seville, 3 miles from stable; HoJo Inn 1 mile from stable.

Rocking H Ranch **Phone/Fax: 330-334-5466**
Christopher and Lori Minnich
3737 Seville Road [44273] Directions: At junction of I-71 and I-76; Exit 2, 15 miles south of I-271. Call for directions. **Facilities:** 12 - 12′ x 12′ stalls, 2 holding pens, 6 pastures, feed/hay and trailer parking available. Fishing pond; dogs welcome. Reservations required. **Rates:** $15 per night. **Accommodations:** Fully furnished room with private bath available.

SPRINGFIELD

Ballentine Stables **Phone: 513-964-8402**
John & Nancy McKeen **Evenings: 513-964-8390**
6049 Ballentine Pike [45502] Directions: Exit 52B off of I-70 and Rt. 41 exit off of I-68. Call for directions. **Facilities:** 33 indoor stalls, 180' x 60' indoor arena, 4 turnout paddocks, 1/2 mile race track, feed/hay & trailer parking. Training of racehorses and breeding of thoroughbreds at stable. Call for reservation. **Rates:** $10 per night. **Accommodations:** Motels 6 miles away.

SPRINGFIELD

High View Stables **Phone: 513-324-4511**
Mike van Dyke, trainer
4711 Springfield-Xenia Road [45502] Directions: 1.5 miles from I-70. Call for directions. **Facilities:** 35 indoor stalls, 40' x 200' turnout paddock, 2 to 3-50' x 50' pens, feed/hay & trailer parking. Call for reservation & availability. **Rates:** $15 per night. **Accommodations:** Holiday Inn & Ramada Ltd. in Springfield, 1.5 miles from stable.

STREETSBORO

Sahbra Farms **Phone: 216-626-2040**
David Gross
8261 Diagonal Road [44241] Directions: Exit 13 off of I-80, 1.5 miles to farm. Call for directions. **Facilities:** 215 indoor stalls, 30 outside paddocks, 1/2 mile race track. This is a beautiful breeding & training center for standardbreds on 400 acres. Call for availability & reservation & rates for use of training center. **Rates:** $14 per night for stall and turnout. **Accommodations:** 1 mile away.

ALL OF OUR STABLES REQUIRE CURRENT NEG. COGGINS, CURRENT HEALTH PAPERS, & OWNERSHIP PAPERS.

SWANTON

Post and Rail Stables **Phone: 419-826-9934**

Bonnie Cicora

10362 State Route 64 [43558] Directions: Call for directions. **Facilities:** 43 indoor stalls, many large pasture areas and round pens, feed/hay & trailer parking available. Arrival must be before midnight and reservations required. This is a complete training facility. **Rates:** $25 per night; $100 per week. **Accommodations:** Cross Country in Toledo and Fairfield Inn in Toledo, 10 minutes away.

WEST ALEXANDRIA

Surecare Farm **Phone: 937-839-4186**

Scott and Barb Stockslaver

7089 State Route 35 East (45381) Directions: Call for Directions. **Facilities:** 4- 10′ x 20′ indoor box stalls. Feed/Hay, trailer parking available. Bed and continental breakfast for guests *and* horses! Easy access for truck and trailer. Located 8 miles from I70 on State Route 35. **Rates:** $90/night for horses, owner, and dog. **Accommodations:** Small apartment with full bath and queen size bed.

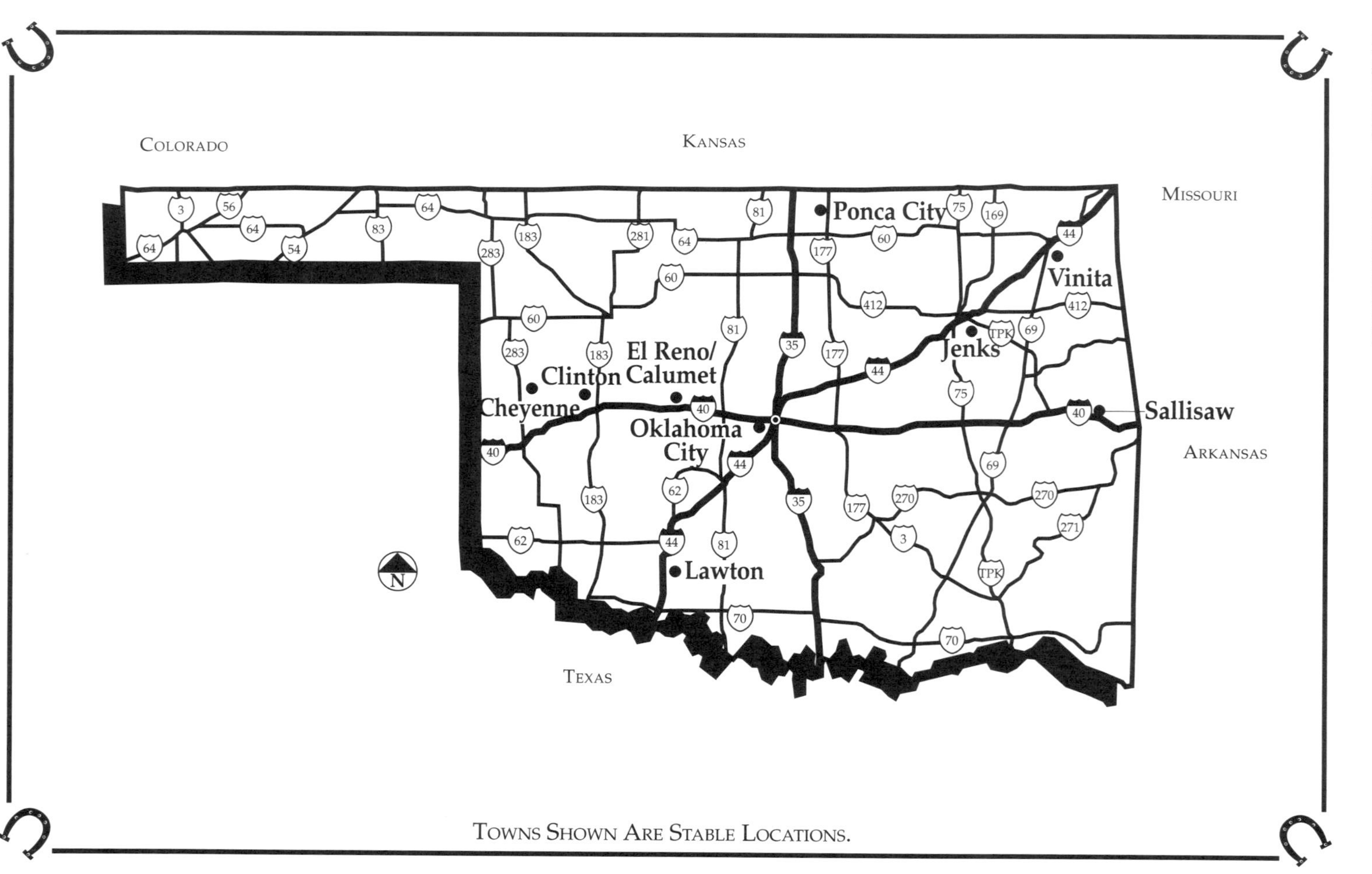

TOWNS SHOWN ARE STABLE LOCATIONS.

ALL OF OUR STABLES REQUIRE CURRENT NEG. COGGINS, CURRENT HEALTH PAPERS, & OWNERSHIP PAPERS.

CHEYENNE
Coyote Hills Ranch **Phone: 580-497-3931**
Kass Nickels
P.O. Box 99 [73628] Directions: Exit 20 off I-40 at Sayre, Hwy 283 north 28 miles to Cheyenne; take Hwy 47 west 4 miles, then 2 miles north, then 2 miles west. **Facilities:** 8 indoor 12' x 12' stalls, 24' x 12' outside barns with 32' x 32' runs, 150' x 60' paddocks, full size outdoor arena and round pen, feed/hay, trailer parking. Horseback riding. Located in the beautiful red hills of Roger Mills County. **Rates:** $15 per night. **Accommodations:** Hotel on ranch; private rooms with bath, heated and air conditioned. RV hookups.

CLINTON
JR'S Western Horse Motel **Phone: 580-323-5888**
M. L. (JR.) Richardson **Phone: 580-323-1588**
1710 Neptune Drive [73601-9722] Directions: Turn south on Neptune Drive at 65A exit. 3/4 mile to 1710 Neptune Drive. **Facilities:** Seven pens approximately 40' X 60', NO Box Stalls. No feed or hay available. Trailer parking available. Back yard facilities, friendly "at home" atmosphere. **Rates:** $10/$15 per night. **Accommodations:** Four nice motels within a mile of stable and R.V. park 1/4 mile.

EL RENO
Little Bit Farm **Phone: 405-262-7504**
David & T.J. Meschberger
Route 2, Box 43, Calumet [73014] Directions: From I-40: Take Exit 119; go about 1/2 mile north to Ft. Reno crossover; take left & go one mile west; take right & go 1/2 mile north; white fence on left is stable. **Facilities:** 20 indoor stalls, 150' x 150' turnout, no feed/hay, trailer parking on premises and can accommodate big rigs. Electric hook-up for camper. Vet nearby. Reservations preferred. **Rates:** $15 per night. **Accommodations:** Best Western Inn in El Reno, 5 miles from stable.

JENKS
Citation Stables **Phone: 918-298-3700**
Pat and Joe Berardi
302 W. 111th Street S. [74037] Directions: From Hwy 75: Go 2.2 miles east on 111th St. South. Stable on right. **Facilities:** 58 indoor 14' x 14' cement/steel stalls, turnout pastures, round pen or paddock, feed/hay & trailer parking on premises. Riding instruction available in English, Western, jumping, & dressage. Bring your own feed or purchase on site. Electric hook-up for campers. Tack shop on premises. Call ahead for reservation. **Rates:** $15 per night, bedding included; $80 per week. **Accommodations:** Best Western in Glenpool, 3 miles from stable.

ALL OF OUR STABLES REQUIRE CURRENT NEG. COGGINS, CURRENT HEALTH PAPERS, & OWNERSHIP PAPERS.

LAWTON

Pool Arena & Stables **Phone: 405-492-4856**
Dodge Pool
HC 30, Box 853 [73505] Directions: I-44 to Hwy 49; go west 4 miles to Hwy 58 & go north 6 miles; big yellow metal building on east side of Hwy 58. (Call for map.) **Facilities:** 24 indoor stalls at least 12' x 12', one 10-acre, three 2-acre, & three 1-acre pasture/turnout areas; 100' x 225' indoor arena; outdoor arena; indoor heated wash rack, hot walker, 80' x 45' round pens; large riding area available; less than 2 miles from mountains & lake. Oats, omolene, & prairie hay available. Trailer parking. Request that horse owners provide own feed buckets. Call for reservation. **Rates:** $12 per night; $60 per week. **Accommodations:** Howard Johnson's 17 miles away in Lawton (1-800-446-4656).

OKLAHOMA CITY

AAA Horse Motel & Thoroughbred Training **Phone: 405-771-3472**
R.L. Odom
6920 N.E. 63rd [73141] Directions: Call for directions 30 minutes to 1 hour in advance. **Facilities:** 20 indoor stalls, five separate 2-6 acre paddocks, 1/2-mile racetrack, feed/hay & trailer parking available. Open 24 hours. Training, racing, & sales of thoroughbred horses. **Rates:** $15 per night; $10 for 2nd and more nights. **Accommodations:** Holiday Inn & Best Western 3 miles away.

Ed Cook **Phone: 405-634-1787**
8400 S. Walker [73139] Directions: Call for directions. **Facilities:** 7 indoor 14' x 14' stalls, four holding pens, 5-acre pasture, feed/hay & trailer parking available. **Rates:** $15 per night; $75 per week. **Accommodations:** Many motels 1/2 mile east of stable.

OKLAHOMA CITY

Robinson's Barn-Bed-Breakfast **Phone: 877-733-2443**
Kenn & Donna Robinson **Web: http://home.earthlink.net/~horseldy**
11200 SE 44 Place [73150] Directions: I-40 & Anderson Road. **Facilities:** 8 indoor concrete 10' x 12' stalls with pipe runs, 100' x 100' turnout, large riding arena, feed/hay, trailer parking; kennel and RV hook-ups ($20 per night) available. **Rates:** $15 per night. Medicinal cob webs, no charge. **Accommodations:** Two bed & continental breakfast in barn, rooms with full bath, heat/air in each room. One room with twin beds, one room with queen bed, $45 per night. 20 minutes from State Fair grounds.

ALL OF OUR STABLES REQUIRE CURRENT NEG. COGGINS, CURRENT HEALTH PAPERS, & OWNERSHIP PAPERS.

PONCA CITY

Sam Smith Quarterhorses **Phone: 405-762-6014**

Sam Smith

Route 3, Box 716 [74604] Directions: From I-35, go 18 miles east on Hwy 60. **Facilities:** 10 wood 11' x 15' stalls, round pen, walker, 150' x 300' paddocks, feed/hay and trailer parking available. Sells and trains quarter horses in performance reining. Stallion standing-at-stud: "Sierra's Notion." **Rates:** $10 per night; weekly rate available. **Accommodations:** Double N Motel two blocks away.

SALLISAW

Back Woods Retreat **Phone: 918-774-0094**

Joyce Keefner **Web: http:kinetic/kedis.com**

Rt. 3, Box 85-1 [74955] Directions: Located 8 miles north of Sallisaw on U.S. Hwy 59 on Brushy Mountain. **Facilities:** Barn, stables and outside pens. **Accommodations:** Suite I - Private bath, king size bed, tree top deck, upper floor. $250.00 per weekend. Suite II & III - Queen size bed, lower floor, bath and shower. **Rates:** Either - $200.00 per weekend. Cabin overlooking lake - two bedroom areas, A/C, refrigerator, linens & towels, bath/shower. $200.00 per weekend, 1 couple, $25.00 per person extra. Call for availability and pricing for your unique situation.

VINITA

Double D Ranch **Phone: 918-256-6268**

Paul & Deanne Brown

Rt. 2, Box 88 [74301] Directions: 7 miles off I-44. Take Afton Exit. go west/south on Highway 60/90 for 7 miles. Or take Vinita Exit and go east/north on Hwy 60/90 for 7 miles. **Facilities:** 80-acre ranch: 40 indoor stalls, turnout pens, indoor riding arena, indoor wash rack, feed/hay & trailer parking on premises. Camper parking okay, if self contained. Ranch breeds and sells Polish Arabians. **Rates:** $15 per night. Extended stays available. **Accomodations:** Motels within 5 miles of ranch.

Hey "cookey", how come the boys are eatin at two differant campfires?
Wellsir, Heres the reglar one. and yonder is the no smokin' section.
FRED LEARY

NOTES AND REMINDERS

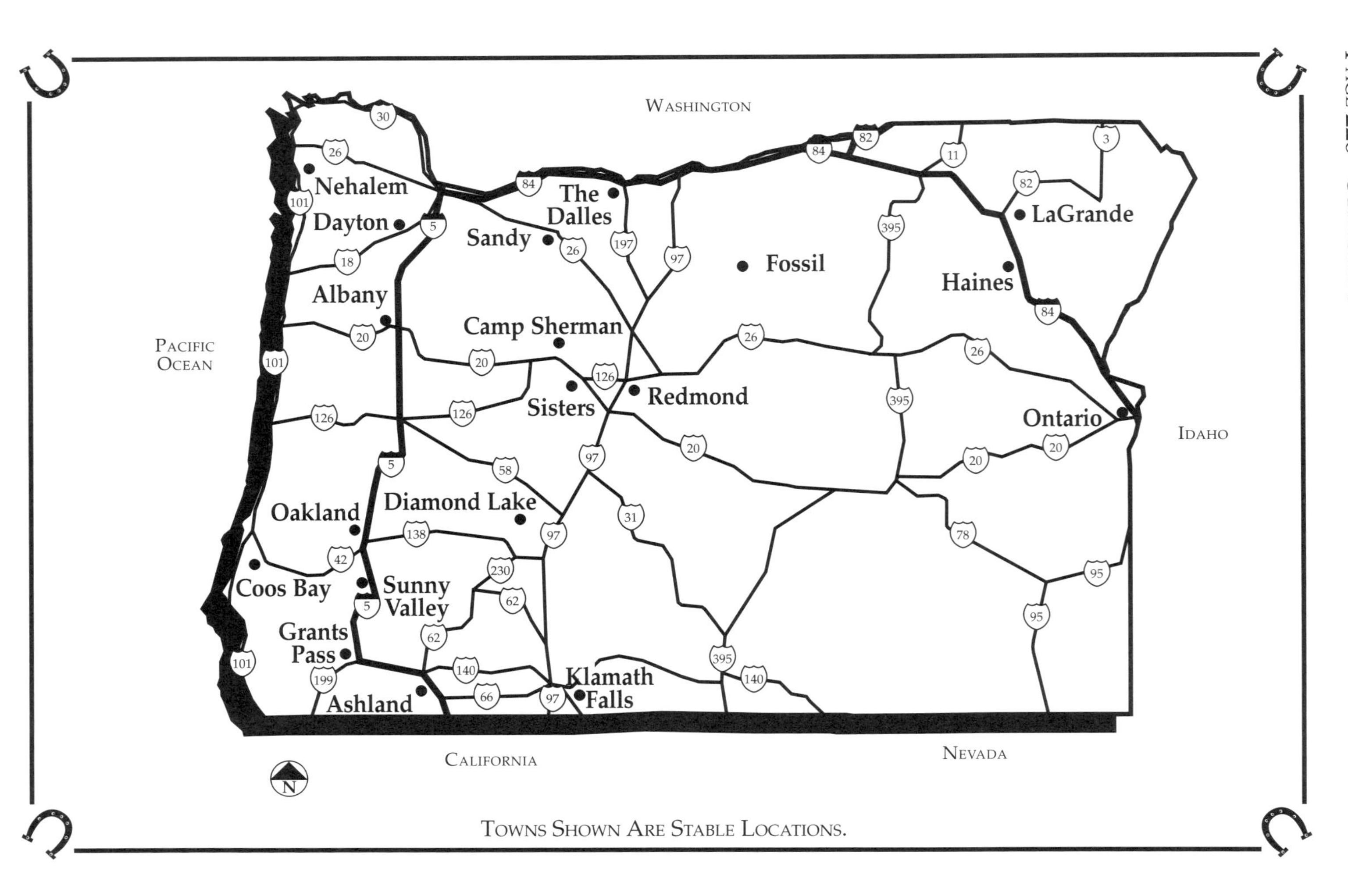

Towns Shown Are Stable Locations.

ALL OF OUR STABLES REQUIRE CURRENT NEG, COGGINS, CURRENT HEALTH PAPERS, & OWNERSHIP PAPERS.

ALBANY

Springhill Boarding Stables **Phone: 541-928-8943**

Liz & Jerry Couzin

5368 Springhill Drive [97321] Directions: Located in North Albany. Call for directions. **Facilities:** 33 indoor stalls, 50' x 150' and 60' x 70' indoor arenas, 80' x 150' outdoor arena, 66' round pen, hot walker, feed/hay & trailer parking on premises. Horses for sale & lease. Training of horses and riders available in all disciplines at all levels. Standing-at-stud: Champion Khemosabi ++++ Son, "Khareen" and Champion Bey Shah Grandson, "SS Shahzahm." Both stallions are throwing beautiful, correct, winning offspring with great minds and fabulous dispositions. Call for reservations or more info. **Rates:** $15 per night with discount for multiple horses. **Accommodations:** Best Western, Holiday Inn, Super 8, Comfort Inn located 5-7 miles from stable.

ASHLAND

B-G Valley View Stables **Phone: 503-482-1772**

Virginia J. Blair

263 Wilson Road [97520] Directions: Exit 19 off of I-5: from the south, exit N. Valley View Rd.; turn right & go 1 mile then turn right onto Wilson Road. Stable is on left, 1,000 ft. from road. S.O.S. sign on driveway. **Facilities:** 8 indoor 12' x 12' stalls, 70' x 120' indoor ring, seven 12' x 32' runs, big turnaround area for trailers, hot & cold horse shower, grain/hay furnished. No smoking. fire sensors. No loose dogs. Must be in by 10 P.M. Please make reservation. Will not return long distance calls. Must have current strangles, rhino, potomac fever, rabies, elw. sleeping sickness, tetanus shots. 42 years experience with horses. Trailer & camper parking available. **Rates:** $15 per night per horse. Full care. **Accommodations:** Best Western Heritage Inn (will take pets) 1 mile away.

CAMP SHERMAN

Sheep Springs Horse Camp **Phone: 503-822-3799**

Hoodoo Ski Area [97730] **For info: 503-549-2111**

Directions: Call for directions. 12 miles west of Sisters on Hwy 20. Look for signs. **Facilities:** 44 - 12' x 12' stalls at 11 sites, trailer parking available. Elevation 3,200 ft. Access to Mt. Jefferson Wilderness Area. Very sensitive area, weather changeable. At one end of 39-mile Metolius Windigo Trail. Call for reservations. **Rates:** $6 per night. **Accommodations:** Camping area with spring water.

COOS BAY

Family Four Stables **Phone: 503-267-5301**

James & Peggie Henriksen

662 Family Four Drive [97420] Directions: 7 miles south of Coos Bay on Hwy 101. Call for directions. **Facilities:** 48 indoor 12' x 12' stalls, 50' x 50' working arena, exercise arena, 100' x 200' covered riding arena, hot walker, round pen, automatic waterer, hay racks, feed/hay, trailer parking available. English and Western riding, instructors available. Call for reservations. **Rates:** $15 per night. **Accommodations:** Motels in Coos Bay, 7 miles from stable.

ALL OF OUR STABLES REQUIRE CURRENT NEG, COGGINS, CURRENT HEALTH PAPERS, & OWNERSHIP PAPERS.

DAYTON

Wine Country Farm **Phone: 503-864-3446**

Joan Davenport

6855 Breyman Orchards Road [97114] Directions: 30 minutes from I-5. From Portland: Take 99W; 3 miles out of Dundee, turn right at McDougal; go 1 block and turn right. Stable is less than 2 miles. **Facilities:** 3 large indoor stalls. Riding trails with magnificent views. Arabian breeding farm and vineyard with own label. **Rates:** $10 per night. **Accommodations:** 5-room Bed & Breakfast with private baths on premises. Rates: $65-$85 in house; $125 for 2-room suite.

DIAMOND LAKE

Diamond Lake Corrals **Phone: 503-793-3337**

Wayne Watson

[97731] Directions: Call for directions; off Rte. 138 to Roseburg. **Facilities:** Five-stall barn, 14 corrals of various sizes, feed/hay, trailer parking available. Horse rentals, pack trips, hunting parties; 20 miles from Crater Lake. Open June-September. Space to pitch a tent or park a camper. **Rates:** $7.50 per night. **Accommodations:** Diamond Lake Resort 1/4 mile away.

FOSSIL

Lightning B Ranch **Phone: 503-489-3367**

Jon & Candy Bowerman

Clarno Road [97830] Directions: Call for directions; off Hwy 218 between Fossil and Antelope. **Facilities:** 5 covered stalls with runs, several pastures, 5 corrals with barn covering, grass and trailer parking available. Natural communication seminar daily; clinics as scheduled, trail riding, training. **Rates:** $10 per night, $50 per week. **Accommodations:** Bed & breakfast on premises.

GRANT'S PASS

Four-Ace's Boarding & Training Stable **Phone: 503-479-1759**

Richard & Effie Ried

1111 Rounds Avenue [97527] Directions: 5.5 miles from fairgrounds. Take Redwood Ave. to Rounds Ave. **Facilities:** 32 indoor stalls, two 1/4 -acre turnouts, covered arena, feed/hay & trailer parking available. Paint sales & breeding at stable. Call for reservation. **Rates:** $15 per night. **Accommodations:** Riverside Motel & Redwood Motel in Grant's Pass, 5 miles from stable.

HAINES

Slash D Arena **Phone: 541-856-3359**

Howard Driggers

Box 337 [97833] Directions: 4 miles from I-84 on Hwy 30. Call for directions. **Facilities:** 25 indoor stalls, outdoor pens, 100' round pen, 90' x 265' indoor arena, feed/hay & trailer parking on premises. Electric & water hook-ups for campers at no charge. Full facilities for rodeos & horse shows. Call for reservation. **Rates:** $15 per night. **Accommodations:** Best Western Sunridge Inn, Eldorado Inn, and Super 8 Motel in Baker City, 8 miles from ranch.

ALL OF OUR STABLES REQUIRE CURRENT NEG. COGGINS, CURRENT HEALTH PAPERS, & OWNERSHIP PAPERS.

KLAMATH FALLS
JT Equestrian Center, Inc. **Phone: 503-882-8299**
Randell & Carolyn Souders **Home: 503-882-9997**
13631 Algoma Road [97601] Directions: Go north from Klamath Falls on Hwy 97 approx. 8 miles from town; take right on Algoma Rd.; go approx. 1.2 miles & turn right on Old Fort Rd.; cross the cattle guard & turn left at the gravel lane that leads into ranch. **Facilities:** 6-8 indoor stalls, 4-6 outdoor stalls, 2 acres of pasture, indoor & outdoor arenas, round pen, pasture land and mountain trail riding. Call ahead. Leave a message and they will return call. **Rates:** $15 per night. **Accommodations:** Oregon Motel 8 (has RV Park) and Super 8 both within 3-4 miles of stable.

LAGRANDE
Mavericks, Inc. **Phone: 503-963-3991**
Maverick Inc. Riding Club
3608 2nd Street N. [97850] Directions: 100 yards off of I-84. Call for directions. **Facilities:** Ten 12' x 12' outdoor pens, large outdoor arena, no feed/hay available, trailer parking on premises. Open horse shows, team-penning, & team-roping events held at facility. Riding trails available. Call ahead. **Rates:** $5 per night. **Accommodations:** Greenwell Motel within 1 mile of stable.

NEHALEM
Pearl Creek Stables **Phone: 503-368-5267**
Janice Woelfle
17150 Camp Four Road [97130] Directions: Call for directions. Hwy 53 off US 101, about 3 miles to second Camp Four Road sign. **Facilities:** 7 indoor 10' x 12' stalls, rubber matted, bedding provided, small corral for turnout, 50' x 60' indoor arena, trailer parking available. Nehalem Bay State Park 5 minutes away, trailer to Manzanita Beach. Dog obedience training. **Rates:** $15 per night. **Accommodations:** Lodging within 5 miles; call for information. Camping available at state park.

NEHALEM
Pegasus **Phone: 503-368-7161**
Karen Joy McCormick
40050 Anderson Road [97131] Directions: Call for directions. **Facilities:** 5 indoor 12' x 12' stalls, 120' x 60' arena, feed/hay available, trailer parking. Riding lessons available. Boarding is only available if the riders stay at Pegasus. **Rates:** $15 per night for stall, $10 per night for arena. **Accommodations:** 2 bedrooms & master suite in cedar shake house, $70-$135 per night, living room, full kitchen. Full-house rental. Ten minutes from ocean and mountain trails.

ALL OF OUR STABLES REQUIRE CURRENT NEG. COGGINS, CURRENT HEALTH PAPERS, & OWNERSHIP PAPERS.

OAKLAND

Dodge Creek Stables **Phone: 541-459-2609**
Geronimo & Mary Bayard **or: 541-459-4682**
E-mail: horses@rosenet.net **Web: www.members.rosenet.net/horses**
3739 Hwy 138W [97462] Directions: Exit 136 (Sutherlin) off I-5, west on Hwy 138 for 3.5 miles, stable on right. **Facilities:** 21 - 12' x 12' and 12' x 14' box stalls with 24' runways, pasture/turnout in various sizes, huge indoor arena, wash rack, trails, jumps, outside lighting, feed/hay, trailer parking.Training and lessons in dressage, eventing, jumping. Outstanding sport horse farrier. Emergency pick-up available. Caretaker on grounds. Electricity, water, shower available for camping. Call for reservations. **Rates:** $20 per night; $100 per week. **Accommodations:** Campers welcome on premises. Town & Country Motel and others in Sutherlin, 3.5 miles from stable.

Ride & Rest Horse Motel **Phone: 503-459-9220**
Ron & Norma Groves
500 Metz Hill Road [97462] Directions: Exit 142 off of I-5. 1/4 mile west on Metz Hill Road. Horse Motel is first driveway on left. **Facilities:** 17 indoor stalls, 160' x 230' outdoor exercise arena, 70' x 85' indoor arena, corral that can hold 2 horses, auto waterers, feed/hay at $2/bale & trailer parking available. Call for reservation. **Rates:** $15 per night with fir shavings; $50 per week. **Accommodations:** Two new motels in Sutherlin and new Best Western in Rice Hill, all six miles north or south with gas stations and 24 hr restaurants.

ONTARIO

Malheur County Fairground **Phone: 541-889-3431 (days)**
Janeen Kressly **Evening: 541-823-2581 (Kevin)**
795 NW 9th Street [97914] Directions: Located 2 miles off of I-84. Call for directions. **Facilities:** 60 outside covered stalls, outdoor riding arena, rodeo-size paddock area, feed store nearby, trailer parking on premises. Open 24 hours with caretaker on site. Fairground hosts rodeos, gun shows, circuses, horse shows & auctions, annual car show and more. **Rates:** $10 per night, includes shavings. **Accommodations:** Close by.

REDMOND

El Dorado Ranch **Phone: 541-548-3356**
Jean Davis
8062 S. Hwy 97 [97756] Directions: 9 miles north of Bend. 4 miles south of Redmond. **Facilities:** 7 box stalls with runouts, paddocks, pastures, 60' round pen, 120' x 220' arena; grass, alfalfa or mix available. Many riding trails on BLM land adjacent to stable. Current vaccinations required. Selling pony & horse carts, wagons, harness. **Rates:** $15 per night. **Accommodations:** Several motels within 4 miles plus 2 RV parks nearby.

ALL OF OUR STABLES REQUIRE CURRENT NEG. COGGINS, CURRENT HEALTH PAPERS, & OWNERSHIP PAPERS.

SANDY

Burnt Spur Ranch **Phone: 503-668-9716**

Linda Keeter & John Keeter

42100 SE Locksmith Lane [97055] Directions: 2 miles east of Sandy: South at Shorty's Corner onto Firwood Road, go 2 miles and turn left onto Locksmith Lane. **Facilities:** 7 indoor stalls, 6 - 12' x 24' covered outdoor pens, hot walker, 110' x 120' outdoor arena, feed/hay, trailer parking. Professional horse trainer on premises. Overnight camper hook-ups. Less than 1 hour from skiing, fishing, trails, etc. Call for reservations. **Rates:** $15 per night. **Accommodations:** B&B within 3 miles. Motel 2 miles. Several Restaurants and Lounges 2 miles.

SISTERS

Black Butte Stables **Phone: 1-800-743-3035**

Sandra Herman, owner; Mike Elmore, manager

P.O. Box 418 [97759] Directions: 7.5 miles west of Sisters on Hwy 20 on Black Butte Ranch, behind the general store. **Facilities:** 10 indoor 12' x 12' stalls, 8 outdoor 10' x 30' paddocks, feed/hay available, trailer parking. Primarily a guided horseback riding and wilderness pack station. Located in foothills of Three Sisters Wilderness Area. Exceptional riding opportunities - waterfalls, creeks, alpine meadows, natural crater-formed lake. Call for reservations 2-3 days in advance in off season; 2 weeks in advance during summer. **Rates:** $7.50 per night self-paddock; $50 per week self-paddock. **Accommodations:** Best Western, Comfort Inn in Sisters, 7.5 miles away.

SUNNY VALLEY

Sunny Valley KOA **Phone: 541-479-0209**

Marty & Les Warren **or: 800-562-7557**

140 Old Stage Road [97497] Directions: From I-5, take Exit 71. Follow signs to KOA Kampground. **Facilities:** 7 indoor stalls with attached paddocks, 3-acre pasture/turnout, trailer parking available. No feed/hay available. **Rates:** $15 per night. **Accommodations:** Campgrounds on site. Motels in Grant's Pass, 12 miles from campgrounds.

THE DALLES

Fort Dalles Days Rodeo Association **Phone: 503-296-6817**

Steve Hunt, President **Night: 503-298-8671**

Bargeway Road [97058] Directions: 1 mile from I-84. Call for directions. **Facilities:** 12 stock pens on 14 acres, 200' x 300' outdoor arena with 3,000 seats & concession stands. Pro & junior rodeos held at facility. Year-round caretaker on site. Hughes Feed & Grain & tack store nearby. K&H Specialties in The Dalles is a full-service repair shop that can fix trucks & trailers with mechanical problems. **Rates:** No charge but donations gladly accepted. **Accommodations:** Quality Inn & Cousins Restaurant 1 mile from rodeo grounds.

NOTES AND REMINDERS

Medical School In The Old West.

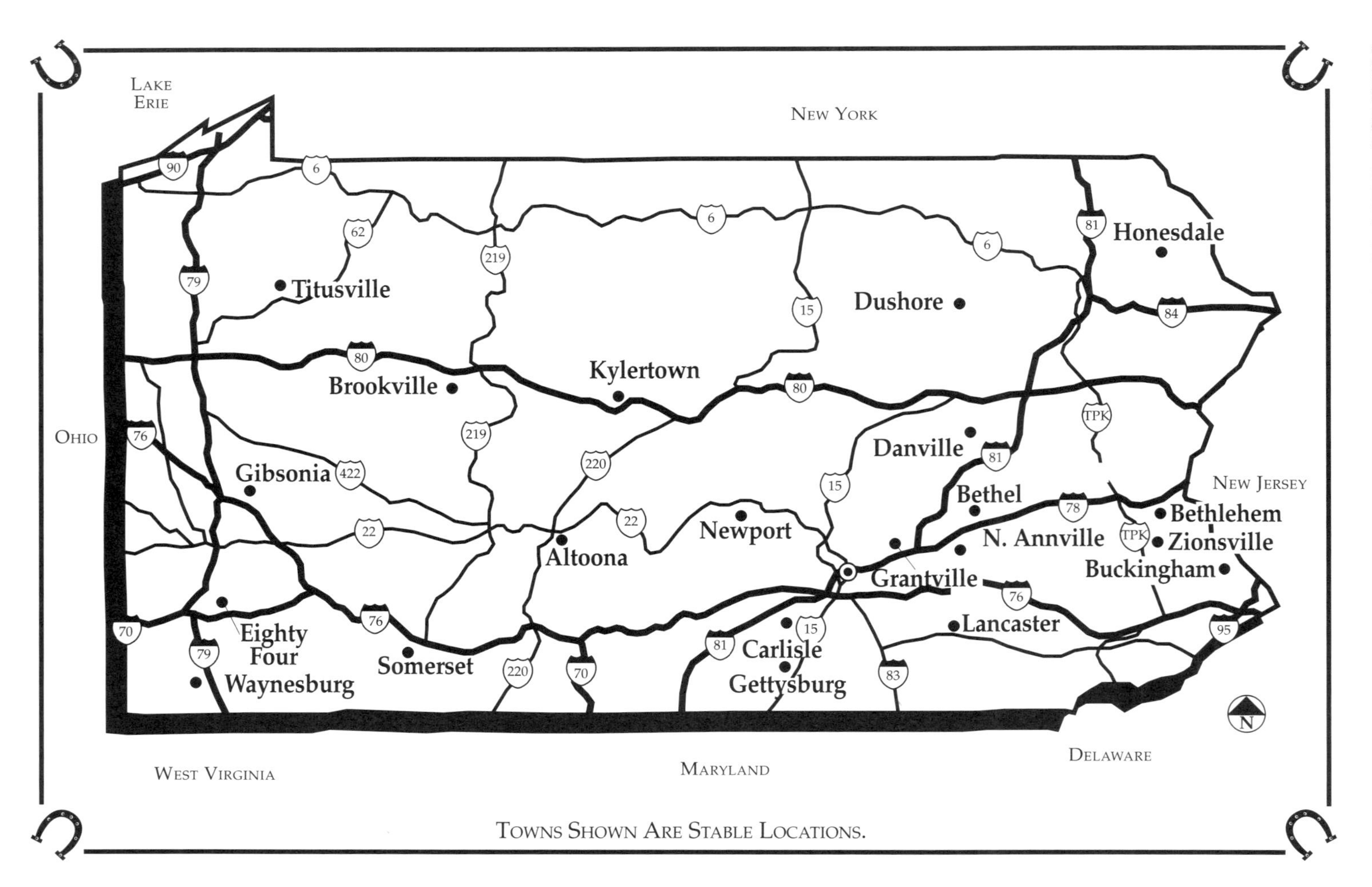
Lake Erie
New York
Ohio
New Jersey
West Virginia
Maryland
Delaware
Titusville
Brookville
Kylertown
Dushore
Honesdale
Gibsonia
Altoona
Newport
Danville
Bethel
N. Annville
Grantville
Bethlehem
Zionsville
Buckingham
Lancaster
Carlisle
Gettysburg
Eighty Four
Waynesburg
Somerset
Towns Shown Are Stable Locations.

ALL OF OUR STABLES REQUIRE CURRENT NEG. COGGINS, CURRENT HEALTH PAPERS, & OWNERSHIP PAPERS.

ALTOONA
Sinking Valley Stables Phone: 814-944-3241
Chris Beaver Barn: 814-944-7063
R. D. 3, Box 255 [16601]
Directions: Rt. 220 N to Altoona from Pittsburgh; 17th St. Exit right at light; go through 1 light & at 2nd light turn right onto Kettle St.; go 7 miles & farm is on left. **Facilities:** 5-18 indoor 10' x 12' stalls; 50' x 50' paddocks; timothy, alfalfa, & orchard grass available. Trailer parking. **Rates:** $10 per night - $8 per night if more than 1 horse; $20 for weekends. **Accommodations:** Holiday Inn, Days Inn, & EconoLodge all in Altoona, 10 minutes from farm.

BETHEL
Windy Ridge Farm Phone: 717-933-5888
Judy Reggio
401 Swope Road [19507] Directions: 2 minutes from I-78, between Exits 2 & 3. 30 minutes from Hershey Park. Located in Amish Country, a vacation destination. Call for directions. **Facilities:** 5 large, airy indoor stalls, three 1/4-acre parcels of grass, one 3-acre board-fenced pasture area with run-in, feed/hay & trailer parking. Farm is a breeding facility for Dutch warmbloods by a national top stallion. Nationally recognized young warmbloods for sale. Devon winners 2 years running. Must have rhino shots. **Rates:** $15 per night if guest at B & B; otherwise $20. **Accommodations:** Guest rooms available in new home on a beautiful farm nestled at the foot of the Blue Mountains.

BETHLEHEM
Koehler's Stable Phone: 610-865-0438
Frank Koehler
3435 Jacksonville Road [18017] Directions: 1/2 mile from Hwy 512 & Hwy 22. Between Hwy 512 & Airport Road. Call for directions. **Facilities:** 5 indoor stalls, four 50' x 50' turnout pens, indoor arena, feed/hay & trailer parking. Buys & sells horses. Call for reservation. **Rates:** $15 per night; $25 per week. **Accommodations:** EconoLodge & Howard Johnson's 1 mile from stable.

BROOKVILLE
Seneca Trial Horses Phone: 814-849-8135
John McAninch & Hollie Nelson
RR 8 P.O. Box 64 (15825) Directions: I-80 exit 13, 2 miles south on Rt 28, turn left on Seneca Trial Rd, farm is 1 mile on left. **Facilities:** 10 stalls, 10′ x 12′ open air / rubber mats. Several pastures available , 1-3 acres. 60′ x 60′ indoor arena, several miles of scenic wooded trails, mare and foal stall available. **Rates:** $15 per night. **Accommodations:** Brookville at exit 13 I-80, 3 miles from stable, Super 8, Days Inn, Holiday Inn. Also B & B, G. Barrier House.

ALL OF OUR STABLES REQUIRE CURRENT NEG. COGGINS, CURRENT HEALTH PAPERS, & OWNERSHIP PAPERS.

BROOKVILLE

Valley View Stables **Phone: 814-849-2407**

Connie & Turk Waterbury

RD 4, Box 345A [15825] Directions: Call from Exit 13 off I-80 and they will come meet you. **Facilities:** 16 indoor 10' x 10' stalls, outdoor arena with pasture, feed/hay, trailer parking available. Travellers with sleeping facilities may sleep on the premises. **Rates:** $10 per night; $12 with electric. **Accommodations:** Days Inn (allows pets), Super 8, Ramada 4 miles away.

BUCKINGHAM

Mill Creek Farm Enterprises, Inc. **Phone: 215-262-7194**

James R. Brame, Sr.

2348 Quarry Road [18912] Directions: I-95 Exit 30: Take Rt. 332 west for 3 miles; take Rt. 413 north 8 miles to property. **Facilities:** 13 large 14' x 14' indoor box stalls, 10 acres of pasture/turnout, feed/hay & trailer parking available. Coggins required. Also swimming pool with spa/jacuzzi, tennis court, etc. **Rates:** $25 per night **Accommodations:** Bed & Breakfast on premises. Rooms are $100 per night, midweek, including breakfast.

CARLISLE

Pheasant Field Bed & Breakfast **Phone: 717-258-0717**

Denise (Dee) Fegan **E-mail: pheasant@pa.net**

150 Hickorytown Road [17013] Directions: From I-76 and I-81: Southeast on S. Middlesex Road, left on Ridge Drive, right on Hickorytown Road. Within 1 hour of Hershey, Harrisburg, Gettysburg. **Facilities:** 12 indoor 15' x 15' stalls, 4 outdoor run-in sheds, 4 acres of pasture, feed/hay available, trailer parking. **Rates:** $25 per night. **Accommodations:** Bed & Breakfast on site, $85-$120 dbl per night, includes full breakfast. Non-smoking.

DANVILLE

Circle G Riding Stable, Inc. **Phone: 717-275-3099**

Charles R. Gordner

2903 Bloom Road [17821] Directions: Call for directions. **Facilities:** 6 indoor stalls, feed/hay & trailer parking available. No smoking in barns. Call for reservation. **Rates:** $15 per night; $75 per week. **Accommodations: Bed and Breakfast on site $85-$115 dbl includes full breakfast.** ``Red Roof Inn & Days Inn in Danville, 20 minutes from stable.

DUSHORE

Drake Hollow Stables **Phone: 570-928-7101**

Robert Brown

RR4, Box 4307 [18614] Directions: Call for directions. Off of Exit 34 on I-80. **Facilities:** 12 indoor 12' x 12' stalls, small pasture, 60' round pen, 100' x 150' arena, 1-acre turnout, feed/hay, trailer parking. Call for reservation. **Rates:** $20 per night. **Accommodations:** Motels in Dushore, 5 miles from stable.

ALL OF OUR STABLES REQUIRE CURRENT NEG. COGGINS, CURRENT HEALTH PAPERS, & OWNERSHIP PAPERS.

EIGHTY FOUR

Gil-Mar Stables **Phone: 412-941-0118**
Gary Stegenga, mgr. **Barn: 412-941-9897**
425 Ross Road [15330] Directions: Canonsburg Exit off of I-79. Stable is 6 miles from exit. **Facilities:** 20-stall barn, 120' x 120' paddock, outside pasture, outside ring, feed/hay & trailer parking available. Call for reservation. **Rates:** $15 per night. **Accommodations:** Motels 10 miles from stable.

GIBSONIA

Sun & Cricket Farm B & B **Phone: 724-444-6300**
John & Tara Bradley-Steck
1 Tara Lane [15044-5507] Directions: 5 miles NE of I-76 (PA turnpike) at Exit 4; 12 miles east of I-79 at Warrendale Exit; 50 miles south of I-80. Call for further directions. **Facilities:** 5 indoor 10' x 12' box stalls, 1 -indoor 12' x 12' box stall, 1/4-acre grassy pasture with split-rail fencing. Extensive trails. Other pets with prior approval only. Deposit required. No arrivals after 10 P.M. No smoking. Guests with horses must stay at B & B. Reservation required. **Rates:** $20 for stall; $15 for pasture per night. **Accommodations:** Carriage house with queen bed, primitive antiques, folk art, quilts, full bath, private porch, $115. Three-room log cabin with king bed or two twins, fireplace, atrium & library, $125. Discounts for two-night stays. Major credit cards accepted.

GRANTVILLE

Centaur Farm **Phone: 717-865-5501**
Edwin Stopherd or Ronald Gerstner
6500 Mountain Road (Rt 443) [17028] Directions: Exit 28 off of I-81: Take left off exit if eastbound - right if westbound; go to "T" & take right; go approx. 2.7 mi. & farm is on right. Green barn in rear. 18 wheelers OK. 1.6 mi. past Penn National Race Course. **Facilities:** 30 indoor stalls in totally enclosed barn, outdoor sheds, 50' x 50' paddocks, five 4-acre paddocks, 24-hr. vet available, lay-up & broodmare care, long- or short-term boarding available. Standing-at-stud: Thoroughbred "Dr. Koch" by "In Reality"; Saddlebred "Centaur's Fame and Fortune" by "Chief of Graystone." Any size trailer welcome. Call for availability. Payment in advance. **Rates:** $15 per night; $100 per week. Any special care is extra charge. **Accommodations:** Holiday Inn & Budget Motel in Grantville, 3 miles from stable.

HONESDALE

Triple W Riding Stable **Phone: 717-226-2620**
Kevin Waller Walter Stolle, mgr.
RR 2, Box 1540 [18431] Directions: 30 miles east of Scranton, 3 miles from Holly. Near Rt. 6. Call for directions. **Facilities:** 20 indoor box stalls, 4 fenced fields, indoor arena, wash rack, viewing room. Corrective shoeing & dentistry work available. Horse training, lessons, & sales, particularly Arabians & Appaloosas. Hay & sleigh rides. Call for reservation. **Rates:** $15-$20 per night. **Accommodations:** Bed & Breakfast on premises has 10 rooms with double occupancy. Also overnight camping.

ALL OF OUR STABLES REQUIRE CURRENT NEG. COGGINS, CURRENT HEALTH PAPERS, & OWNERSHIP PAPERS.

KYLERTOWN
Terri's Stable **Phone: 814-345-6940**
Terri Fisch
N. Second Street [16847] Directions: 1/2 mile north of Exit 21 off I-80. **Facilities:** 25 indoor stalls, hot & cold water, 3 wooden-fence turnout paddocks, indoor arena, feed/hay & trailer parking available. Training & breeding of quarter horses at stable. Call ahead for reservation. **Rates:** $15 per night. **Accommodations:** Stop 21 Motel 1/2 mile.

LANCASTER
Foxfield Farm **Phone: 717-484-2250**
Susan Walmer, owner
230 Holtzman Road [17569] Directions: Exit 21 off of I-76. Call for directions. **Facilities:** 45 box stalls, indoor arena, paddocks, outdoor riding rings, feed/hay & trailer parking available. Call for reservation. **Rates:** $20 per night. **Accommodations:** Motels within 1 mile of stable.

NEWPORT
Windy Ridge Acres **Phone: 717-567-7457**
Laura Martin
RD 2, Box 354C [17074] Directions: Located off of Rt. 322. Call for directions. **Facilities:** 6 indoor box stalls, 4 tie stalls, 100' x 200' outdoor ring, 50' x 100' indoor ring, trail riding on 125 acres of wooded trails. English & Western lessons given at all levels. Also Clydesdale wagon rides available. Call for reservation. **Rates:** $15 per night. **Accommodations:** Motels within 5 miles of stable.

SOMERSET
Lori-M-Stables **Phone: 814-443-6627**
Mina Stutzman
RD 5, Box 72 [15501] Directions: Exit 10 off of I-76. Call for directions. **Facilities:** 42 indoor stalls, 11 acres of pasture/turnout, 60' x 120' indoor arena, 3 paddocks, & jump course. Breeding and training facility for horses & riders. Arabian & Morgan stallions at stud & Arabians showing on the "Class A" circuit. Call for reservation. **Rates:** $20 per night. **Accommodations:** Motels 6 miles from stable.

TITUSVILLE
Knapping Knapp Farm Bed & Breakfast **Phone: 814-827-1092**
Vern & Peg Knapp
43778 Thompson Run Road [16354] Directions: 1 hour from I-80, I-79, I-90; southeast of Erie. **Facilities:** 6 permanent 8' x 10' wooden inside stalls, various-size outdoor stalls (small, medium, large) with Hi-T fence, pasture/turnout in various sizes for short-term stays, feed/hay available, trailer parking but no hook-ups. Various activities on site. New sand-base 120' x 240' outdoor arena with scheduled cowboy events; practice times for roping & penning in arena for beginners to advanced, singles and teams. Pleasure rides on wooded, logging trails on about 1,000 acres of private land. Chuck wagon dinners; can accommodate groups. **Rates:** Varies, depending on accommodations. **Accommodations:** Bed & breakfast on premises.

ALL OF OUR STABLES REQUIRE CURRENT NEG. COGGINS, CURRENT HEALTH PAPERS, & OWNERSHIP PAPERS.

WAYNESBURG

High Moon Stables **Phone: 412-324-2770**

Joann Moon

Rt. 2, Box 175 [15370] Directions: From I-79 S: turn right off of Exit 2 - Kirby; go 1.6 miles to Rt. 19; turn left & go .5 mile & turn right; go .8 mile and see stable sign on right. **Facilities:** 9 indoor stalls, pasture/turnout available, feed/hay & trailer parking also available. Trail rides, overnight riding trips, hayrides, wagon trains, camping on a historic Indian Trail, and riding lessons are all offered at the stable. Please call for reservation. **Rates:** $15 per night; $75 per week. **Accommodations:** Bed & Breakfast on premises plus camping and an RV camper.

ZIONSVILLE

Horseman's Hollow Equestrian Center **Phone: 215-541-4363**

Elizabeth Lafrenz

8300 School House Lane [18092] Directions: Off PA. Turnpike, take Exit 32. Exit toll booth, right turn onto PA. Rt 663. Go 4/10 mile to Spinnerstown Rd. Turn right and go straight for 2 miles. At 2nd stop sign continue straight onto Orchard Rd. Go 6/10 mile to stop sign. Left onto School House Lane. 1 Mile to Horseman's Hollow on right.
Facilities: 8 - 12′ X 12′ stalls. Tongue and groove boards. Hay & feed included. Trailer parking available. 12 acres of pasture. Boarding, lessons and 80′ X 152′ indoor arena.
Rates: $15 per night. **Accommodations:** Econolodge and Rodeway Inn in Quakertown 4 miles away.

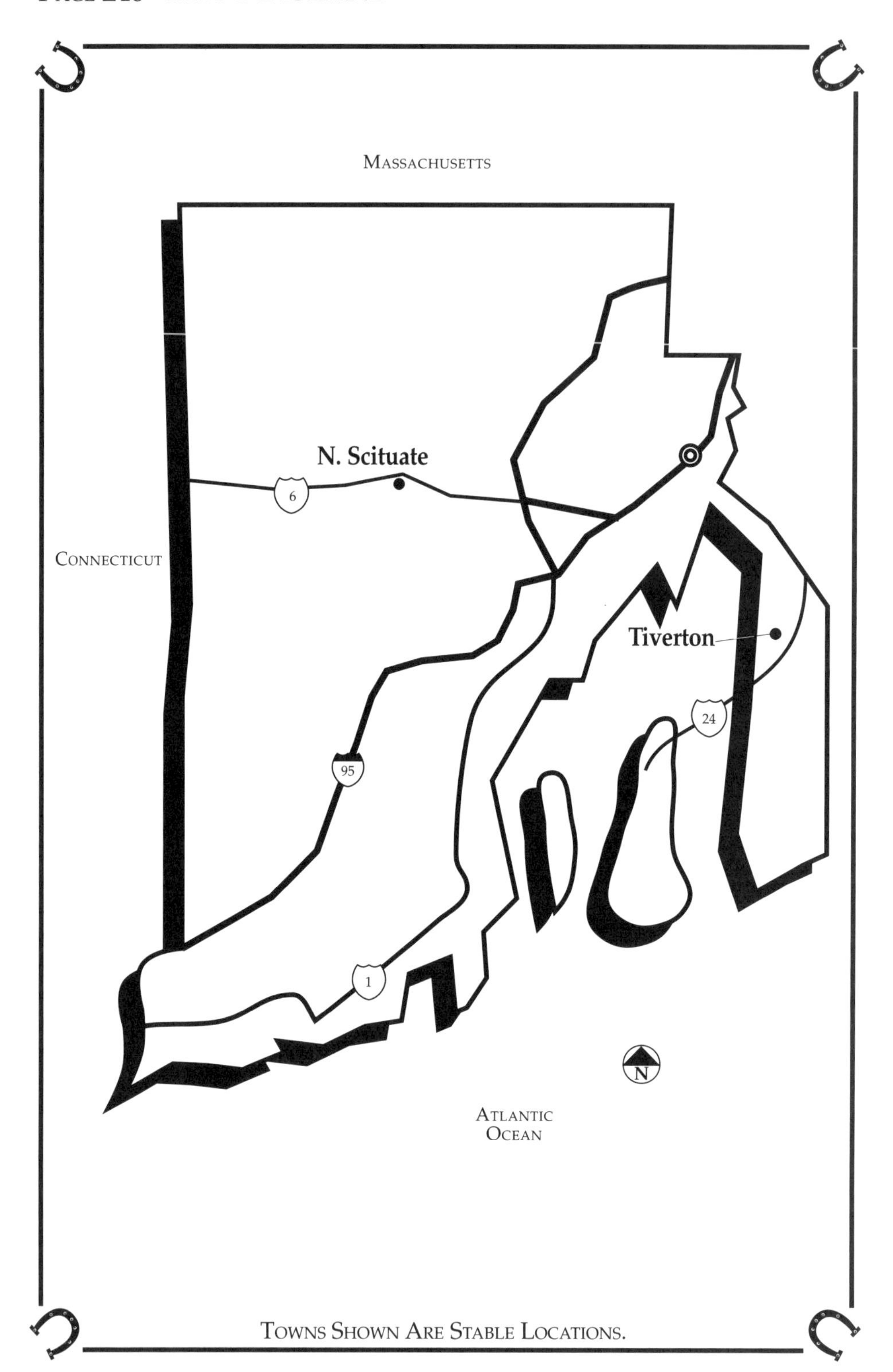
Massachusetts
N. Scituate
6
Connecticut
Tiverton
24
95
1
N
Atlantic
Ocean
Towns Shown Are Stable Locations.

ALL OF OUR STABLES REQUIRE CURRENT NEG. COGGINS, CURRENT HEALTH PAPERS, & OWNERSHIP PAPERS.

NORTH SCITUATE

Stone House Farm **Phone: 401-934-0272**

George Bessette

86 Peeptoad Road [02857] Directions: 15 minutes from Providence off Rt. 6. Call for directions. **Facilities:** 17 indoor stalls, 120' x 80' indoor arena, pasture, and 6 paddocks each over 1 acre. Instruction in hunter/jumper for horses & riders from beginner to Grand Prix. Also buys and sells quality horses on consignment. Call for reservation. **Rates:** $20 per night. **Accommodations:** Motels 5 miles from stable.

TIVERTON

Sakonnet Equestrian Center **Phone: 401-625-1458**

Robert & Helen Bobeoch

3650 Main Road [02878] Directions: 5 miles off Rte. 24. Call for directions. **Facilities:** 48 indoor stalls, indoor and outdoor riding ring, paddocks available, feed/hay, trailer parking. Six horse shows per year, lessons at all levels. Adjacent to 400-acre town-owned park. Call for reservation. **Rates:** $20 per night. Monthly boarding available. **Accommodations:** Motels within 6 miles of stable.

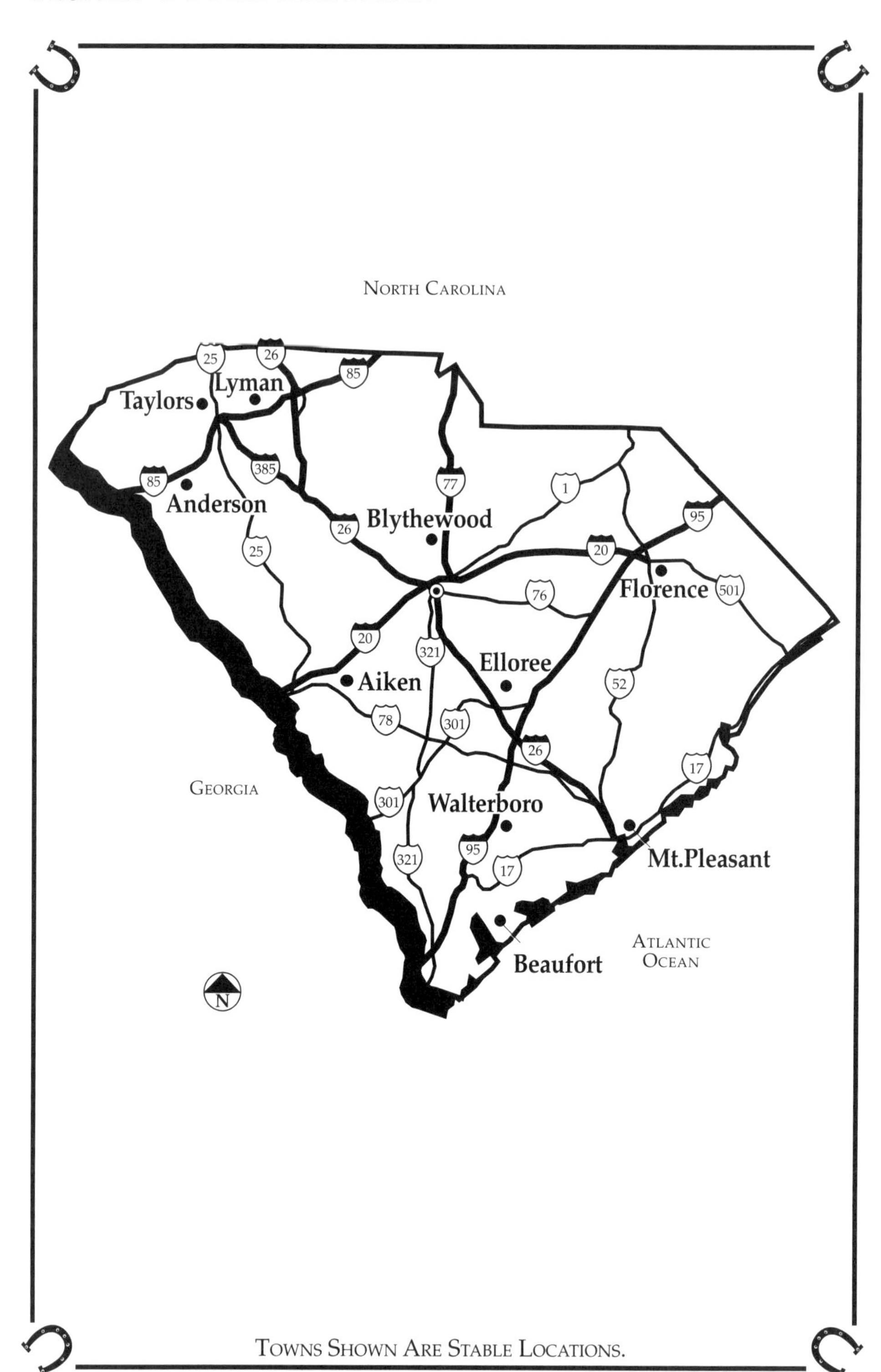

Towns Shown Are Stable Locations.

ALL OF OUR STABLES REQUIRE CURRENT NEG. COGGINS, CURRENT HEALTH PAPERS, & OWNERSHIP PAPERS.

AIKEN

Fulmer International **Phone: 803-649-0505 Fax: 803-649-1200**
Robert N. Hall, D.B.H.S. **E-mail: fulmer@scescape.net**
2500 Dibble Road [29801-3381] **Web: www.equitation.com**
Directions: Aiken-Augusta Hwy 1/78, go South on 118, east on Dibble Road. Go 1/2 mile, just past the natural gas sub-station and turn right up the sandy driveway & continue up to the stable. **Facilities:** 10 indoor stalls, 1 large paddock. Trailer parking available for guests. Please bring your own feed/hay. Stable adjoins the famous 2,000-acre Hitchcock Woods, which has over 40 miles of trails and jumps (hunt fences). Maps available. Boarding facilities, breaking and training through Gran Prix level, student instruction in Dressage, Eventing and Show Jumping. Breeding facilities and home-bred horses for sale. Telephone in advance for reservation. **Rates:** $20 per night. **Accommodations:** Variety available, including B&B.

ANDERSON

Penn's Woods Stable **Phone: 803-261-8476**
Bill Payne
1930 Denver Road [29625] Directions: Exit 19B off of I-85. Call for directions. **Facilities:** 12 box stalls, 4 tie stalls, pasture, 4 turnout paddocks, outdoor riding ring, jumper & cross-country course & fox hunting. Training of hunter/jumper for horses & riders. Complete blacksmith shop on premises. Near Clemson University. Call for reservation. **Rates:** $20 per night. **Accommodations:** Motels less than 2 miles from stable.

BEAUFORT

Beaufort Equestrian Center **Phone: 843-846-4765**
Helen Martz
163 Keanes Neck Rd [29940] Directions: 8 minutes from I-95, 2 minutes from Hwy 17, call for directions. **Facitilies:** 22 12'x12' matted floor, modern stalls, 6 2 and 3 acres pastures. Feed/hay avialable, trailer parking. Modern, clean barn with excellent turn-out, wash racks, lessons in Hunter Jumper and Western Pleasure. **Rates:** $15 per day, $90 per week, lower rates for monthly. **Accommodations:** Howard Johnson Express Inn, Beaufort 4 miles. Hampton Inn, 5 miles. Many more available.

BLYTHEWOOD

Rice Creek Stables **Phone: 803-786-0649**
Carl Blaylock & Melissa Rice
1046 Mickle Road [29016] Directions: From I-77N from Columbia: Take Exit 27. Call for directions. **Facilities:** 2 indoor stalls, 60 acres of turnout, 1/2 mile oval track, 10-acre jumping course, feed/hay & trailer parking available. Call for reservation. **Rates:** $18 per night. **Accommodations:** Ramada Inn & Best Western 6 miles from stable.

ALL OF OUR STABLES REQUIRE CURRENT NEG. COGGINS, CURRENT HEALTH PAPERS, & OWNERSHIP PAPERS.

ELLOREE

R.V. Shirer Thoroughbreds **Phone: 803-897-2238**
R.V. Shirer, owner
East Harlin Street [29047] Directions: From I-95: Take Exit 98 towards Elloree & go 7 miles; 2 streets past only traffic light, turn right. Stable is on right side. **Facilities:** 8 indoor stalls, 5 outdoor stalls, 5 paddocks, hot walker, feed/ hay & trailer parking. Campers OK if self-contained. **Rates:** $15 per night; $75 per week. **Accommodations:** Motels in Santee, 7 miles from stable.

FLORENCE

Florence Horse Center **Phone: 843-679-5502 or: 843-667-0951**
Jack Belew **E-mail: jack@easternx.com**
3508 Cherrywood Road [29501] **Fax: 843-667-3504**
Directions: 5 minutes from I-95 and US 52. Call for directions. **Facilities:** 37 indoor stalls, ring, paddocks, dressage ring, and jumps. Western & English dressage & hunter training for horses and riders. Horses for sale. Truck & trailer parking available. Also, home of Eastern Equine Express Horse Transportation, serving all of the lower 48 states. Reservation required. **Rates:** $25 per night. **Accommodations:** Ramada Inn, Days Inn, Hampton, & EconoLodge all within 5 minutes of stable.

LYMAN

Scotsgrove Farm **Phone: 864-877-9392**
Mr. & Mrs. Robert C. Scott **Barn: 864-968-9401**
39 Hillcrest Street [29365] Directions: Call for directions. **Facilities:** 12 indoor 12' x 12' stalls in main barn & 3 stalls in smaller barn, 6 paddocks up to 3 acres each, 3 pasture area, 150' x 250' sand ring, dressage arena, and cross country course. Fescue & coastal Bermuda hay available, & trailer parking on site. Lessons, training, foxhunting available. **Rates:** $20 per night. **Accommodations:** Bed & Breakfast on premises with large private suites and baths.

MT. PLEASANT

Pelican's Roost Stables, Inc. **Phone: 803-856-0444**
Liza Antley, owner **Evenings: 803-856-0731**
Jane Watson, mgr.
1351 Venning Road [29464] Directions: Georgetown Exit off of I-526. Call for directions. **Facilities:** 28 indoor stalls, 9 turnout paddocks, 3 rings, jump course, tractor-trailer parking, & feed/hay available. Training of riders in hunter/jumper. 24-hr security. Call for reservation. **Rates:** $25 per night. **Accommodations:** Motels 3 miles from stable.

ALL OF OUR STABLES REQUIRE CURRENT NEG. COGGINS, CURRENT HEALTH PAPERS, & OWNERSHIP PAPERS.

TAYLORS

Foxcroft Farms **Phone: 803-244-2636**

Andre D'eridder

175 McConnell Road [29687] Directions: Located off of I-29 near Greenville. Call for directions. **Facilities:** 5 indoor stalls, pasture/turnout, lessons & trail rides available, trailer parking on site, and feed/hay available at additional charge. Call for reservation. **Rates:** $15 per night. **Accommodations:** Motels 3 miles away.

WALTERBORO

Double D Stable **Phone: 843-893-3894**

Tommie Derry

1256 Rodeo Drive. (29488) Directions: Call for Directions. **Facilities:** 16 12X12 indoor stalls, 4 multi-acre pastures, 2 barns with 8 stalls each, hook-ups for overnight, hot walker, round pen, arena with lights and bleachers, stalls have auto water and overhead lights, wash room with hot and cold water, small pets must be on leash and contained, prefer reservations. **Rates:** $25 per night, $150 per week. **Accommodations:** Holiday Inn, Econo Lodge and many other major hotels nearby.

Mt. Carmel Farm Bed & Breakfast **Phone: 843-538-5770**

Maureen Macknee

Rt. 2, Box 580A [29488] Directions: 3-1/2 miles off I-95. Reservations required. Call for directions. **Facilities:** 8 stalls, paddocks, turnout, layups, round pen. Trailer parking on site. This is a bed & breakfast for people traveling with horses and/or other animals. Overnight boarding is only for guests of the B & B. **Rates:** Call for reservations. **Accommodations:** B & B with 2 guest rooms with private baths.

Walterboro

Double D Stable

- 16 12X12 indoor stalls
- 4 multi-acre pastures
- two barns with 8 stalls each
- hook-ups for overnight
- hot walker
- round pen
- 175′ x 250′ arena with lights and bleachers
- stalls have auto water & overhead lights
- wash room with hot & cold water

Note: small pets must be on leash and contained, prefer reservations. **Rates:** $25 per night, $150 per week. Holiday Inn, Econo Lodge and many other major Hotels nearby.

Tommie Derry
1256 Rodeo Drive. Walterboro, SC 29488
Phone: 843-893-3894

NOTES AND REMINDERS

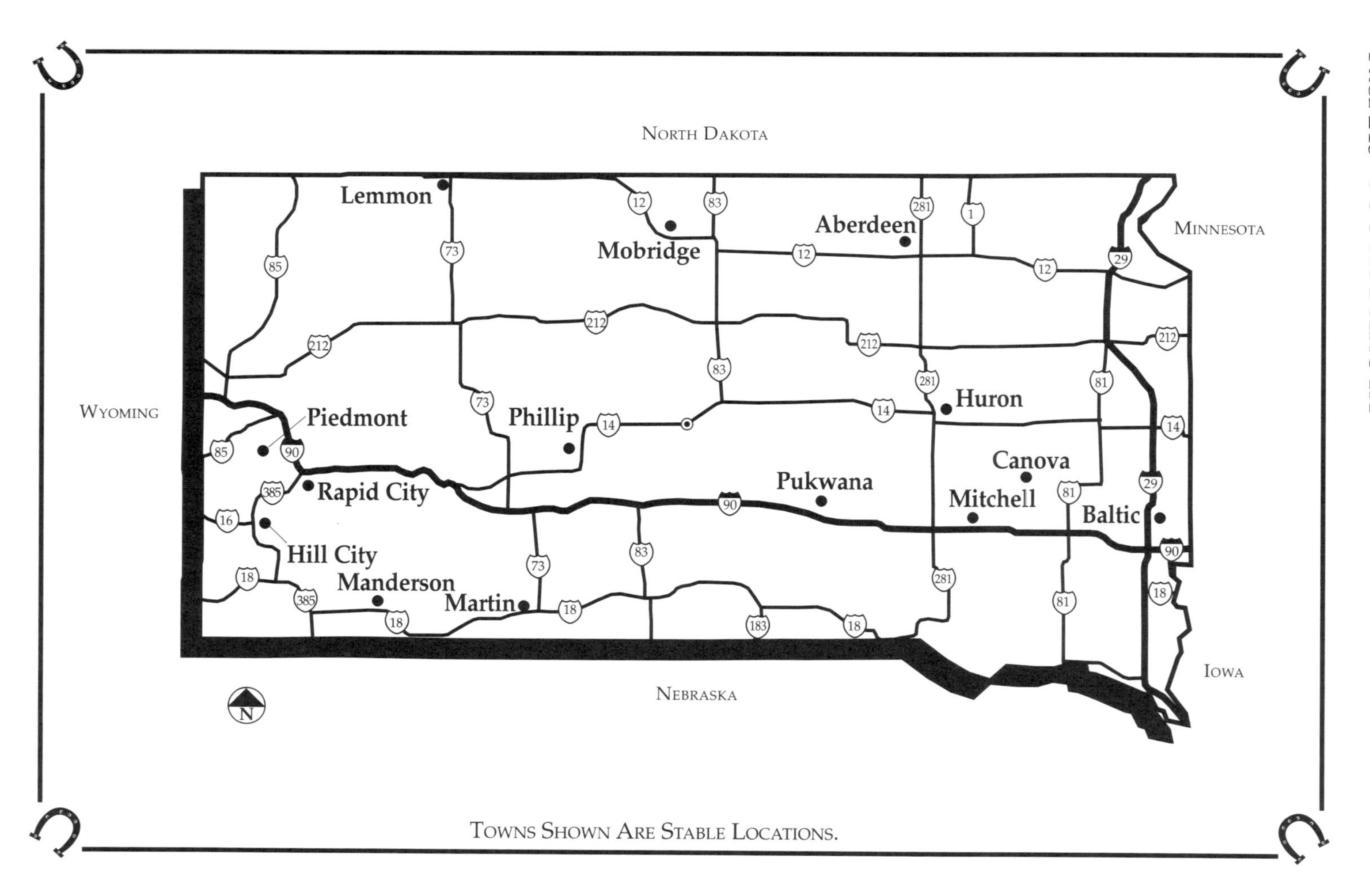
NORTH DAKOTA
MINNESOTA
WYOMING
IOWA
NEBRASKA
Lemmon
Mobridge
Aberdeen
Huron
Piedmont
Phillip
Rapid City
Pukwana
Canova
Mitchell
Baltic
Hill City
Manderson
Martin
TOWNS SHOWN ARE STABLE LOCATIONS.

ALL OF OUR STABLES REQUIRE CURRENT NEG. COGGINS, CURRENT HEALTH PAPERS, & OWNERSHIP PAPERS.

ABERDEEN

EL-JO-MAR Arabian Ranch **Phone: 605-229-0022**

Margaret Forseth

RR 1, Box 14 [57401] Directions: Call for directions. **Facilities:** 15 indoor & outdoor stalls, 4 turnout runs, 2 pens, 130' x 60' indoor arena, 200' x 100' outdoor arena, feed/hay & trailer parking. This is an Arabian breeding & training facility with Ferzon, Basque, & Padron bloodlines available. Standing-at-stud: "Firzak ++," Legion of Merit & National Champion & "Gallant Fashion++." Lessons given and horses for sale. Easy to find location. Call for reservation. **Rates:** $15 per night; weekly rate available. **Accommodations:** Motel 5 minutes from ranch.

BALTIC

Heartland Arabian Farm **Phone: 605-543-5900**

Jane & Lloyd Solberg, owners

25467 473rd Avenue [57003] Directions: Exit 399 at I-29 & 1-90 junction. Call for directions. **Facilities:** 64 indoor heated 10' x 12' stalls, 3 outside paddocks on a total of 190 acres. Feed/hay & trailer parking available. An Arabian breeding & training farm. 5 stallions standing-at-stud including "The Chief Justice." Horses for sale. Call for reservation. **Rates:** $20 per night; $100 per week. **Accommodations:** Ramkota Inn in N. Sioux Falls, 4 miles from stable.

CANOVA

Skoglund Farm Bed & Breakfast **Phone: 605-247-3445**

Alden Skoglund

Rt. 1, Box 45 [57321] Directions: Call for directions. **Facilities:** 3 outdoor pens, pasture/turnout area, no feed/hay, overnight horse trailer parking available. Boarding only for guests at B & B. **Rates:** No charge. **Accommodations:** Bed & Breakfast on premises: Adults $30, teens $20, children $15, 5 & under, free.

HILL CITY

Happy Hill Ranch & Tack Shop **Phone: 605-574-2326**

Marlen & Grayce Larson

23845 Penalua Gulch Road [57745] Directions: 3 miles east of Hill City; 1/8 mile off Rtes. 385 & 16. **Facilities:** 8 - 10' x 10' and 10' x 12' stalls, 40' x 60' grassy paddocks, feed/hay, trailer parking. Black Trakehner standing-at-stud; Warm Blood Crosses for sale. Tack shop, training. **Rates:** $10 per night; $60 per week. **Accommodations:** Motels within 3 miles of stable.

ALL OF OUR STABLES REQUIRE CURRENT NEG. COGGINS, CURRENT HEALTH PAPERS, & OWNERSHIP PAPERS.

HURON

Huron Veterinary Hospital **Phone: 605-352-6063**

Dr. Brunswig

RR 2, Box 149 [57350] Directions: Hwy 281 to Hwy 14 to Huron. Call for further directions. **Facilities:** 9 indoor stalls, 30' x 30' pasture/turnout area, feed/hay & trailer parking available. Medical services available on premises. Call for reservation. **Rates:** $20 per night. **Accommodations:** Motels 1 mile.

LEMMON

Johnson Racing Stables **Phone: 605-374-5733**

Bob Johnson

HCR 82, Box 83 [57638] Directions: 9 miles south of Hwy 12 on White Butte Road. Call for directions. **Facilities:** 2 indoor stalls, runs, feed/hay & trailer parking available. Advance notice requested. **Rates:** $15 per night; weekly rate available. **Accommodations:** Motels in Lemmon 20 miles from stable.

MANDERSON

Ecoffey Stables **Phone: 605-867-5698**

Gilbert Ecoffey

Box 345 [57756] Directions: Call for directions. Short distance from Wounded Knee. **Facilities:** 8 indoor stalls, 100' x 150' arena, feed/hay & trailer parking available. Can assist anyone in area if they have broken down. Will get horses & trailer. Call for reservation. **Rates:** $15 per night. **Accommodations:** 3 motels in Gordon, NE 30 miles from stable.

MARTIN

T - (T Bar) **Phone: 605-685-6900**

Thomas H. Loomis

HC #2, Box 5A [57551] Directions: From I-90: South on 73 to Rt. 18. Straight at stop onto "Old 18" 3.5 miles heading west past Deadman's Lake on south side of road to top of hill turn south, 2 miles on dirt road then east 1/4 mile. **Facilities:** 6 - 7' x 10' box stalls plus corral, 26+ acres of pasture/turnout, feed/hay available, trailer parking. Horse owners care & maintain their own animals. Trail maps for riding available. A confirmed reservation is mandatory! **Rates:** $12 per night; $70 per week. **Accommodations:** 6-bed bunkhouse with 2 showers and outside shower on premises; $35 per person per night.

MITCHELL

Mitchell Livestock Auction Co. **Phone: 605-996-6543**

Tim Moody

P.O. Box 516 [57301] Directions: Exit 332 off of I-90: go 1/4 mile south on Hwy 37. Look for big sign & go 1/4 mile east. **Facilities:** 200 indoor & outdoor pens. Feed/hay & trailer parking available. **Rates:** $3 per night. **Accommodations:** Super 8 & Best Western 1/4 mile away.

ALL OF OUR STABLES REQUIRE CURRENT NEG. COGGINS, CURRENT HEALTH PAPERS, & OWNERSHIP PAPERS.

MOBRIDGE

Oahe Veterinary Hospital **Phone: 605-845-3634**

Tami Schanzenbach

North of City Airport Road, Box 547 [57601] Directions: Hwy 12 to Mobridge. Located on right side. See wood sign. **Facilities:** 2 indoor stalls, numerous 200' x 50' pens, feed/hay & trailer parking. 24-hr veterinary care available at hospital. Located in a beautiful area on Lake Oahe next to the Missouri River. Call for reservation. **Rates:** $15 per horse including feed; weekly rate available. **Accommodations:** Super 8 and Wrangler 1 mile away.

PHILIP

Triangle Ranch Bed & Breakfast **Phone: 605-857-2122**

Kenny & Lyndy Ireland

HCR 1 Box 62 [57567] Directions: Call for directions. **Facilities:** 4 single and 2 double indoor tie stalls, 2 indoor box stalls, 9 outdoor plank & pole corrals, alfalfa/grass hay available, trailer parking. Working cattle ranch, seasonal riding to check cows, checking fences, scenic rolling prairie and river breaks, possible trail rides, guests ride their horses. **Rates:** $10-12 per night. **Accommodations:** B&B on ranch, historic home with 4 bedrooms/2 shared baths, $50-$60 per night.

PIEDMONT

Six Mile Ranch **Phone: 605-787-9631**

Mike & Jackie Stahly

2853 Elk Creek Rd. (57769) Directions: Nine miles from Rapid City. I-90, exit 46. Go East on Elk Creek Rd. Go six Miles, Ranch on the right. **Facilities:** 19 12X10 indoor stalls, 30x50 and larger turnouts,Grass/Alfalfa, trailer parking , Private guided tours in the heart of the Black Hills. Old railroad beds and game trails. Riding lessons/training/outdoor roping arena. **Rates:** $15 per night. **Accommodations:** Campsites on ranch, Elk Creek Resort-6 miles.

PUKWANA

Diamond A Cattle Co **Phone: 605-778-6885**

Crystal & Tucker Ashley **605-730-1074**

35540 250th St [57370] Directions: Kimball Exit off of I-90: Go 6 miles west on paved road; 1 mile north on gravel, & 3/4 mile west to indoor arena. **Facilities:** 2-10 large pens to turnout, can make other pens with portable panels, large pasture areas, indoor & outdoor roping arenas, feed/hay & trailer parking. Stock available for roping practice. **Rates:** $10 per night. **Accommodations:** Cabin available to rent.

ALL OF OUR STABLES REQUIRE CURRENT NEG. COGGINS, CURRENT HEALTH PAPERS, & OWNERSHIP PAPERS.

RAPID CITY

Bunkhouse Bed & Breakfast & Working Ranch **Phone: 605-342-5462**
Carol Hendrickson **Toll Free: 1-888-756-5462**
Web: www.bbonline.com/sd/bunkhouse
14630 Lower Spring Creek Road, Hermosa [57744] Directions: Exit 60 off of I-90 into Rapid City; left at first light onto Cambell St.; go thru 1 more light; at second light, read odometer and go 8 miles south on Hwy 79 to Lower Spring Creek Rd; take left and stable is 4.3 miles. **Facilities:** 3 indoor 10' x 12' & 14' x 14' stalls, 5 outdoor pens with wind break, no pasture/turnout, feed/hay available at $3 per horse, trailer parking available. No stallions. Riding trails on ranch plus trail maps of great trails in area. B & B not open Jan 1 - May 1. **Rates:** $7 or $10 ($3 for hay) per horse per night if guest at B & B; $20 or $23 ($3 for hay) per horse per night if not. **Accommodations:** Bed & Breakfast on premises with 3 guest rooms. Full breakfast, great accommodations. Motels nearby in Rapid City.

NOTES AND REMINDERS

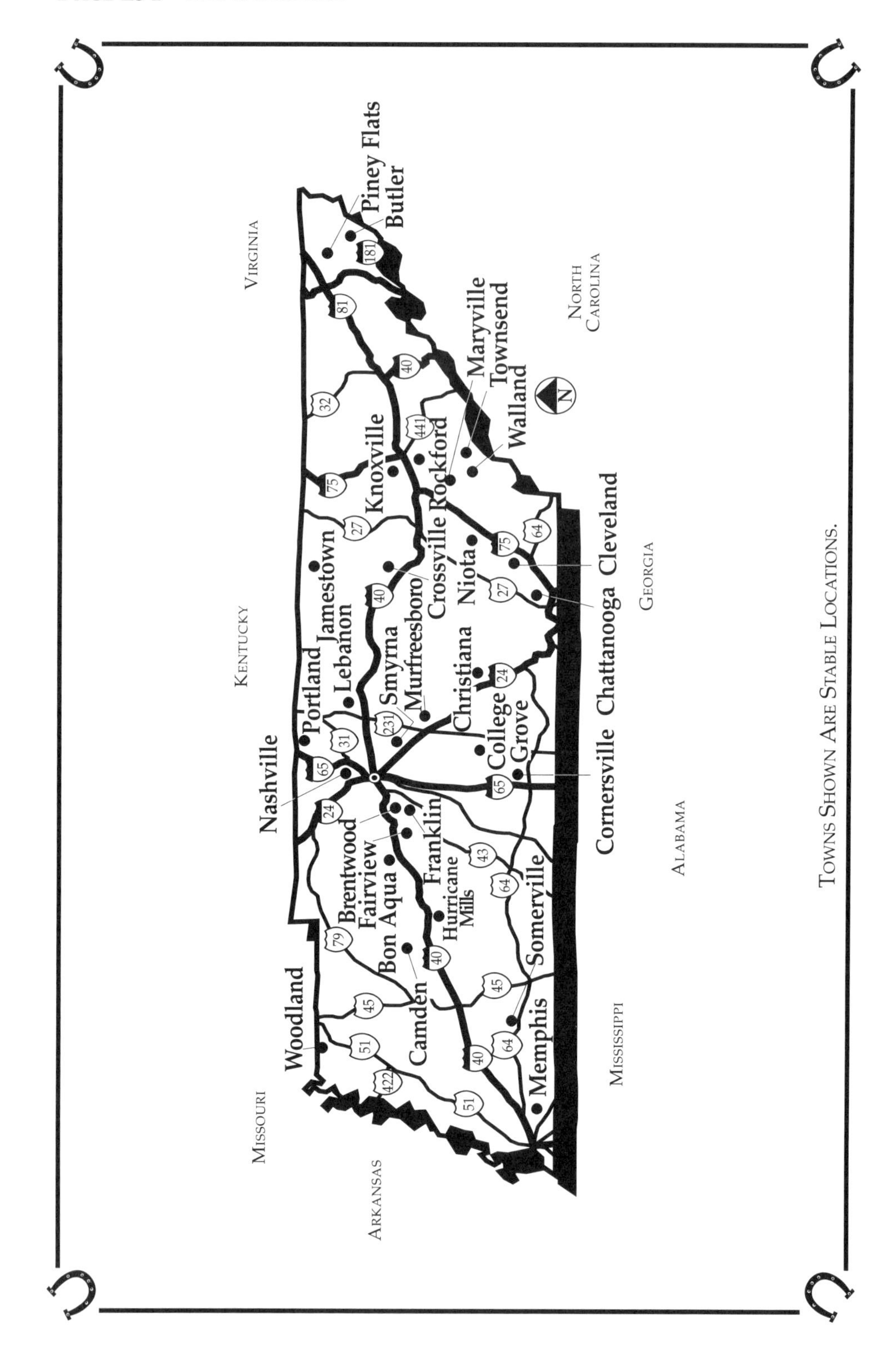

Towns Shown Are Stable Locations.

<u>ALL OF OUR STABLES REQUIRE CURRENT NEG. COGGINS, CURRENT HEALTH PAPERS, & OWNERSHIP PAPERS.</u>

BON AQUA
<u>Harmony Farms</u> **Phone: 615-670-4737**
Joann & Randy Jackowski **or: 615-670-6560**
10726 Harmony Farm Lane [37025] Directions: Exit 172 off of I-40. Go left 4 miles - Harmony Farms on left. **Facilities:** 10+ box stalls, 10+ slip stalls, 40' & 60' round pens, feed store & saddle shop on premises. Trailer parking on premises, up to & including, tractor trailers. Mares with foals & stallions welcome. Thoroughbred Sporthorse breeding program. Standing-at-stud: "EVENING CZAR," "MAKE IT ALL," "SEBASTIAN," "JESTCINO." **Rates:** $15 for box stall, $10 for slip stall per night; $95 per week. **Accommodations:** Comfort Inn & Days Inn nearby.

BRENTWOOD
<u>English Manor Bed & Breakfast Inn</u> **Phone: 800-332-4640**
Willia Dean English **or: 615-373-4627**
6304 Murray Lane [37027] Directions: 10 miles south of Nashville off I-65 South. **Facilities:** 3 stalls, 5 acres of pasture, 40' barn and tack room, feed/hay, trailer parking. Call for reservations. **Rates:** Call for rate. **Accommodations:** B&B on premises, 7 bedrooms with private baths, full breakfast.

BUTLER
<u>Iron Mountain Inn</u> **Phone: 423-768-2446**
Vikki Woods **Web: www.ironmountaininn.com**
138 Moreland Drive [37640] Directions: Call for directions. **Facilities:** Two stalls on premises with additional nearby, 10'x12' run-in-sheds, various pasture/turnout areas. Feed/hay available, trailer parking. Located near hundreds of miles of riding trails in Tennessee, Virginia and North Carolina. **Rates:** $15 per night, $60 per week. **Accommodations:** Iron Mountain Inn, which is a B & B that encourages its guests to enjoy the trails, andthe Inn.

CAMDEN
<u>Bird Song Trail Ride</u> **Phone: 901-584-9206**
Pat & Norman Fowler **or: 901-584-4280**
4700 Old Hwy 69 [38320] Directions: 85 miles west of Nashville. Exit 126 off I-40, 8 miles north towards Camden, right on Shiloh Church Road for 1 mile, right on Little Birdsong Road. Stable is first place on right. **Facilities:** 300 covered boxed stalls, feed/hay, trailer parking, camper hook-ups. Riding trails along creeks and the Tennessee River; trail rides, ride alone or in groups. **Rates:** $15 per night, weekly rate on request. **Accommodations:** Campsites on premises, electric available. Colonial Inn, Guest House Inn. Brochures available on Trail riding and dates.

ALL OF OUR STABLES REQUIRE CURRENT NEG. COGGINS, CURRENT HEALTH PAPERS, & OWNERSHIP PAPERS.

CHATTANOOGA
Cedar Creek Farm **Phone: 706-673-4040**
Judy Noel
500 Old Hwy 2, Varnell, GA [30756] Directions: 13 miles south of Chattanooga, 12 miles north of Dalton, Ga. From I-75: At Exit 139, go east 1 mile, cross bridge, and turn right, go 3.5 miles, sign/mailbox/gravel drive on right. **Facilities:** 15 indoor stalls, paddocks, 4-board fencing, feed/hay & trailer parking available. Large tack store on site. As much advance notice as possible, please. **Rates:** $10 per horse per night. **Accommodations:** Many motels in area.

CHRISTIANA
Kanawha Farm **Phone: 615-895-9262**
Janet E. Stevens
9996 Manchester Hwy [37037] Directions: .8 mile off of I-24 at Exit 89 on Hwy 41E. Call for further directions. **Facilities:** 6 indoor 10' x 10' stalls . One 1/2-acre turnouts with run-ins, feed/hay, trailer parking. Standing-at-stud: "Sombreado" (TB) by "Raja Baba" and "Dam Straight" (black & white Paint) by "Dam Yankee," sport type horses. **Rates:** $15 per night. **Accommodations:** Holiday Inn, Best Western, Howard Johnson in Murfreesboro, 9 miles away.

CLEVELAND
BJ's Stables **Barn Phone: 423-472-2523**
B. J. Owens Dan DeFriese, manager **House: 423-559-9253**
655 Urbana Road NE [37312] Directions: Just off I-75, Exit 27. 5 minutes from downtown Cleveland. Call for further directions. **Facilities:** 22 - 12' x 12' stalls, 30 acres of pasture/turnout, feed/hay available, parking area large enough for tractor-trailers. Two wash racks, hot & cold water, lighted round pen and arena. Large animal veterinarian clinic on premises with 24-hour call. **Rates:** $20 per night; $100 per week. **Accommodations:** Holiday Inn, Red Carpet Inn, Scottish Inn, and others 5 minutes from stable.

COLLEGE GROVE
BPeacock Hill Country Inn **Phone: 615-368-7727**
Anita & Walter Ogilvie **or: 800-327-6663**
6994 Giles Hill Road [37046] Directions: Call for directions. **Facilities:** 8 new indoor 12' x 12' stalls, pasture/turnouts, feed/hay, trailer parking. 650-acre working cattle farm, new barn, miles of trails for riding, hiking. Reservations required. **Rates:** $12 per night. **Accommodations:** Luxury Country Inn on premises, $95-$125 per night includes breakfast.

ALL OF OUR STABLES REQUIRE CURRENT NEG. COGGINS, CURRENT HEALTH PAPERS, & OWNERSHIP PAPERS.

CORNERSVILLE

Lairdland Farm **Phone: 615-363-9080**

Jim Blackburn

3174 Blackburn Hollow Road [37047] Directions: From I-65, take Exit 22, south on 31-A 1 mile to Blackburn Hollow Road, 1 mile to farm. **Facilities:** 22 indoor 12' x 12' stalls, feed/hay available, trailer parking. Miles of trails to ride, hiking, bicycling & gold available in area. **Rates:** $15 per night; ask for weekly rate. **Accommodations:** Bed & Breakfast on premises; log cabin circa 1830s. $85 per night. Privacy is their specialty.

FAIRVIEW

Best Little Horse House **Phone: 615-799-8833**

Gennette S. Norman **Cell: 615-500-8812**

7201 Cumberland Dr (37062) Directions: 5 miles from I-40, exit 182. 25 miles SW of Nashville. **Facilities:** 6 indoor stall, 2 holding pens and pasture. Close to fishing, camping, hiking, horse trails and Nashville attractions. (Formerly Sweet Annies).

Lazy Susan Appaloosas **Phone: 615-799-0991**

Rick & Susan Morrison

7250 Northwest Hwy [37062] Directions: Call for directions. Located 20 miles west of Nashville; 3 miles off of I-40, Exit #182. **Facilities:** 5 - 10' x 12' stalls, 1 - 20' x 12' stall, 4 acres of turnout, 60' round pen, outdoor riding area, wash rack. Grain & hay available; trailer parking. **Rates:** $15 per night per stall. **Accommodations:** Motels in Dickson, 10 miles away or will rent up to 2 double occupancy bedrooms on premises. Hook-ups for campers and trailers available. Ideal for overnight equine travelers and their owners.

Pick up ad from last edition
page # 257
Combs Crest Farm

DANDRIDGE

Combs Crest Farm

Loyal & Hillary Combs, Owners

Phone: 865-397-3045 /Barn: 865-397-1212

Directions: Call for directions. Farm 2 miles off I-40; 25 minutes east of Knoxville. **Facilities:** Modern, clean block barn with 12' x 12' box stalls, mats, shavings, hay and hot water wash rack. Outdoor grass paddock with shelter available. RV hook-up & laundry, 24-hour security & vet services. **Rates:** $25 per night for stalls with shavings, $15 per night for paddock. **Accommodations:** Charming suite in barn; sleeps 4, breakfast, kitchen, A/C, phone, TV, VCR; 24-hour restaurants nearby. **Reservations:** please, but will accommodate emergencies.

ALL OF OUR STABLES REQUIRE CURRENT NEG, COGGINS, CURRENT HEALTH PAPERS, & OWNERSHIP PAPERS.

FRANKLIN

Namaste Country Ranch Inn — **Phone: 615-791-0333**
Lisa Winters — **E-mail: namastebb@aol.com**
5436 Leipers Creek Road [37064] — **Web: namastacres.com**
Directions: Call for directions. **Facilities:** Arena, round pen, walker. Quiet valley setting, 26 miles of scenic horse trails, swimming pool & hot tub. Open year round. AAA approved. Horse owners must be guests of B & B. 1 mile off scenic Natchez Trace Parkway, 11 miles from Franklin. **Rates:** $10 stalls. **Accommodations:** Country home offers 3 private suites, in-room coffee, phone, fridge, TV/VCR, firelplace, private entrance and bath. $75-$85.

HURRICANE MILLS (BUFFALO)

Seraphim Stables (Isaiah 6-2:3) — **Phone: 931-296-1967**
Shirley Owens
125 Owens Cove [37078] Directions: Call for reservations. I'll send a brochure with map. **Facilities:** New barn. 13 indoor stalls 12' X 12'. 11 outdoor stalls with 12' X 12' area covered. Horses MUST be well behaved, NO dangerous animals accepted. Hay, feed and trailer parking available. NO pasture boarding. Boarding facility with indoor and outdoor arenas. Also raising registered Quarterhorses for sale. Riding lessons available. Miles of quiet country lanes to ride or walk. **Rates:** $15 per horse per day (foals at side free). $10 a day up to six days. $8 a day for seven or more days. **Accommodations:** Holiday Inn, Super 8, Buffalo Inn & Best Western, all within 1.6 mi. at I-40 Exit 143, Hwy 13.

JAMESTOWN

Big South Fork Ridge Cabin Rental — **Phone: 931-879-9413**
Roy and Margaret Smith — **E-mail: equinestar@aol.com**
576 Evergreen Lane (38556) Directions: Off Rte 297, between Jamestown and Oneida, call for Directions. **Facilities:** 8 indoor stalls, trailer parking and RV hookups. Turnout Pen. Over 100 miles of riding trails in the nearby Big South Fork National River and Recreational Area. **Rates:** $10 per night, $50 per week. **Accommodations:** Cabin available for rent.

KNOXVILLE

Cumberland Springs Ranch — **Phone: 865-584-5857**
Gene & Anne French — **or: 865-558-0914**
4105 Sullivan Road [37921] Directions: From I-640: Take Western Ave. Exit (West); turn right on Sullivan Rd.; go about 1 mile & turn right at natural wood fence. **Facilities:** 5 indoor 12' x 20' & 12' x 14' stalls, 2 large outdoor arenas, 60' x 120' indoor arena, grass hay & 12% grain or oats available. Trailer parking. Also offers local and long-haul horse transportation. "Your horse's safety and comfort is our goal." **Rates:** $20 per night. **Accommodations:** Several top-name motels less than 2 miles away.

ALL OF OUR STABLES REQUIRE CURRENT NEG, COGGINS, CURRENT HEALTH PAPERS, & OWNERSHIP PAPERS.

Hunter Valley Farm **Phone: 615-690-6661**
Becky Elmore
9111 Hunter Valley Lane [37922] Directions: I-40 West to Maloney Hood Exit: Go left & go to first light (Kingston Pike) & turn right; at next light turn left (Pellessippi Parkway); go to Northshore Drive Exit & turn left at bottom of ramp; go 1/2 mile to Keller Bend; turn right & go .1 mile & turn left on Hunter Valley Lane. Farm is .1 mile on left. **Facilities:** 10 indoor stalls, large paddocks, feed/hay & trailer parking. **Rates:** $25 per night. **Accommodations:** Red Roof Inn, LaQuinta, & others 10 minutes away.

LEBANON
Cedars of Lebanon Stables **Phone: 615-444-5465**
Don L. Heil
Cedar Forest Drive [37090] Directions: From I-40, Exit 238 to Hwy 231 south, 6.5 mi. south of Lebanon to Cedars of Lebanon Park. Follow signs to stables. **Facilities:** 19 indoor stalls, 10 acres of pasture/turnout, feed/hay available, trailer parking. Overnight rides, plus hayrides; 1-hour, 2-hour, and half-day rides. **Rates:** $15 per night; $75 per week. **Accommodations:** Best Western in Lebanon, 6 miles from stables.

Cedar Stable **Phone: 615-286-2635**
Gerald & Dana Chapman
7930 Murfreesboro Road [37090] Directions: Hwy 231 S Exit off of I-40. Call for directions. **Facilities:** 15 indoor stalls, outdoor 12' x 10' stalls, indoor arena, two 8-acre pastures, one 15-acre pasture, feed/hay & trailer parking available. Call for reservation. Monthly boarding and all levels of lessons & training available. **Rates:** $20 per night; weekly stays with advance notice. **Accommodations:** Days Inn & Shoney's 12 miles from stable.

Cool Breeze Ranch **Phone: 615-443-0347**
J.R. & Juli Kelley
1400 Peyton Road [37087] Directions: Located 1.25 miles off of I-40. Exit 239 B off of I-40 East, Exit 239 off of I-40 West. Turn right on Peyton Rd 1/4 mile from I-40. Ranch is on right. **Facilities:** 17 indoor stalls, 75' x 125' round pen, 2-acre pasture, feed/hay & trailer parking available. **Rates:** $10 per night; $100 per week. **Accommodations:** Eight motels 2 miles from stable.

Shannondale Boarding Stables **Phone: 615-983-7197**
Edie Murnane
2543 Tuckaleechee Pike [37803] Directions: I-40 to 321, Lenoir City; 129 S. to Alcoa Hwy to Maryville to Tuckaleechee Pike; stable is 1.6 miles on left. **Facilities:** 27 indoor stalls, 160' ring, 200' ring, lighted arena, feed/hay at additional cost, trailer parking on premises. Bathrooms with showers available and air conditioned lounge. Full board, partial board, & lessons available. Call for reservation. **Rates:** $20 per night; $15 per day weekly rate. **Accommodations:** 3 rooms with bath available on premises.

ALL OF OUR STABLES REQUIRE CURRENT NEG, COGGINS, CURRENT HEALTH PAPERS, & OWNERSHIP PAPERS.

MEMPHIS

The Shelby Farms Show Place Arena **Phone: 901-756-7433**
Shelby County Government **(see ad on opposite page)**
105 S. Germantown Road [38018] Directions: Take Exit 13 East (Walnut Grove Rd.) off of I-240 or take Exit 16 South (Germantown Rd.) off of I-40. Arena located at Walnut Grove & Germantown. **Facilities:** 632 covered 10' x 10' stalls, 6' high cattle holding pen, overnight trailer parking & RV hook-ups on grounds (electric & water only - dump station on grounds) no feed/hay . Hook-ups: $15 per space, per night. Shavings available at $5 per 50lb bag. **RESERVATIONS REQUIRED**. No security on grounds. Rates: $12 per night. **Accommodations:** Wellesley Inn & Suites (901-386-1525, Winngate Inn (901-386-1110), Best Inn & Suites (901-757-7800).

MURFREESBORO

Hunters Court Stable **Phone: 615-896-4189**
David Q. Wright
410 DeJarnett Lane [37130] Directions: Located between I-24 and I-40, 1/2 mile east of Hwy 231 North. **Facilities:** 30 stalls, several small paddocks, feed/hay, trailer parking. Instruction in hunter seat equitation; several hunters and jumpers for sale at all times. **Rates:** $18 per night. **Accommodations:** Many motels in Murfreesboro within 5 miles.

NASHVILLE

Apple Brook Bed, Breakfast, and Barn **Phone: 615-646-5082**
Donald & Cynthia Van Ryen
9127 Hwy 100 [37221-4502] Directions: Exit 199 off I-40 West, south to Hwy 100, right 6.5 miles, stable on left. **Facilities:** 4 indoor stalls, feed/hay on request, trailer parking. Nearby attractions include Natchez Trace, Fairview Nature Park, and many others. **Rates:** $10 per night. **Accommodations:** B&B on premises, 4 rooms, 2 w/private baths.

NIOTA

Sweetwater Equestrian Center, Inc. **Phone: 800-662-4042**
Grady V. & Charlotte J. Maraman **or: 423-337-2674**
1065 County Road 316 [37826] Directions: Exit 60 off of I-75: Go north on Hwy 68 for 3 miles; turn left on McMinn County Rd. 316; located approx. 1/2 mile on right. **Facilities:** 20 indoor & 8 outdoor stalls, one acre of pasture/turnout, feed/hay & parking for trailer. Electric hook-up for self-contained campers. **Rates:** $20 per night. **Accommodations:** Days Inn, Comfort Inn, Quality Inn & others located at Exit 60, 3.5 miles from Center.

Strip in Shelby Farms Showplace Ad
Position 'Patch' over old bullet copy

- ☛ Central location!
- ☛ 75% of US Population lives within a 600-mile radius.
- ☛ Over 600 stalls available.
- ☛ RESERVATIONS REQUIRED.
- ☛ Bedding available for purchase on site.
- ☛ RV/Trailer hook-ups available on site.
- ☛ All barns equipped with complete fly spray system.
- ☛ Full service restaurant located right next door.
- ☛ Host hotels within 5 miles.

IMPORTANT: Original 12-month coggins test is required before entry onto Show Place grounds. No exceptions.

ALL OF OUR STABLES REQUIRE CURRENT NEG. COGGINS, CURRENT HEALTH PAPERS, & OWNERSHIP PAPERS.

PINEY FLATS
Walnut Creek Farm **Phone: 423-538-8931**
Jane and Phil Elsea, D.V.M. **Fax: 423-915-0371**
2490 Enterprise Road [37686] Directions: 9 miles south of I-81, scenic upper northeast Tennessee area. Call for detailed directions. **Facilities:** 4 indoor 12′ x 12′ box stalls, 2 indoor 12′ x 24′ box stalls (could be divided), several turn-out areas/pastures, feed/hay & trailer parking available. Riding ring, draft horse shoeing stocks, walker, video camera monitoring system in two box stalls, Owner is equine veterinarian. Clinic located close by in Johnson City. **Rates:** $15 per night. **Accommodations:** Two bedroom apartment w/ kitchen and bath adjoining barn or numerous motels in Johnson City and Bristol, each about 10 miles from farm.

PORTLAND
Fox Acres Ranch **Phone: 615-325-5114**
A.C. & Alicia Fox
310 Brandy Hollow Road [37148] Directions: Approx. 10 miles off of I-65: take Exit 108, Hwy 76E; turn right on Brandy Hollow Road; ranch is 3/4 mile on right. Located 45 min. from Nashville & Opryland. "Leave your horses while you visit." **Facilities:** 10 indoor stalls, up to 10 outdoor holding pens, 5 acres of pasture, outdoor arena, 43' x 75' indoor arena, hot & cold water for horse showers, feed/hay & trailer parking. 3-5 water & electric hook-ups for campers. **Rates:** $15 per night. **Accommodations:** Possible B & B arrangements on premises. Also, Hawk's Motel in Portland 3 miles from stable.

ROCKFORD
Porter Brakebill Farm **Phone: 423-982-0200**
Don Brakebill
803 Martin Mill [37853] Directions: Call for directions. **Facilities:** 6 indoor 12' x 12' stalls, 1/2 to 5 acres of pasture/turnout, riding area, trails, round riding pen, feed/hay available, trailer parking. **Rates:** $15 per night. **Accommodations:** Bed & Breakfast with 5 rooms available on premises; $90 per night includes "bountiful" continental breakfast. Motels in Maryville and Knoxville, 10-15 miles.

SMYRNA
CKW Tanner Stables **Phone: 615-355-1776**
Bill & Kathy Tanner
3886 Rock Springs Rd. [37167] Directions: 1.5 miles off of I-24. Call for specific directions. **Facilities:** 7-13 indoor 12' x 12' stalls, 5 acres of pasture, water & elec. in barn wIth concrete pad for washing, feed/hay & trailer parking. Vet, farrier, & tack repair available. Grain store nearby. Call for reservations. **Rates:** $20 per night; ask for weekly rate. **Accommodations:** Comfort Inn in Lavergne less than 5 miles away. Also, KOA campgrounds nearby.

ALL OF OUR STABLES REQUIRE CURRENT NEG. COGGINS, CURRENT HEALTH PAPERS, & OWNERSHIP PAPERS.

SOMERVILLE

Lucky D Ranch **Phone: 901-465-2066**
Vicki Duffy
2950 Tomlin Road [38068] Directions: I-40 to Hwy 64E to Tomlin Road just east of Oakland. North on Tomlin .6 mile; stable is on west side. **Facilities:** 2 indoor 12' x 22' stalls, 6 indoor 12' x 12' stalls, 1-acre pasture/turnout with loafing shed, feed/hay, trailer parking. Standing-at-stud: PHBA Champion "Palo Buddy" 1988 Palomino QH, "Hankins Glo" AQHA Grandson of "King P-234." Specializing in palomino quarter horses. Weanlings, yearlings, & 2-year-olds for sale. **Rates:** $20 per night. **Accommodations:** Days Inn in Lakeland (15 miles), Somerville Inn in Somerville (6 miles), Wilson Inn in Memphis (17 miles).

TOWNSEND

Davy Crockett Stables **Phone: 615-448-6411**
J.C. Morgan
234 Stables Drive [37882] Directions: 30 miles from I-40 and I-75. Call for directions. **Facilities:** 10 indoor stalls, no pasture/turnout, sweet feed/hay available. Trailer parking on premises. No stallions. Stable property joins Great Smoky Mountain National Park. **Rates:** $10 per night. **Accommodations:** Tremont Campground 200 yards away. Also eight motels 3 miles from stable.

Gilbertson's Lazy Horse Retreat **Phone: 865-448-6810**
Melody H. Gilbertson **Web: www.thesmokies.com/lazy_horse/**
938 Schoolhouse Gap Road [37882] Directions: Entering Knoxville on I-75, do not take I-75 east exit, but continue south on 275. Take Hwy 129 to Alcoa/ Maryville (past airport), then 321 north to townsend. **Facilities:** 12 indoor 10' x 10' stalls with dutch doors, 4 corrals approximately 80' x 100', 1/2-acre pasture, feed available, trailer parking. **Rates:** $10 per night in stalls, $5 per night in corral; weekly rate, $60 in stalls, $30 in corrals. **Accommodations:** Two-bedroom cabins on premises with jacuzzi, fireplace, TV-VCR, central heat & air, equipped kitchen, phone, washer/dryer. Also 1-bedroom efficiency cabin. 1.5 miles from Great Smoky Mountain National Park.

Packs Stables and Cabin Rentals **Phone: 423-448-6318**
Greg Pack
7760 Cedar Creek Road [37882] Directions: 30 miles from I-40 & I-75. Call for directions. **Facilities:** 12 indoor stalls with pasture/turnout, outside paddocks, sweet feed & hay furnished. Trailer parking on premises. Stallions allowed. 4 miles from Great Smoky Mountain National Park. **Rates:** $10 per night; $60 per week. **Accommodations:** Cabin rentals on premises. Campground 1/2 mile away; motels 1 mile away.

ALL OF OUR STABLES REQUIRE CURRENT NEG. COGGINS, CURRENT HEALTH PAPERS, & OWNERSHIP PAPERS.

WALLAND

Twin Valley Bed & Breakfast Horse Ranch **Phone: 423-984-0980**
Janice Tipton
2848 Old Chilhowee Road [37886] Directions: Call for directions.
Facilities: 14 indoor & outdoor covered stalls, 1/2-acre corral with creek & shelter, paddock with shelter, wash rack, pond for swimming & fishing, hiking & riding trails, trailer parking available. 12 miles from Great Smoky Mountain National Park. Must sign a liability release form. Must call for reservation.
Rates: $10 per night including hay. **Accommodations:** Bed & Breakfast & cabins on premises. Ranch offers many activities including riding lessons, fishing, BBQs, etc.

WOODLAWN

Creek Top Stables **Phone: 931-906-4077**
Bill & Sylvia Prettyman
3001 Cooper Creek Road (37191) Directions: West on I-24, 10 miles west of Clarksville in Woodland. Off US 79 South. Call for directions. **Facilities:** 4 12′ x 12′ stalls and 2 10′ x 10′ stalls (covered). Feed available and trailer parking. 6 acres of pastures. Large tack room, washrack, farrier sevices available. Miles of horse trails. **Rates:** $10 per night and $50 per week. **Accommodations:** Quality Inn, River View Inn, Clarksville all 20 minutes away.

Dang Fool Tried To Ride Into The Sunset

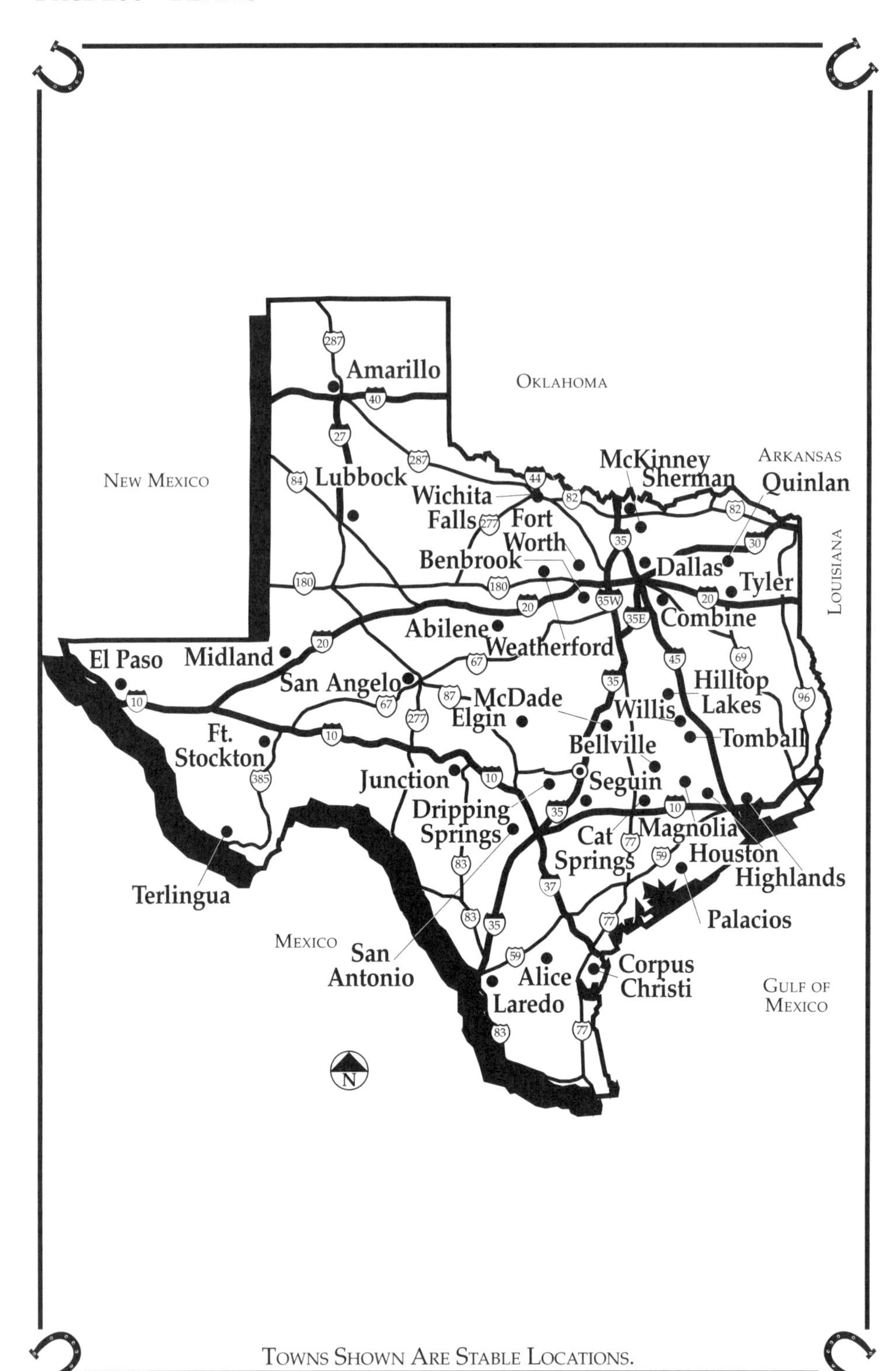

Towns Shown Are Stable Locations.

ALL OF OUR STABLES REQUIRE CURRENT NEG. COGGINS, CURRENT HEALTH PAPERS, & OWNERSHIP PAPERS.

ABILENE

Expo Center of Taylor County, Inc. **Phone: 915-677-4376**
1700 Hwy 36 [79602] Directions: From IH 20, take Loop 322 south to Hwy 36. The Expo Center is located on the corner of Loop 322 and Hwy 36. **Facilities:** 750 indoor stalls, limited pasture/turnout. Trailer parking on the 117-acre facility. Feed stores 2 miles from Center. Reservation required. **Rates:** $12 per night. **Accommodations:** Travelodge & Garden Inn 1-3 miles.

ALICE

Running Bear Ranch **Phone: 361-668-0996**
Dianne Jackson **Or: 361-660-4743**
3244CR 170 North Directions: Call for Directions. **Facilities:** 10 available stalls. 4 15x30 with shelter and 6 stalls with runs. 4 150x100 turnouts with shelter. Lighted arena(250x300). Round pen with 200 acres of beautiful riding areas. Oat and coastal hay and trailer parking available. Vet on Call. No alcohol on premises. 4 star facilities and care. **Rates:** $15 with feed and hay per night, $12 with own feed. **Accommodations:** Kings Inn Holiday Inn and Executive Inn within 15 minutes of ranch.

AMARILLO

Tascosa Stables, Inc. **Phone: 806-342-9061**
Cheryl Rhoderick **Toll Free: 866-272-3996**
e-mail: tascosastables@earthlink.net **www.tascosastables.com**
3513 N. Western (79106) Directions: Approximately 4 miles north of I-40 and 1/2 mile south of Loop 335. **Facilities:** 35 12x15 indoor with automatic waste disposal, and organic fly spray system. Dutex doors, outside runs, full arena, 2 round pens, 8 horse walker, feed/hay, trailer parking, security gates. Must have health papers and current Coggins test. **Rates:** $20 per night **Accommodations:** Discount motel reservations arranged for Tascosa Stables customers.

BELLVILLE

Banner Farm Bed & Breakfast **Phone: 800-865-8534**
Toni Trimble
681 Farm Road 331 [77418] Directions: Exit 720 off I-10, go 4.5 miles north on Hwy 36, turn right at Farm Road 331, facility is 5.5 miles on right. **Facilities:** 9 indoor 10' x 10' and 10' x 12' stalls, 1/3-acre turnout paddocks, lighted riding arena, hay, lighted trailer parking. Call for reservations. **Rates:** $15 per night. **Accommodations:** B&B on premises in turn-of-century farmhouse, 3 guest rooms with shared bath, swimming pool, $60 per night.

ALL OF OUR STABLES REQUIRE CURRENT NEG. COGGINS, CURRENT HEALTH PAPERS, & OWNERSHIP PAPERS.

BENBROOK
Shepherds Valley **Phone: 817-249-8400**
Gary Reynolds
10180 Rolling Hills Drive [76126] Directions: Off FM 2871, 2 miles off I-20. 10 miles west of downtown Fort Worth. **Facilities:** 70 indoor stalls, shavings over rubber mats, 60,000-sq-ft indoor arena, turnout in indoor and outdoor arena if available, trailer parking available. No feed/hay provided. **Rates:** $15 per night; call for weekly rate. **Accommodations:** Hampton Inn in Fort Worth, 5 miles from stable; special Shepherds Valley rate.

CAT SPRING
Southwind Horse Farm **Phone: 409-992-3270**
Sunny & John Snyder
Rt. 1, Box 15-C [78933] Directions: Just west of Houston & 18 minutes north of I-10. Take Sealy/Hwy 36 North (#720) Exit. Call for detailed directions. **Facilities:** 11 large indoor stalls, large turnout area, 4 holding/exercise pens, pasture, 15 acres for riding. Can safely accommodate mare with foal or stallions. Camper hook-up and vet nearby if needed. Kennel for small dogs. Reservations not required but recommended. **Rates:** $15 per night, discounts for 3 or more head. **Accommodations:** Bed & Breakfast on premises with private baths. Discounted B & B/stable rate available.

COMBINE
Double B Farms **Phone: 972-476-6667**
Bill & Laura Brakefield
220 F.M. 3039 [75159] Directions: 25 miles S.E. of Dallas, Call for directions. **Facilities:** 10 - 12' X 16' indoor stalls. Hay and trailer parking available. 1 - 1/2 acre pasture, 1 - 3 acre pasture and 1 - 10 acre pasture. Round pen, hot walker and wash rack available. Call for reservations. **Rates:** $15 per night or $75 per week. **Accommodations:** Countryside Inn - Kaufman, TX - 15 mi., Holiday Inn - Terrell, TX - 20 mi., Seagoville Inn - Seagoville, TX - 8 mi.

CORPUS CHRISTI
Golden Gait Farm, **Phone: 512-939-7828**
Dana Riley **or: 512-939-8330**
1616 Ramfield Road [78418]
Directions: Call for directions. **Facilities:** Four 100' x 40' and one 200' x 40' paddocks, lighted arena, feed/hay & trailer parking available. Hunter/jumper & dressage training. Reservation required. No arrivals after 11 P.M. **Rates:** $15 per night. **Accommodations:** Motel 6 & LaQuinta about 10 minutes away.

ALL OF OUR STABLES REQUIRE CURRENT NEG. COGGINS, CURRENT HEALTH PAPERS, & OWNERSHIP PAPERS.

DALLAS

Bettina Stable — **Home phone: 214-343-4747**
Irmgard Christina Pomper — **Office phone: 214-361-8300**
E-mail: pompbarn@flash.net
7921 Goforth Road at White Rock Trail [75238] Directions: Located in NE Dallas: From LBJ Freeway go south on Audelia, right on Kingsley, left on White Rock Trail & then right on Goforth Road. Stable is on corner, big white barn with red roof. **Facilities:** Indoor stalls. Would prefer 2 days notice. Located in the city of Dallas. **Rates:** $25 with own feed; short & long term rates negotiable. **Accommodations:** Comfort Inn, Doubletree Hotel, and others & within 5 min.

DRIPPING SPRINGS

Grosvenor Farm — **Phone: 512-894-0815**
Carol & David Grosvenor — **Fax: 512-894-4313**
HC 01 Box 1954 (Procknow Rd, CR 195) — **E-mail: grosvenr@bga.com**
Directions: 4.5 miles from 290 West. Call for directions or to get map.
Facilities: 6 indoor 12' x 12' stalls with rubber mats and semi-private paddock, 1 indoor 10' x 10' stall with no paddock, 2 arenas, small dressage area, jumps, 1-acre paddock, round pen. Bring own feed/hay. Trailer parking available. Farm is close to 450-acre ranch that offers guided rides; also close to 5,000-acre state park open to equestrians. Many local attractions for swimmers, golfers, shoppers, diners. **Rates:** $15-$18 per night; ask for weekly rate. **Accommodations:** 2 B&B's within 5 miles; Dabney House 1.5 miles away, Short Mama's 4.5 miles away, $65-$85 per night. Many more in Wimberly, 15 miles away.

EL PASO

Coachlight Stables — **Phone: 915-860-8199**
Lloyd & Tina McNeil
10949 E. Burt Road [79927] Directions: I-10 to Horizon Blvd. Exit 36; go south 1 mile, left on Burt Road. Stables on left. **Facilities:** 7 indoor stalls, 6 inside/outside stalls, feed/hay available. **Rates:** $15 per night. **Accommodations:** Americana Motel 1 mile from stable.

Johnny Bean Horse Farm — **Phone: 915-877-3788**
Jim Bean
6201 South Strahan Road [79932] Directions: Call for directions. **Facilities:** 47 box stalls 14' x 16' to 20' x 22', 1-acre pastures, feed/hay, trailer parking. "First-class horse motel." **Rates:** $17.50 per night. **Accommodations:** Motels and restaurants nearby.

ALL OF OUR STABLES REQUIRE CURRENT NEG. COGGINS, CURRENT HEALTH PAPERS, & OWNERSHIP PAPERS.

ELGIN

Ragtime Ranch Inn **Phone: 512-285-9599**
Roberta Butler & Debbie Jameson **800-800-9743**
P.O. Box 575 [78621] **Fax: 512-285-9651**
Directions: 23 miles east of Austin on Hwy 290. Call for directions.
Facilities: 2 - 12' x 12' stalls with 12' x 40' runs, 2 private pastures. Training, indoor and outdoor arenas available next door. Ample parking. Fishing pond, walking trails. **Rates:** No charge for ranch guests. **Accommodations:** B&B on premises, 4 rooms with private baths, private porch/deck, pool.

FT. STOCKTON

Merrell and Ann Daggett **Phone: 915-336-5490**
E. 53rd Lane [79735] Directions: From I-10: Exit 259 & go north to Hwy 1053; go north about 5 miles; turn right on Stone Rd.; go 1 block then turn left on 53rd Lane. Stable is approx. 1/2 mile on right. **Facilities:** 4 indoor & 3 outdoor stalls, 7 large holding pens. Standing-at-stud: "Winter's Star" APHA 21770. Paints for sale. Room for large transport vehicles. Electric & water hook-ups available for trailers. Reservations preferred. Daggett Trucking: livestock hauling available. **Rates:** $15 per night. **Accommodations:** Motels 6 miles away.

HILLTOP LAKES

Hilltop Lakes Equestrian Center **Phone: 409-855-2842**
Noah Powell
P.O. Box 1794 [77871] Directions: From I-45: Hwy 79 Exit South, go 20 miles. Take Hwy 3 East 10 miles to Center on left. **Facilities:** 20 indoor stalls, five 50' x 100' paddocks, rodeo arena on premises, horse walker, racetrack on premises. Restaurant on site. Reservations 24 hours in advance. **Rates:** $20 per night; weekly rate negotiable. **Accommodations:** Hilltop Lakes Resort Motel within walking distance.

HOUSTON

Magic Moments Stable **Phone: 713-461-1228**
Granger & Jean Durdin, owners
Peter Smit, Manager
1726 Upland Drive (77043) Directions: I-10 to Wilcrest exit, Go North, immediate left onto Old Katy(west), immediate right onto Upland Drive(North). Second stop sign, Barn is at Upland and Chatterton. **Facilities:** 12x12 indoor stalls with rubber matted floors, Individual 100x100 paddocks. Trailer parking, feed/hay, 24 hour security, lessons and full board. Mostly Arabians and 1/2 Arabians. Appaloosa stud "Thunder Bay May" with references. Vet and farrier on call. Call ahead for reservations. **Rates:** $15 per night. **Accommodations:** La Quinta, Houston (1 mi).

ALL OF OUR STABLES REQUIRE CURRENT NEG. COGGINS, CURRENT HEALTH PAPERS, & OWNERSHIP PAPERS.

HOUSTON/CAT SPRING

Rancho Texcelente IXL **Phone: 979-865-3636**
Nancy Flick **Fax: 979-865-1929**
E-mail: ixl@paso.net **Web: www.paso.net**
14012 Paso Fino Rd P. O. Box 55, FM 1094 [78933] Directions: From I-10: 50 miles west of Houston. Take Exit 720 & go north on 36 for 2 miles, take left on FM 1094 & go 12 miles. Ranch is on left side. **Facilities:** 32 indoor 12' x 12' stalls, 2 holding pens, 250 acres of pasture, feed, 1 open and 1 covered arena, covered round pen, walker, escorted or unescorted trails, camper hookup. One kennel. Leases Paso Fino horses. Local horse transportation available. Advance reservation preferred. **Rates:** $15 per night. **Accommodations:** 2 bedroom guest house. Also, Hotel Wayne in Bellville & Best Western in Sealy, about 12 miles from stable.

HOUSTON (HIGHLANDS)

Texas Star Stables **Phone: 281-843-3042**
Glen Weeks, manager
230 Highlands Shores [77562] Directions: I- 10E, Exit 787, Go north 3 miles to Highlands Shores. Turn left and go to first gate on right. 20 minutes east from Houston. **Facilities:** Newest and biggest barn in the area. Can pull truck and trailer into barn for loading in bad weather. 15 - 12′ X 12′ stalls. Feed, hay and trailer parking available. Hot walker, round pen. Private paddocks with tack rooms. Standing studs, "Paint Me Special" and others. **Rates:** $20 per night, $75 per week. **Accommodations:** Best Western, Holiday Inn & Ramada Inn near by.

JUNCTION

Caverhill Ranch **Phone: 915-446-2448**
Roxanne & John Fargason
Hwy 377 North [76849] Directions: Exit 456 off of I-10; go north on Hwy 377 towards Mason; turn right at fork at Junction Stockyards; go 4 miles; on left hand side is County Rd.; 1/4 mile on right is Caverhill. **Facilities:** 2 outdoor stalls, round pen, 50 acre turnout, no feed/hay, miles of ranch roads for riding, & trailer parking. Reservation required 24 hours in advance. **Rates:** $10 per night with use of guest house; $15 per night if not. **Accommodations:** 3-bedroom guest house on premises. $40 minimum charge for rental.

LAREDO

El Primero Training Center, Inc. **Phone: 210-723-5436**
Keith Asmussen **or: 210-723-9451 or: 210-722-4532**
Box 1861 [78044] Directions: I-35 to Hwy 59E. Call for directions. **Facilities:** 392 indoor box stalls, 57 holding pens, round pens, 5/8-mile racetrack, feed/hay & trailer parking available. Call for reservation. **Rates:** $12 per night. **Accommodations:** Motels 1 mile from stable.

ALL OF OUR STABLES REQUIRE CURRENT NEG. COGGINS, CURRENT HEALTH PAPERS, & OWNERSHIP PAPERS.

LUBBOCK
Four Bar K Ranch **Phone: 806-789-8682**
Chuck Kershner
2811 98th Street [79423] Directions: I-27 Exit on 98th St. Go west 2 miles. Cross over University. Go four blocks. Gate on left. **Facilities:** 8 indoor & 8 outdoor stalls, large riding pasture, turnout areas, roping arena, round pen, 2 outdoor arenas. Summer horse camp for kids/adults. Private lessons year round. Overnight camping. Neg. coggins papers. Call for reservations. **Rates:** $15 per night, $20 with feed, $75 weekly rate. $20 RV hookup, Free trailer parking. **Accommodations:** Holiday Inn, Lubbock Plaza, 2 miles from stable. Marriott 1.5 miles from stable.s

MAGNOLIA
Whisper Breeze Farm **Phone: 409-273-2398**
Denise Jones, John Coleman **Beeper: 281-490-1730**
1826 Cattle Drive [77354] Directions: Approx. 10 miles west of I-45 on FM 1488. Call for exact directions. **Facilities:** 3 indoor 12′ x 12′ stalls, 2 indoor 12′ x 18′ stalls, 3 pastures, 1 paddock for turnout, feed/hay available, trailer parking and RV hook-ups available. Many miles of riding trails available. Call for reservations. **Rates:** $15 per night; weekly rate negotiable. $15 for RV hook-up. **Accommodations:** Many motels in Conroe, The Woodlands, and Tomball, all within 12 miles of farm.
′

McDADE
Martin Ranch **Phone: 512-273-9027**
Gary W. Martin
R.R. #1 Box 19 [78650] Directions: 9 miles east of Elgin on 290. Turn north on Marlin; 3.7 miles to road's end and ranch. **Facilities:** 33 stalls, 24' x 24' or 2.5 acres of pasture/ turnout, coastal alfalfa and grains available, trailer parking. RV hook-up. Riding & training areas available. Standing-at-stud: B&W Overo "Sonny's Creation," Chestnut Overo "Sir Teddy Clue," B&W Tobiano "Stormin King." **Rates:** $14 per night; $70 per week. **Accommodations:** Motels in Elgin, 12 miles from ranch.

McKINNEY
McKinney Stables **Phone: 972-562-9302**
Terri & Bryan Collins
807 Hwy 380 East [75069] Directions: Exactly 2 miles east of Hwy 75 on Hwy 380, next to Mobil station. **Facilities:** 65 indoor oversize stalls, no pasture/turnout, oats/coastal hay, and parking for any size truck & trailer. Visa, MC, Discover, & Am Ex welcome. **Rates:** $15 per night; $45 per week. **Accommodations:** Holiday Inn, Comfort Inn, & Super 8 Motel all within 2 miles of stable.

MIDLAND
Rebelee Kennels & Stables **Phone: 915-682-5032**
Cindi & Jack Bates
4200 N. Fairgrounds Road [79705] Directions: Within a few miles of I-20. Call for directions. **Facilities:** 29 indoor & outdoor stalls, arenas, pens, & trailer parking available. Monthly boarding available for $175. Call for reservation. **Rates:** $10 per night; $70 per week. **Accommodations:** Plaza Inn 1 mile & Hilton 5 miles from stable.

ALL OF OUR STABLES REQUIRE CURRENT NEG. COGGINS, CURRENT HEALTH PAPERS, & OWNERSHIP PAPERS.

PALACIOS

Elbow Creek Station — **Phone (day): 361-972-0202**
Mr. & Mrs. R.J. Travers, owners — **Phone (night): 361-972-2088**
Max Travers, mgr.
Rt.1, Box 440 [77465] Directions: From State Hwy 35, go east on Farm Road (FM) 521 for 3miles to FM 2853. Go North on FM 2853 for 1/2 miles. Elbow Creek Station is on left. Sign out front. **Facilities:** Up to 5 stalls, 12′ x 12′ round pen, arena, various turnout arenas. Trailer parking, but no camping available. Easy access to Matagorda Beach and Gulf of Mexico for beach riding or fishing. Reservations required. No personal checks or credit cards accepted **Rates:** $15 per day $75 per week. **Accommodations:** In Palacios (8 miles); Luther Hotel, The Main B & B, Deluxe Inn.

QUINLAN

Davie Tatum Stables — **Phone: 903-883-0486**
Davie Tatum — **Voice-mail: 972-661-2361**
6135 Hwy 34 South [75474] — **E-mail: DavieTatum@aol.com**
Directions: Exit Hwy 34 South off of I-30. Go 8 miles on left in Casli Community . **Facilities:** 12-10′ x 12′stalls, 8 large runs with shelter, round pen, feed, arenas wash rack and camper hook-up. RATES: $20 for stalls, $15 for runs per night. Weekly rates $70 for run , $100 for stalls (7 nights). **Accommodations:** Best Western, Ramada Inn and Super 8 located in Greenville, TX.

SAN ANGELO

San Angelo Horse Center, Inc. — **Phone: 915-374-5391**
185 W. FM 2105 (76901) — **or: 915-658-6613**
Trish L. Hutchinson
Directions: Between HWY 87 North and S.H. 208. Facilities: 12 12x12 box stalls, 10 16x20 outdoor pens, 3-20 acre turnout: cross fenced, lighted arena, feed/hay, trailer parking (no hookups), secure property with keypad gate entry, private 20 acre riding area. Owners live on premises. 1 minute from Collisseum, 5 minutes from State Park with equestrian center. Negative Coggina and ownership papers. Call ahead for reservations. Rates: Inside-$25 per night, $150 per week.Outside-$20 per night. $120 per week. Accommodations: Limited Express of San Angelo (92mi), Inn of the Conchos (2mi), Motel 6 (5mi).

SAN ANTONIO

T-Slash-Bar Ranch — **Phone: 210-677-0502**
Tommy Mayhue
13901 Hwy 90 West [78245] Directions: Entrance is located on US Hwy 90 at state Hwy 211. Call with questions. **Facilities:** At least 5 indoor box stalls, outdoor with runs, outdoor pens, numerous pasture/turnout, feed/hay and trailer parking available. Lighted arena, team roping/penning, full boarding/care, access to vet/farrier. Located near Sea World of Texas, Hyatt Resort, Historic Castroville, Kelly/Lackland AFB. **Rates:** $10 per night; weekly rate negotiable. **Accommodations:** Bed & breakfast available at ranch in two wonderful suites, with full kitchen.

ALL OF OUR STABLES REQUIRE CURRENT NEG. COGGINS, CURRENT HEALTH PAPERS, & OWNERSHIP PAPERS.

SEGUIN

Buzzard Creek D Ranch — **Phone: 210-914-3343**

Danny & Mary Davis

2985 Gin Road [78155] Directions: Located 26 miles east of San Antonio, 10 miles west of Seguin. Take Exit 599 off I-10, go south on FM 465 for 1 mile, right on Gin Road, first gate on left. **Facilities:** 8 outdoor 8' x 12' stalls, 4 - 50' x 75' lots, 10-acre pasture, feed/hay, trailer parking. **Rates:** $12 per night. **Accommodations:** Holiday Inn, Best Western, EconoLodge in Seguin (10-12 miles); Winfield's Motel and Restaurant in San Antonio (11 miles).

PGL Ranch — **Phone: 210-303-4949**

Paul Garcia — **or: 210-372-3435**

4001 Hwy 90E [78155] Directions: Exit 612 (US Hwy 90) off I-10, ranch is one mile on right. **Facilities:** 13 indoor 12' x 20' and outdoor 14' x 30' stalls, pasture sometimes available, hay, trailer parking. Horse Transportation: Alamo Equine Service. **Rates:** $15 per night. **Accommodations:** Holiday Inn and EconoLodge at next exit (west) off I-10.

SHERMAN

TLC Quarter Horse Ranch & Equestrian Center — **Phone: 903-786-2484**

4158 Refuge Road [75092] Directions: From I-35 go east on US 82 or from US 75 go west on US 82. North on FM 1417 for 4.5 miles, left on Refuge Road 4 miles. Blue barns on left. **Facilities:** 12 indoor 12' x 12' box stalls, 5 indoor 13' x 20' covered 3-sided shed, 4 outdoor stalls (12' x 12', 12' x 16', 12' x 36', 12' x 24'); 5-40 acres of pasture/turnout, outdoor arena, round pen, indoor working area, horse wash facilities; alfalfa and grass, grains available; trailer parking and RV hook-ups. Horses for sale, all ages and levels of training. **Rates:** $10-$15 per night. **Accommodations:** Comfort Inn, Best Western in Sherman, 12 miles away.

TERLINGUA

Turquoise Trail Rides — **Phone: 915-371-2212**

Barbara Russell — **or: 800-887-4331**

P.O. Box 178 [79852] Directions: At Y of Hwys 118 & 170, next to Big Bend National Park, 80 miles south of Alpine. Call for further directions. **Facilities:** 1-10 corrals with Shades metal panels, alfalfa hay and trailer parking available. Horse outfitters, guided trail rides, "your horse or ours." **Rates:** $10 per night, $60 per week. **Accommodations:** Big Bend Motor Inn in same block; Oasis RV Park nearby.

TOMBALL

Circle G Stables — **Phone: 713-698-3456**

Sim & Pat Gounarides — **(voice pager)**

19406 Lindsey Lane [77375] Directions: From I-290 West: Go to Telge Road & go north to Self Road; left on Self to dead-end; asphalt road to the right & stable is first on right. **Facilities:** 11 indoor stalls, 7 paddocks, 12' x 24' birthing stall, feed/hay & trailer parking available. Feed store nearby. Please call ahead. "Come and feel at home." **Rates:** $15 per night; $75 per week. **Accommodations:** Best Western on Rt. 249, 6 miles from stable.

ALL OF OUR STABLES REQUIRE CURRENT NEG. COGGINS, CURRENT HEALTH PAPERS, & OWNERSHIP PAPERS.

TYLER

Pine Lake Stables **Phone: 903-592-8075**
Joanne Casmo **Web: www.pinelakestables.com**
11015 Pine Lake Blvd. [75709] Directions: Only 15 miles off of I-20. Call for exact directions. **Facilities:** 34 indoor 12' x 12' stalls, wash rack, walker, outdoor arena, round pen, hay, chips, trailer parking, camper/RV hook-ups available. Reservations preferred, but not necessary. Several studds available for breeding (App, Qtr & Paints) **Rates:** $15 per night, bedding extra. **Accommodations:** $15 per night. Motels within 7 miles in Tyler.

WEATHERFORD

Hay USA and Horse Hotel **Phone: 817-599-0200**
Liz Blitzer
1714 Blair Drive [76086] Directions: 25 miles east of Ft. Worth right off I-20. Exit Hwy 180 and go 2 miles. Sale Barn is on right; pawn shop is on left. Hay USA behind pawn shop. **Facilities:** 12 half-covered outdoor 10' x 22' stalls, 60' round pen, 265' calf-roping arena, 2 holding pens, feed/hay, trailer parking. Can handle one stallion and/or one mare & foal. Hay dealer on premises, NHA member. Horsemanship, calf-roping, barrel-racing lessons & clinics. Reservations requested, but drop-ins welcome. Bring own water buckets. **Rates:** $15-20 per night, $30 per stallion. **Accommodations:** Best Western, Santa Fe Inn in Weatherford, 3 miles away.

Hidden Lakes Riding Stable **Phone: 800-935-0397**
Chris Willingham **or: 817-448-9910**
5400 White Settlement Road [76087] Directions: 5 minutes from intersection of I-20, I-30, & FM 1187. Call for directions. **Facilities:** 30 indoor 15' x 15' insulated stalls, 30 acres of pasture/turnout, feed/hay & trailer parking available. **Rates:** $10 per night; $65 per week. **Accommodations:** Small furnished apartments $20. Ramada Ltd. in Willow Park, 5 miles from stable.

WICHITA FALLS

Turtle Creek Stables & Arena **Phone: 817-692-8130**
Tambra Holcomb **Barn: 817-691-6291**
2110 Turtle Creek [76304] Directions: Call for directions. Only facility in Wichita Falls. **Facilities:** 12 indoor 18' x 15' treated pine stalls with 40' private runs, pasture and turnout in various sizes, 2 lighted arenas, 100' round lighted pen, hot walker, feed/hay, trailer parking, RV hook-up. Call for reservations. **Rates:** $15 per night; variable rates for multiple horses. Shavings available.
Accommodations: Motels within 5 miles of stable.

ALL OF OUR STABLES REQUIRE CURRENT NEG. COGGINS, CURRENT HEALTH PAPERS, & OWNERSHIP PAPERS.

WILLIS

Diamond T Farms — **Phone: 409-856-7709**
Ron & Linda Tullis — **or 800-687-0944**

12110 Maggie Lane [77378] Directions: From I-45: Take Exit 92 west toward lake; go 5 miles, turn right on Cude Cemetary Rd.; go approx. 1 city block & take right on Maggie Lane; go to end of cul-de-sac; stable is last house on left next to the blue/white barn. **Facilities:** 4 indoor 12' x 12' stalls with auto waterers, 2 pastures, round pen, indoor wash rack, tack room, riding area. Space for self-contained campers. Vet/farrier on call. 1 mile from Lake Conroe. Reservations requested. Home of Diamond T Transportation, equine transportation service. **Rates:** $5 - $15 per night; weekly rate negotiable. **Accommodations:** Ramada Inn & Woodlands Inn 10 miles from stable.

NOTES AND REMINDERS

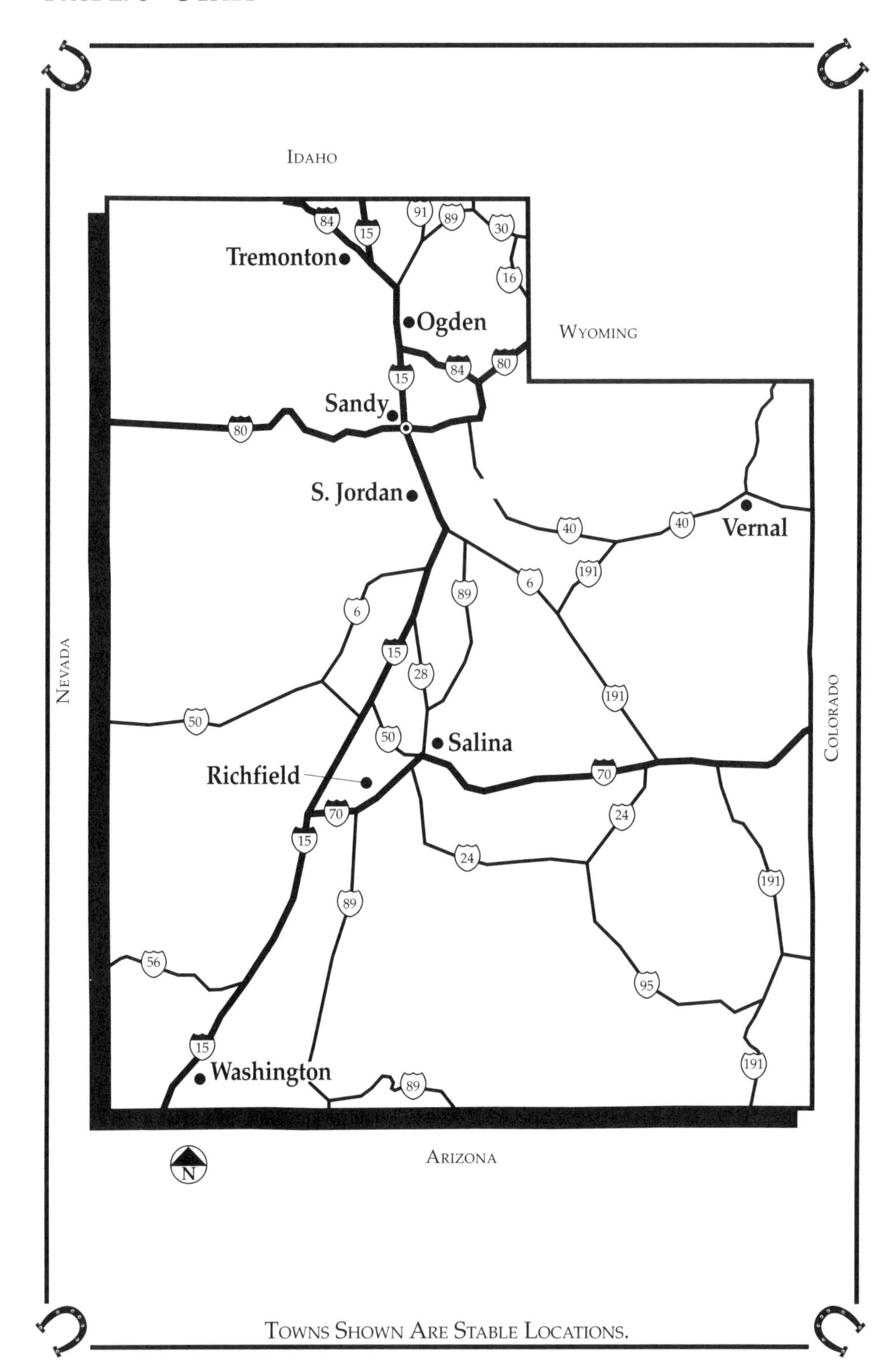

TOWNS SHOWN ARE STABLE LOCATIONS.

ALL OF OUR STABLES REQUIRE CURRENT NEG. COGGINS, CURRENT HEALTH PAPERS, & OWNERSHIP PAPERS.

OGDEN

Golden Spike Arena Events Center **Phone: 800-44-ARENA**

Weber County

1000 No. 1200 West [84404] Directions: Take Exit 349 off I-15 and follow signs to fairgrounds. **Facilities:** 383 covered outdoor 10' x 10' stalls, trailer parking available. Call for reservations. **Rates:** $12 per night. **Accommodations:** High Country Inn within 2 miles.

RICHFIELD

Stan Gleave **Phone: 435-893-8600**

1625 North Main Street [84701] Directions: Exit 40 off I-70. Halfway between Denver and Los Angeles, also half way between Phoenix and Salt Lake City. Call for directions. **Facilities:** 10 outdoor, 10 indoor stalls, walker, outdoor exercise area, feed available. Trailer parking available. Miles of scenic trails. Vet and farrier on call. Frontier Village, motel and restaurant. **Rates:** $10 per night. **Accommodations:** Super 8 Motel.

SALINA

Best Western Shaheen Equestrian Motel **Phone: 801-529-7455**

Larry Shaheen

1225 S. State Street [84654] Directions: Exit 54 off of I-70. Motel is 1,000 yards north of exit. Located on or near US 50 & US 89. **Facilities:** 61 Horse stalls, trailer parking available. Adjacent to Blackhawk Arena, an equestrian event center. Reservations preferred. **Rates:** $10 per night; $60 per week. **Accommodations:** Best Western hotel and restaurant on premises.

SANDY

Alta Hills Farm **Phone: 801-571-1712**

C. Diane Knight

10852 South 20th E [84092] Directions: I-15 to 10600 south, turn east 20 blocks, turn right 1-1/2 blocks south. **Facilities:** 35 indoor stalls, feed/hay, trailer parking. **Rates:** $15 per night. **Accommodations:** Motels nearby.

SOUTH JORDAN

Terry Teeples Horse Boarding **Phone: 801-446-8343**

Terry Teeples

11040 South 2700 West [84095] Directions: Call for directions. **Facilities:** 32 indoor stalls, 15 outdoor pipe pens, 2-acre pasture, hot walker, round corral, beautiful riding trails. Stallions welcome. Trailer parking & room for big rigs. Horse transportation company that services the inter-mountain area of Utah, Idaho, Wyoming, & Montana and also has weekly runs to southern California, Oklahoma, & Texas, located here. Call for information and reservations. **Rates:** $10 per day. **Accommodations:** Many motels in South Jordon.

ALL OF OUR STABLES REQUIRE CURRENT NEG. COGGINS, CURRENT HEALTH PAPERS, & OWNERSHIP PAPERS.

TREMONTON

Box Elder County Fairgrounds **Phone: 801-257-5366**
Bill Smoot, manager **or: 744-2600 or: 257-5828**
400 N. 1000 West [84337] Directions: Take Exit 40 off of I-15. You will see the fairgrounds. **Facilities:** 120 outdoor, completely covered stalls with bedding, large indoor arena, round pen, large open corrals, water accessible. Call 24 hours a day for information. No need to call in advance but it would be helpful. Horse shows, reining & rodeos, year-round team penning held at fairgrounds. **Rates:** $10 w/ bedding for stall, $10 for corrals per night. **Accommodations:** Western Inn next to fairgrounds, Sandman Motel 2 miles away.

VERNAL

Western Park **Phone: 801-789-7396**
Derk Hatch
300 E. 200 S [84078] Directions: 2 blocks from Hwy 40. Call for directions. **Facilities:** 400 covered stalls, 102' x 203' indoor arena, 160' x 270' outdoor arena, 5/8-mile racetrack, convention center, Old West Museum, amphitheatre, playground, & trailer parking. **Rates:** $10 per night. **Accommodations:** EconoLodge & Best Western 1 mile from stable.

WASHINGTON

Harmony Horse Haven **Phone: 435-673-3991**
Steven L. Hafen **or: 435-680-2650**
2321 So. Washington Field Rd. [84780] Directions: I-15 to exit 10, in Washington to 300 East (only light), turn right or South, go 2 1/4 miles across bridge by church on left, around hill, white fence, pine trees every 25′. 2 1/2 miles from Washington light. **Facilities:** Twenty-four 12′x36′ outdoor stalls w/cover, feed/hay available, plenty of trailer parking, arena. Pretty Wilkin, Standing at stud, foundation quarter horse. V.E.W.T. Flu & Rino shots required. **Rates:** $10/$12 per day, ask for weekly rates. **Accommodations:** Red Cliffs Inn, 3 miles from stable.

TUMBLEWEED
DRIP-DRY
WEED

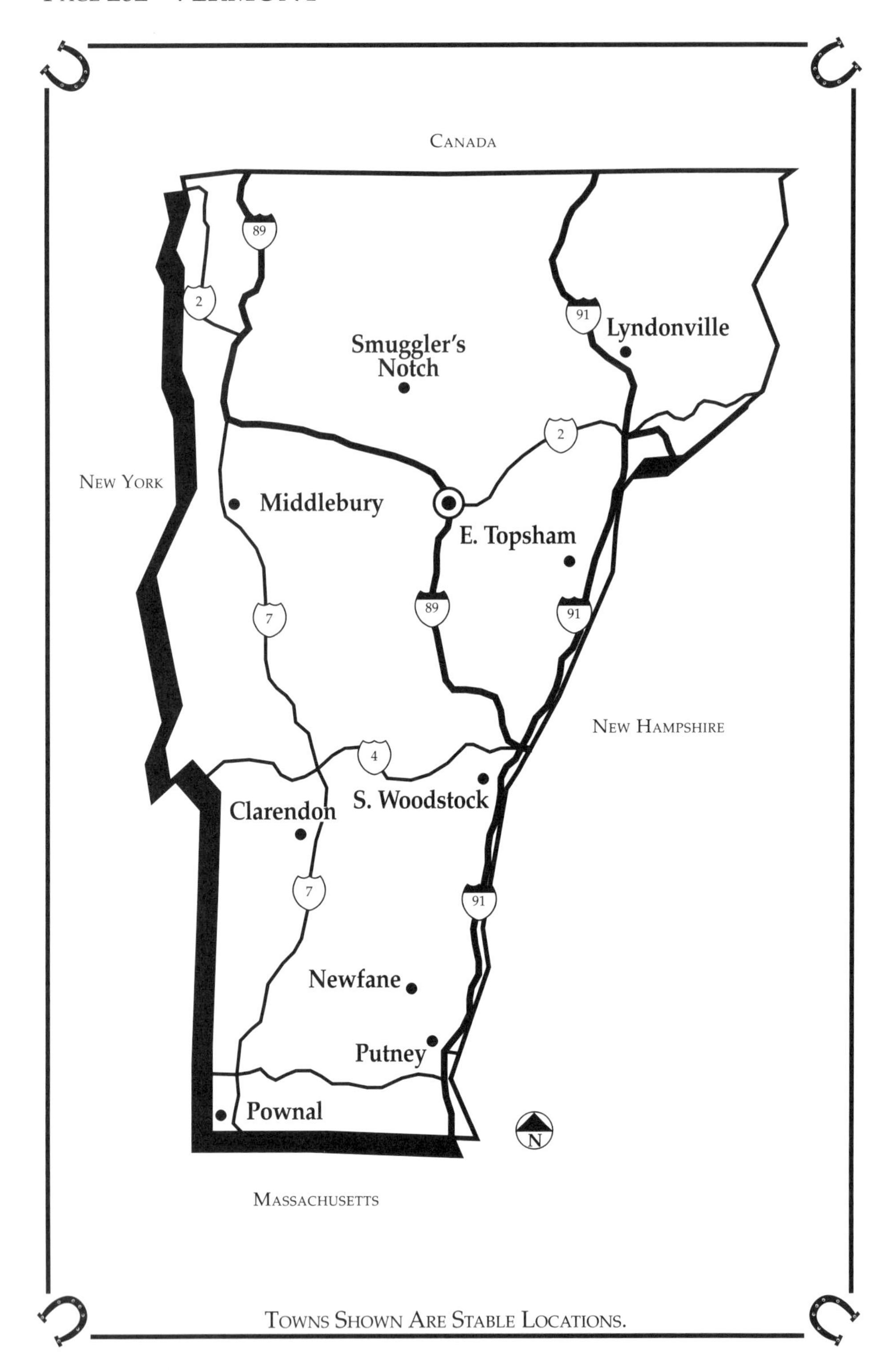

TOWNS SHOWN ARE STABLE LOCATIONS.

ALL OF OUR STABLES REQUIRE CURRENT NEG. COGGINS, CURRENT HEALTH PAPERS, & OWNERSHIP PAPERS.

CLARENDON
Meadowbrook Stables **Phone: 802-438-5799**
Ellen Franklin
1706 Walker Mountain Road [05777] Directions: Call for directions. **Facilities:** 12 indoor stalls, outdoor arena, feed/hay & shavings available. 24-hr advance notice requested. **Rates:** $10 per night; $30 per week. **Accommodations:** Holiday Inn & Howard Johnson's both in Rutland, located 10 miles from stable.

EAST TOPSHAM
Back-in-Time "Horse Resort" **Phone: 802-439-5448**
Glenn & Burnice Dow
Main Street [05076] Directions: I-91 to Exit 16, Rte. 25 west for 7.1 miles, turn right at East Corinth General Store, continue 5 miles to Old Millers Store. **Facilities:** Three 10' x 12' stalls, 1 - 8'x 10' stall, 1 - 24' x 24' stall, two 2-acre pastures, wooden paddock, feed/hay, trailer parking. Historic store/warehouse. Guided trail rides on old logging roads/dirt roads. **Rates:** $20 per night, $100 per week. **Accommodations:** Completely furnished apartment on premises, $75 per day per person/horse. Motels in Bradford, 12 miles away.

LYNDONVILLE
Breezy Knoll Stable **Phone: 802-626-9685**
Harold & Nancy Dresser
RFD 1, Box 361 A [05851] Directions: 2.5 miles north of Rt. 5 off of I-91. Call for directions. **Facilities:** 30 indoor box stalls, 70' x 150' indoor arena, 4 paddocks, 2 large fenced pasture areas. Trail riding. Trains Standardbreds for racing. Call for reservations. **Rates:** $15 per night; $75 per week. **Accommodations:** Motel 1 mile from stable.

MIDDLEBURY
Cobble Hill Farm **Phone: 802-388-7027**
Peggy Ward
RD 3, Painter Road [05753] Directions: Call for directions. **Facilities:** 23 indoor stalls, no pasture/turnout, feed/hay & trailer parking available. Call for availability & reservations. **Rates:** $15 per day. **Accommodations:** Many nearby.

NEWFANE
West River Stables **Phone: 802-365-7745**
Roger Poitras
RR 1, Box 695 [05345] Directions: Exit 2 off of I-91 on Rt. 30N. Call for directions. **Facilities:** 24 indoor stalls, 3 paddocks, 160' x 200' outdoor ring. Jumping & dressage lessons up to Grand Prix level based on centered riding. Call for reservations. **Rates:** $20 per night. **Accommodations:** Bed & Breakfast on premises.

ALL OF OUR STABLES REQUIRE CURRENT NEG. COGGINS, CURRENT HEALTH PAPERS, & OWNERSHIP PAPERS.

POWNAL

Valleyview Horses & Tack Shop, Inc. **Phone: 802-823-4649**

Shelley & Kati Porter; Cindy Legge

Box 48A Northwest Hill Road [05261] Directions: Located off of Rt. 7. Call for further directions. **Facilities:** 7 box stalls, 9 straight stalls, 100 acres fenced pasture, outdoor run-in sheds with 2 horses to each paddock, 80' x 100' outdoor ring, miles of trails on 250 acres. Full tack shop with Western apparel on premises. Buys & sells horses and lessons in English & Western. Trail rides, pony rides, & parties. Call for reservations. **Rates:** $20 per night; $80 per week. **Accommodations:** Many motels 2 miles from stable.

PUTNEY

The Horse Farm **Phone: 802-387-2782**

Patsy Alexander

Westminster West Road [05346] Directions: Exit 4 off of I-91. Call for directions. **Facilities:** 52 indoor 12' x 12' & 12' x 20' stalls, 10 paddock areas, 60' x 150' outdoor arena, 75' x 160' indoor arena. Horses for sale. Training of horses & riders. Trail rides available. Call for reservations. **Rates:** $15 per night; ask for weekly rate. **Accommodations:** Motels within 1 mile.

SMUGGLER'S NOTCH

Vermont Horse Park **Phone: 802-644-5347**

Suzanne Learned

Rt. 108, Mountain Road [05464] Directions: Call for directions. **Facilities:** 14 standing stalls, run-in shed, 3-20 acre pastures, 100' x 150' outdoor arena, guided trail rides. Located across from Smuggler's Notch Ski Resort. Call for reservations. **Rates:** $20 per day. **Accommodations:** Motel located 2 miles from stable.

SOUTH WOODSTOCK

Green Mountain Horse Association **Phone: 802-457-1509**

Jack Dortch, General Manager

Rt. 106 South [05071] Directions: Call for directions. **Facilities:** 136 indoor stalls, dressage arena, cross-country course with jumps. Unlimited trail riding. Sponsor of yearly events. Call for reservations. **Rates:** $15 per night for members; $25 per night for non-members. **Accommodations:** Many motels in Woodstock, which is a beautiful & historic town with many sites & activities.

NOTES AND REMINDERS

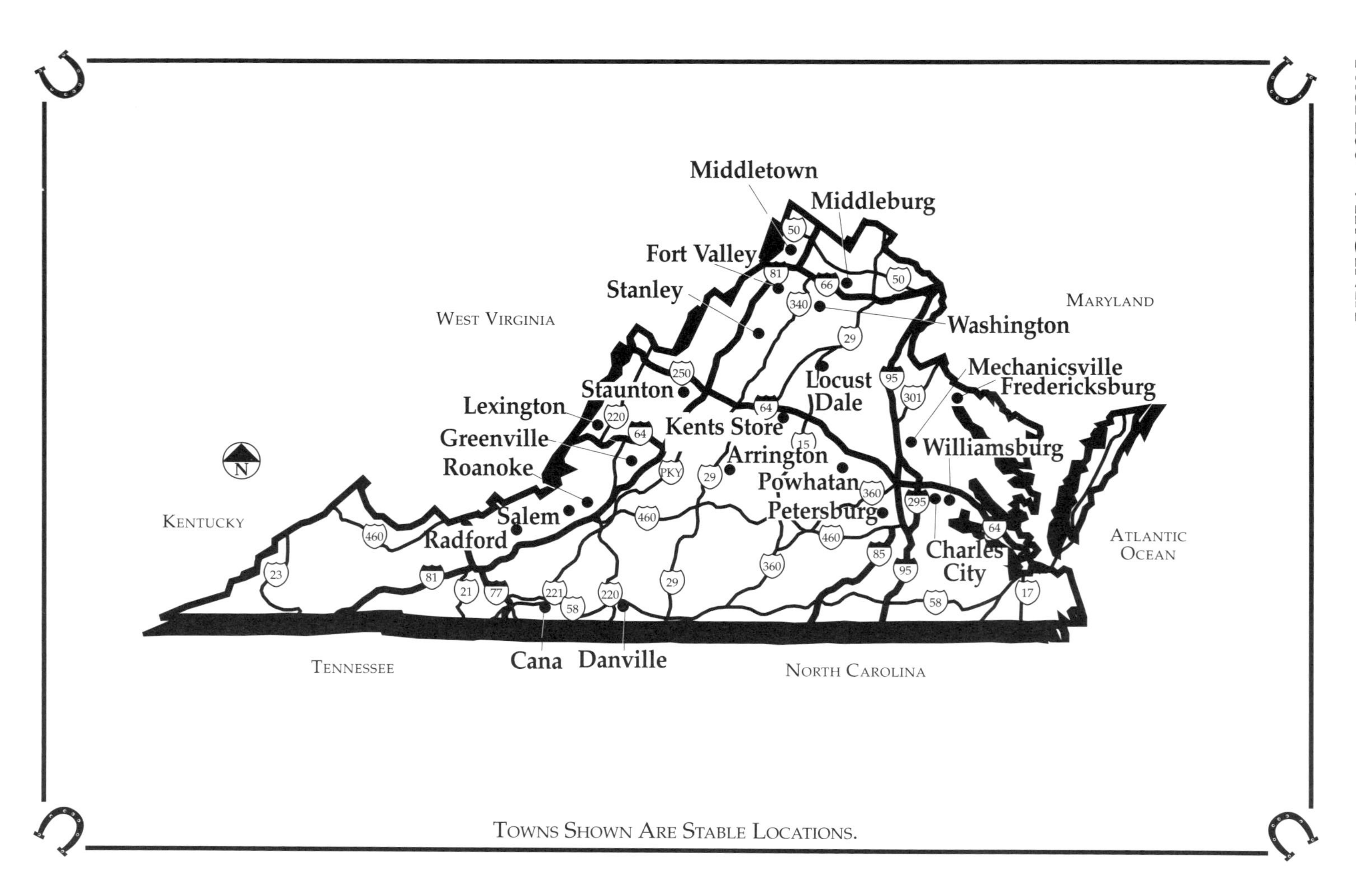
Middletown
Middleburg
Fort Valley
Stanley
Washington
Maryland
West Virginia
Staunton
Locust Dale
Mechanicsville
Fredericksburg
Lexington
Kents Store
Greenville
Williamsburg
Arrington
Roanoke
Powhatan
Petersburg
Salem
Radford
Kentucky
Charles City
Atlantic Ocean
N
Cana
Danville
Tennessee
North Carolina
Towns Shown Are Stable Locations.

ALL OF OUR STABLES REQUIRE CURRENT NEG. COGGINS, CURRENT HEALTH PAPERS, & OWNERSHIP PAPERS.

ARRINGTON

Twin Rivers Farm **Phone: 804-277-9435**

Vivienne Skidmore

298 Bellevette Place [22922] Directions: 2 miles east of Hwy 29 halfway between Charlottesville & Lynchburg VA. Call for details. **Facilities:** 6 - 16′ X 16′ indoor stalls, 50′ X 50′ turnout pasture available. Feed, hay & trailer parking available. Dogs welcome but must be kept on leash. **Rates:** $25 per day or $140 per week. NO Credit Cards. **Accommodations:** Motels available 10 miles away. Small apartment in barn with private bath, $25 per night.

CANA

Tanbark Acres **Phone: 540-755-5191**

Carlton and Dee Everhart Fax: 540-755-2739 Web: www. tanbarkacres.com

240 Tanbark Trail (24317) Directions: 4 miles from I-77, 1 1/10 from US #52. Call for Directions. **Facilities:** 10 12x12 rubber-matted indoor stalls. 4 (small but grassy) paddocks, trailer parking, feed/hay available, Friesian breeding farm, new Morton Barn, negative Coggins, health records, proof of ownership, New River Trail; 157 miles of walking/riding trails **Rates:** $20 per night. **Accommodations:** Several Hotels and Motels within a few miles of the farm.

CHARLES CITY

Pleasant Springs Farm **Phone: 804-829-5541**

Gale Bazzichi

6700 Old Union Road [23030] Directions: Talleysville Exit off of I-64. Call for directions. **Facilities:** 4 indoor 12' x 12' stalls, total of 40 acres of separate pasture/turnout areas, show size outdoor riding ring, feed/hay & trailer parking. Full-service boarding facility with twice daily turnout. James River Plantation nearby. Call for reservation. **Rates:** $20, including feed, per night. **Accommodations:** Nellie Custis Motel 7 miles from stable.

DANVILLE

Grenadier Farm **Phone: 804-685-7568**

Virginia Wiseman

Rt. 6, 786 Pine Lake Rd, Box 2756 [24541] Directions: Located on Rt. 58 West. Call for specific directions. **Facilities:** 4 indoor stalls available in a 45-stall barn, 5 fenced pastures, feed/hay & trailer parking available. Fox hunting on Wednesdays of every week. Hunt seat & general equitation lessons offered for both horses & riders. Call for reservation. **Rates:** $15 per night; ask for weekly rate. **Accommodations:** Stroft Inn 4 miles from farm.

ALL OF OUR STABLES REQUIRE CURRENT NEG. COGGINS, CURRENT HEALTH PAPERS, & OWNERSHIP PAPERS.

FORT VALLEY

Fort Valley Stables **Phone: 540-933-6633**
Rick & Sandy Deshenes **Toll Free Phone: 888-754-5771**
Web: www.fortvalleystable.com
299 South Fort Valley Road [22652] Directions: Exit 279 off I-81. east 1 mile to Hwy 11 - Left. 1/2 mile to Hwy 675 - Right. 5-1/2 miles to Kings Crossing - Right. Hwy 678, 1-1/2 miles to stable entrance. **Facilities:** 8 indoor box stalls, 2 - 10' X 16' & 6 - 8' X 12'. 20 corral pens 16' x 12', 18 water & ewlectic trailer sites, 26 no hook up sites. Swimming pond, 3 fishing ponds. Over 80 miles of mountain trails. Feed, hay and trailer parking available. 3 pastures about 1 acre each. **Rates:** $5 to $15 per horse per night. **Accommodations:** 2 cabins on site. Trailer or tent camping with water and electric hookups. Ramada Inn & Budget Host, both in Woodstock 9 miles away.

FREDERICKSBURG

Cedar Crest Farm & Stables **Phone: 540-752-7302**
314 Poplar Road (22406) Directions: I-95 to rt. 17. Take 4 miles to Poplar Rd. on right(rt. 616). 1 mile to farm sign, take right. Glendie Beside Drive, Follow road to barn. **Facilities:** 8 10x16, 16x16 indoor stables. trailer parking, electric hookups, feed/hay available, 4 separate paddocks, full size riding ring, pasture, year-round full care boarding available. **Rates:** stall-$20 pasture-$10 per night. weekly rates negotiable. **Accommodations:** Holiday Inn, Motel 6, Court Yard, Ramada, EconoLodge within 5 miles of barn.

GREENVILLE

Penmerryl Farm/The Equestrian Centre **Phone: 1-800-808-6617**
Ken Pittkin, owner; Karen Evans, mgr. **or: 540-337-0622**
662 Greenville School Road, Box 402 [24440] **Fax: 540-337-0282**
Directions: 10 minutes from I-81, Exit 213A Greenville. Call for directions, reservations and rates. **Facilities:** 20-34 indoor 12' x 12' stalls, paddocks, large pastures, horse walker, trailer parking. Farm is working breeding and training center. Also on premises tennis court and two lakes for swimming, fishing, sailing. Must make reservations. **Rates:** $25 per night. **Accommodations:** B&B lodge & cabins with pool and hot tub.

KENTS STORE

Another Bay Farm **Phone: 804-457-3408**
John Pearsall Blants
4375 Hickory Hill Road [23084] Directions: Take Hadensville Exit off off I-64. Go .5 mile & cross over Rte. 250. Turn left on 606/629 & go 1.5 miles. Turn right on Rt. 609 (Hickory Hill Rd). Go exactly 1 mile then turn left onto farm road. **Facilities:** 7 indoor 12' x 12' box stalls plus 10 outdoor run-ins. 6 pastures on 400-acre farm, riding ring, round pen and miles of trails. Feed/hay and ample trailer parking available. Miles of riding trails. Quarter horses and Paints for sale. **Rates:** $18 per night; $100 per week. **Accommodations:** Motels in Charlottesville area, 20 miles away.

ALL OF OUR STABLES REQUIRE CURRENT NEG. COGGINS, CURRENT HEALTH PAPERS, & OWNERSHIP PAPERS.

LEXINGTON

Fancy Hill Farm **Phone: 540-291-1000**
Patricia A. Magner, manager **Fax: 540-291-4057**
100 Equus Route [24578] Directions: From I-81 South, take Exit 180B (Route 11, Fancy Hill), turn right on Rte 11, 6/10 mile to entrance on left. From I-81 North, take Exit 180 to Rte 11 North, left on 11, 1.2 miles to entrance on left. **Facilities:** Barn has 19 indoor 12' x 12' stalls with individual windows, indoor arena has 14 indoor 12' x 10' stalls, 8 paddocks 3-6 acres, 17-acre pasture; indoor ring, 3 large outdoor rings, training ring. Trailer parking. Boarding, lessons, clinics, training, shows. **Rates:** $20 per night, plus bedding. **Accommodations:** Westmoreland Budget Motel (1/2 mile), Natural Bridge Motel (4 miles). Several motels in Lexington, 7 miles away.

Virginia Horse Center **Phone: 540-463-2194**
Frank Bierman, Director of Facilities Programs
Rt. 39 West [24450] Directions: 3 miles from both I-64 & I-81. Call for directions. **Facilities:** 500 indoor stalls, 80 outdoor covered stalls, 4 dressage outdoor rings, cross-country course, two 200' x 95' warm-up rings, 130' x 200' outdoor oval ring, 300' x 150' indoor coliseum, feed/hay & trailer parking on premises. Center hosts all breed horse shows, hunter/jumper shows, dressage events, horse sales, & rodeos. Seats 4,000 with concession stands. Must call ahead for reservation. **Rates:** $20 per night plus necessary security. **Accommodations:** Super 8, Hunt Ridge Motel, Comfort Inn all 1 mile from Center.

LOCUST DALE

The Inn at Meander Plantation **Phone: 703-672-4912**
Suzanne Thomas, Suzie Blanchard, & Bob Blanchard
James Madison Hwy, US Route 15 [22948] Directions: I-81 to I-64 to Rt. 15. 9 miles south. Adjacent to Robinson River. **Facilities:** 5 to 10 indoor stalls in 3 buildings, 60 acres of pasture/turnout, outside riding ring, riding trails, feed/hay at add'l cost, trailer parking. Kennels available for dogs. This is a vacation and sightseeing area. Overnight boarding only for guests of inn. Advance notice preferred. Innkeepers will assist you in finding local trails & areas to ride in the mountains. **Rates:** $20 per night; weekly rate upon request. **Accommodations:** Inn on premises in a stately Colonial mansion. Five guest rooms with private baths and full breakfast. Only a short drive to Skyline Drive, Charlottesville & Monticello, & Montpelier.

MECHANICSVILLE

Rose Hill Stable **Phone: 804-746-5906**
Neal Blair
8138 Rose Hill Drive [23111] Directions: Call for directions. 3 minutes off I-295, Richmond bypass; 10 minutes from I-64 and I-95. **Facilities:** 33 indoor 10' x 12' stalls, pasture/turnout, feed/hay, tractor/trailer parking. Lessons, showing, buying, selling, training. All weather footing, sand ring, 2 grass rings. Call for availability. **Rates:** $20 per night. **Accommodations:** Motels 5 minutes from stable

ALL OF OUR STABLES REQUIRE CURRENT NEG. COGGINS, CURRENT HEALTH PAPERS, & OWNERSHIP PAPERS.

MIDDLEBURG

Middleburg Equine Swim Center **Phone: 540-687-6816**

Roger Collins & Laura Hayward

35469 Millville Road [22117] Directions: 3 miles west of Middleburg on Rte. 50, turn right on Rte. 611, after 1 mile turn right onto Millville Road. Swim Center is 1 mile on right. **Facilities:** 42 indoor 12' x 12' stalls, pasture/turnout of varying sizes, 100' x 200' outdoor riding ring. feed/hay, trailer parking. Unique facility specializing in swimming horses for rehab and conditioning. In the heart of Virginia, foxhunting on 42 acres, miles of trails and cross-country riding. **Rates:** $15 per night. **Accommodations:** Numerous B&Bs and country inns within 3 miles.

MIDDLETOWN

Monte Vista Stable **Phone: 540-869-4621**

Dr. N. Lee Newman **Fax: 540-869-0979**

8183 Valley Pike [22645] Directions: Call for directions. Located 2 miles from I-81 and I-66 intersection. **Facilities:** 6 indoor stalls minimum 10' x 14', small paddocks, feed/hay, trailer parking. National historic register property; veterinarian on premises; Rowdi Arabians, foals for sale; near National Forest trails, Old Dominion 100 trails. Reservations required. **Rates:** $20 per night, $100 per week. **Accommodations:** Bed & Breakfast available on site. Call for prices. Wayside Inn in Middletown, 1 mile; Comfort Inn in Stephens City, 5 miles; Battle of Cedar Creek Campground in Middletown, 1 mile. Super 8 Motel 1.5 miles.

PETERSBURG

Idle Moment Farm **Phone: 804-862-4463**

Garry G. & Bobbie L. Moretz

7724 Vaughan Road [23805] Directions: Call for directions. Facilities: 15 indoor stalls, 66′ x 166′ arena, 40 x 40′ paddock, dressage arena, feed/hay & trailer parking. Recent negative coggins and health certificate "We'll be happy to accommodate almost any request." **Rates:** $20 per night. **Accommodations:** Several motels & restaurants within 5 miles.

POWHATAN

Allengeny Stables **Phone: 804-379-2970**

Ron Ervin

1735 Old Powhatan [23139] Directions: Located 22 miles from the center of Richmond on Rt. 60W. Call for directions. **Facilities:** 4 indoor stalls, 4 acres of pasture/turnout, feed/hay & trailer parking available. Limited trail riding and pony rides available. Call for reservation. **Rates:** $20 per night; ask for weekly rate. **Accommodations:** Days Inn 7 miles from stable.

ALL OF OUR STABLES REQUIRE CURRENT NEG. COGGINS, CURRENT HEALTH PAPERS, & OWNERSHIP PAPERS.

POWHATAN

Windsor Farm Stables **Phone: 804-598-2679**

Joe Hairfield

2600 Huguenot Trails [23139] Directions: Located off of Rt. 60. Call for directions. **Facilities:** 4 indoor stalls, 240 acres of pasture/turnout, feed/hay & trailer parking available. 2 days notice if possible. Major tourist attractions nearby. **Rates:** $15 per night includes feed. **Accommodations:** Motels within 15 miles of stable.

RADFORD

Bedlam Manor Farm **Phone: 540-639-4150**

Rebecca Thompson **or: 540-639-9756**

6363 Belspring Rd [24141] Directions: From I-81 S: Take Exit 109; bear right onto Rt. 177/Tyler Ave. & continue to Norwood St.; turn left on Rt. 11 (Norwood St.) & follow Rt. 11 S, turning right to cross Memorial Bridge; turn right onto Rt. 114 at 2nd stop light; turn left onto Rt. 600 at first stop light. Farm is one mile on left. **Facilities:** 4 large box stalls, turnout paddocks available with shelter, feed/hay available with prior notice, trailer parking on farm. Located reasonable distance from trails in Jefferson National Forest and the New River Trail. **Rates:** $20 per night for one horse; $10 per night each for more than one. **Accommodations:** Dogwood Motel & Executive Motel both 5 miles from stable.

ROANOKE

Slocum's Appaloosa Ranch **Phone: 540-473-3778**

Marilyn Slocum **Barn: 540-977-4432**

4860 Glade Creek Road; Mailing Address: P.O. Box 433 [24090] Directions: Exit 150A off of I-81. Call for directions. **Facilities:** 4 indoor stalls, paddocks, feed/hay & trailer parking available. Located near Dixie Caverns & Lexington Horse Center. Trail riding, lessons (pleasure, English, Western), all trails guided, beginners and large groups welcome. Boarding, sales. Black & white Appaloosa stud on premises. Call for reservations. **Rates:** $20, including feed, per night; weekly rate negotiable. **Accommodations:** Campground & many motels nearby.

Walking M Stables **Phone: 703-774-4663**

Charles & Sarah Morris

5701 Bridlewood Drive [24018] Directions: Exit 581 off of I-81. Call for directions. **Facilities:** 20 indoor stalls, 70' x 140' indoor arena, 110' x 240' outdoor arena, 40' round pen, feed/hay. 24-hr security. Training of horses & riders in English & hunt seat at all levels. Horses for sale. Call for reservation. **Rates:** $15 per night. **Accommodations:** Motels 3 miles from stable.

ALL OF OUR STABLES REQUIRE CURRENT NEG. COGGINS, CURRENT HEALTH PAPERS, & OWNERSHIP PAPERS.

SALEM

Sundance Manor **Phone: 703-380-4001**

LaClaire Dantzler, trainer

5091 Glenvar Heights Blvd. [24153] Directions: Dixie Caverns Exit off of I-81. Call for directions. **Facilities:** 11-15 indoor stalls, outdoor paddocks, pasture/turnout area, feed/hay & trailer parking at stable. Teaching, training, & breeding done at stable. Specializes in American Saddlebred. As much advance notice as possible. **Rates:** $30 for stall, $15 for paddock per night. **Accommodations:** Blue Jay & Super 8 less than 10 minutes from stable.

STANLEY

Jordan Hollow Farm Inn **Phone: 540-778-2285**

Web: www.jordanhollow.com **or: 540-778-2209**

326 Hawksbill Park Road (SR 626) [22851] Directions: From I-81 at New Market: Take Rt. 211 east to Luray; turn on 340S Business & go 6.5 miles to left on Rt. 624 to stop sign; take left on Rt. 689 & go approx 1/2 mile; take right on Rt. 626. Farm is .3 mile on right. **Facilities:** 10 indoor stalls, 2 small paddocks, riding ring, small field, timothy/grass mix at $2.50 per bale. **Rates:** $20 for non-guests of inn; $15 for guests. **Accommodations:** Jordan Hollow Farm Inn is a beautifully restored colonial horse farm that has been converted to a country inn. It has 20 guest rooms and a restaurant on the property. $110 or $154 per night for 2 people including breakfast.

STAUNTON

Cabin Creek Stables **Phone: 703-337-7636**

Judy & Donald Cromer

Rt. 5, Box 92 [24401] Directions: 3 miles off of I-81 & 2 miles off I-64. Call for directions. **Facilities:** 40 indoor stalls, 29-acre pasture, 1.5 & 2.5-acre paddocks, 75' x 50' outdoor ring, inside riding available. English & Western lessons offered. Horses for sale. Call for reservation. **Rates:** $20 per night. **Accommodations:** Holiday Inn, Days Inn & Master Host Motel 2 miles from stable.

Westwood Animal Hospital **Phone: 703-337-6200**

Susan Trout, mgr.

Rt. 6, Box 453A [24401] Directions: 5 miles from I-64 & I-81. Call for directions. **Facilities:** 6 indoor stalls, 3.5 acre pasture, turnout facilities, outdoor riding ring, feed/hay & trailer parking on premises. Large animal care available. Call for reservation. **Rates:** $20 per night. **Accommodations:** Holiday Inn & Master Host Hotel 5 miles from hospital.

ALL OF OUR STABLES REQUIRE CURRENT NEG. COGGINS, CURRENT HEALTH PAPERS, & OWNERSHIP PAPERS.

WASHINGTON

Caledonia Farm - 1812 **Phone: 540-675-3693**
Phil Irwin **Reservations: 800-BNB-1812**
47 Dearing Road [22627] Directions: Call for directions. 4 miles north of Washington, Virginia, and 68 miles southwest of Washington, D.C. Close to I-66, I-81, I-95, and I-64. **Facilities:** 2 indoor barn stalls, 50' x 50' pasture/turnout on 52 acres, trailer parking. Adjacent to Shenandoah National Park with its 500 miles of trails. Western & English stables/studs nearby. **Rates:** Free to B&B guests. **Accommodations:** B&B on premises.

WILLIAMSBURG

Carlton Farms **Phone: 804-220-3553**
C. Lewis Waltrip
3516 Mott Lane [23185] Directions: I-95 to I-64 E. Call for further directions. **Facilities:** 10 indoor stalls, huge indoor ring, pasture/turnout area, feed/hay & trailer parking available. A boarding & lessons facility that also sponsors horse shows. 1 day notice if possible. **Rates:** $20, including feed, per night. **Accommodations:** Motels in Colonial Williamsburg 10 minutes.

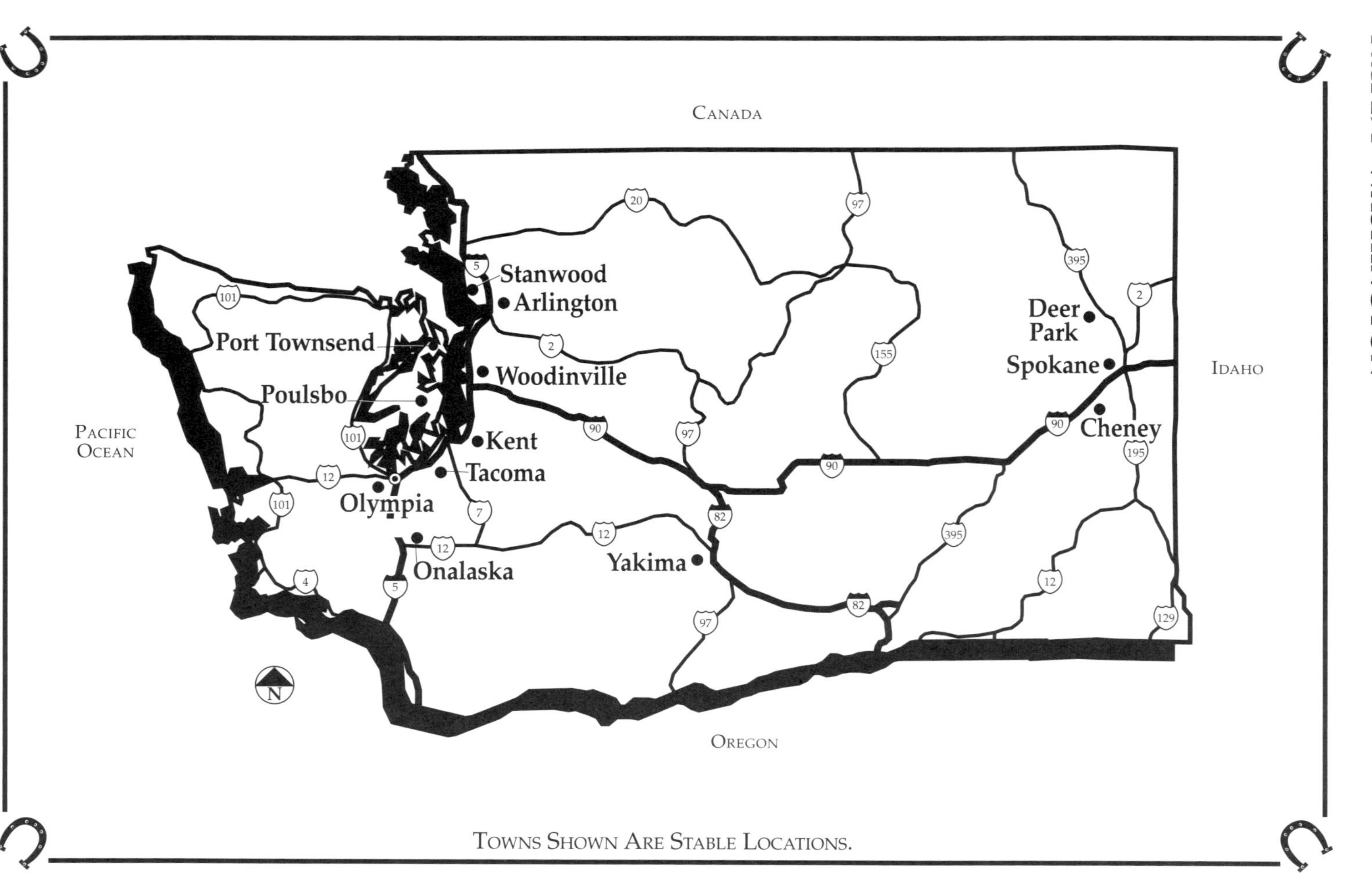
Canada
Idaho
Pacific Ocean
Oregon
Stanwood
Arlington
Port Townsend
Poulsbo
Woodinville
Kent
Tacoma
Olympia
Onalaska
Yakima
Deer Park
Spokane
Cheney
Towns Shown Are Stable Locations.

ALL OF OUR STABLES REQUIRE CURRENT NEG. COGGINS, CURRENT HEALTH PAPERS, & OWNERSHIP PAPERS.

ARLINGTON

Bill and Stevie Somes **Phone: 360-435-3374**

6007 267th Place N.E. [98223] Directions: From I-5 north of Everett, WA: Take Exit 212; go east on Stanwood-Bryant Rd.; go 4.5 miles to stop sign on Hwy 9; go thru stop sign onto Grand View for 1/2 mile; go right on 59th St.; stable is first driveway on left - red house & white barn. **Facilities:** 2 indoor stalls, 4 outdoor stalls. All stalls have 300' x 40' paddocks. Feed/hay & trailer parking available. 14,000 acres of riding trails adjacent to stable & 64,000 acres available within 3 miles. Access to trailhead for Pacific Crest Trail 90 minutes away. 30 minutes from Puget Sound. **Rates:** $12 per night; $52.50 per week. **Accommodations:** Arlington Motel on I-5 & Hwy 530 about 3.5 miles away.

CHENEY

Marshland Equestrian Center **Phone: 509-448-0681**
Carolynn Bohlman, Owner **or: 509-448-0466**

12711 S. Gardner Road [99004] Directions: Call for directions & availability. **Facilities:** 40 stalls, covered outdoor pens & paddocks. 1 official outdoor Dressage Arena, 1 outdoor Jumping Ring & 1 indoor arena with heated lounge. Quality hay & grain. Trailer parking on premises. Boarding, training, lessons, schooling horses & sales **Rates:** $20 per night. **Accommodations:** Hampton Inn & Quality Inn 10 minutes from stable.

DEER PARK

Blue Haven Stables **Phone: 509-276-7968**
Randy & Pamela Heiman

W. 8516 Staley Road [99006] Directions: Located 4 miles west of Hwy 395 at Staley Road Exit. Call for directions. **Facilities:** 19 indoor stalls, indoor arena, 40 acres of pastures & paddocks, 1/4-mile outdoor track, feed/hay & trailer parking available. Has jogging machine for horses that goes up to 18 mph for race horses. Breeds, trains, & sells American Saddlebreds. Farrier on premises. Call for reservation. **Rates:** $15 per night; $75 per week. **Accommodations:** Motels 6 miles from stable.

KENT

Reber Ranch **Phone: 206-630-3330**
Todd Reber

28436 132nd Avenue SE [98042] Directions: From Hwy 18: take 272nd St. exit; go west 2 miles; turn left onto 132nd Ave.; go 1 mile & ranch is on left. **Facilities:** 50 indoor stalls available out of a total of 225 stalls, pasture & turnout, trailer parking on ranch. Feed & tack store on premises. One day notice if possible but not necessary. **Rates:** $12 per night. **Accommodations:** Value Inn & Nendels within 3 miles of ranch.

ALL OF OUR STABLES REQUIRE CURRENT NEG. COGGINS, CURRENT HEALTH PAPERS, & OWNERSHIP PAPERS.

OLYMPIA

James Gang Ranch **Phone: 360-491-3216**
Linda James
8935 Mullen Road SE [98513] Directions: Take Exit 111 off of I-5. Call for further directions. **Facilities:** 30-35 indoor stalls, turnouts, feed/hay included, trailer parking on premises. Boarding, training, & breeding facility specializing in Pintos, American Saddlebreds, & Arabians. 1 day notice if possible. **Rates:** $10 per night. **Accommodations:** Motel 8, Olympic Motel, & Quality Inn 3 miles from stable.

Weeping Willow Ranch **Phone: 360-491-3217**
4437 Shincke Road NE [98506] Directions: Located off of I-5. Call for directions. **Facilities:** 2 indoor stalls, paddocks, indoor & outdoor arenas, feed/hay, trailer parking on site, riding trails. Call for reservation. **Rates:** $20 per night; weekly rate available. **Accommodations:** Many motels in Olympia, 4 to 5 miles from stable.

ONALASKA

Cashel Farm **Phone: 360-978-4330**
Kay & Carlos Sabich **or: 1-800-333-2202**
446 Gore Road [98570] Directions: Exit 68 off of I-5, east 8 miles on Hwy 12. Call for specific directions. 1/2 hour north of Portland; 1 hour south of Olympia. **Facilities:** 8 indoor 12' x 13' stalls with rubber mats, 2 acres electric-fenced pasture/turnout, local grass hay available, trailer parking, water & electric for RV hook-up. 300 acres of trail riding adjacent to property, 1 hr to Mt. St. Helen's and Mount Rainier trails, 10 min. to lake and river fishing, 25 min. to 60 outlet stores & 12 antique stores. No smoking facility. Dogs must be on leashes. **Rates:** $20 per night. **Accommodations:** Bed & Breakfast facilities available on site.

PORT TOWNSEND

Jefferson County Fairgrounds **Phone: 360-385-1013**
Bob Bates
49th & Kuhn [98368] Directions: Take Hwy 20 into Port Townsend. Look for signs to fairgrounds, 1 mile but in city limits. **Facilities:** 75 covered all-wood 12' x 12' stalls, 50' x 150' outside riding arena, small inside 70' x 70' arena, feed/hay, trailer parking. Caretaker 24 hours a day. Camping, travel trailer park, dance hall, dining room & kitchen available for rent. Must clean own stalls. **Rates:** $5 per night, weekly rate available.

POULSBO

Sandamar Farm **Phone: 360-779-9861**
Reg & Julie Gelderman
4499 NE Ganderson Road [98370] Directions: Call for directions. 45 minutes north of Takoma, 1 hour west of Seattle. **Facilities:** 10 indoor 12' x 12' stalls, pasture/turnout, feed/hay, trailer and RV camper parking available. Close to Olympic National Park, guided trail rides available. Call for reservations. **Rates:** $15 per night. **Accommodations:** Motels within 3 miles.

ALL OF OUR STABLES REQUIRE CURRENT NEG. COGGINS, CURRENT HEALTH PAPERS, & OWNERSHIP PAPERS.

RIDGEFIELD

J M J Ranch **Phone: 360-887-1826**

Jeff & Julie Meyer

24909 NE 29th Avenue [98642] Directions: Extremely convenient to Clark County Fairgrounds; less than 10 minutes from I-5 freeway. Call for additional directions. **Facilities:** 3-4 indoor 12' x 12' box stalls with mats, 2 with turnouts, 60' round pen, alfalfa/grass available, trailer parking. **Rates:** $15 per night. **Accommodations:** Studio apartment with bathroom is attached to barn, $35 per night.

SPOKANE

Spokane Sport Horse Farm **Phone: 509-448-3722**

Christel Carlson **Advance Res/Leave message - Barn: 509-448-5064**

Message/Fax: 509-448-2658 **Cell/ Day of Stay Res: 509-993-6786**

E-mail: ccarlson@spokanesporthorse.com Web: www.spokanesporthorse.com

10710 S. Sherman Road [99224] Directions: From I-90, take Pullman Hwy (195) south approximately 3 miles, right on Cheney-Spokane Road, keep right at Y in road, 1 mile past cemetery take left on Sherman Road. Farm is 1.8 miles on right. **Facilities:** 60 indoor stalls with runs, 3 covered stalls with runs, 70' round pen for turnout, 150' x 250' and 100' x 230' outdoor arenas, 216' x 122' indoor arena, automatic heated waterers, feed/hay, trailer parking. Miles of trails on 150 acres. Raise Warmbloods, especially for dressage. Dressage shows and instruction available. Lipizzan Stallion Conversano ll Natasha at stud. "Deep Creek" feed and tack store on farm premises. **Rates:** $20 per night. **Accommodations:** Electric hook-up for campers. Hampton Inn, Best Western, others within 10 minutes.

STANWOOD

Foggy Hollow Farm **Phone: 360-629-3937**

Sharon & Al Gileck

5031 - 324th Street N.W. [98292] Directions: From I-5 North: Take Exit 215, 300th St. NW; from ramp, take left; go under freeway to stop; go right on Old 99; follow that for about 2 miles; at 324th St. NW, take left & farm is 2nd house on right. Call for directions from I-5 South. **Facilities:** 6 indoor 12' x 12' stalls, 60' round pen, 50' x 150' & 150' x 200' pasture/turnout areas; grain & hay available, trailer parking on premises & electric & water hook-up for RV. Vet on call 24 hours. **Rates:** $15 per night; $85 per week. **Accommodations:** Hill Side Motel in Conway, about 4 miles from stable.

TACOMA

Horseland Farms, Inc. **Phone: 206-584-3781**

6420 150th Street SW [98439] Directions: Exit 123 off of I-5. Call for directions. **Facilities:** 5 indoor stalls, paddocks, lighted indoor arena & outdoor arena, feed/hay included, trailer parking on premises. At least 24 hours notice preferred, 1 week notice if possible. **Rates:** $15 per night. **Accommodations:** Howard Johnson's & Lakewood Motor Inn within 10 minutes of stable.

ALL OF OUR STABLES REQUIRE CURRENT NEG. COGGINS, CURRENT HEALTH PAPERS, & OWNERSHIP PAPERS.

WOODINVILLE

Derby Farms **Phone: 425-483-9583**
Pamela Pentz
17720 N.E. 185 [98072] Directions: Approximately 2 miles off of I-405. Call for directions. **Facilities:** 23 indoor 12' x 12' box stalls with auto waterers, 60' x 180' indoor arena, 60' x 180' outdoor arena, pasture, feed/hay & trailer parking available. 10-acre dressage training facility. Hanoverian stallion standing-at-stud: "Weisswein," approved Oldenburg. Ms. Pentz is a dressage trainer at all levels and an AHSA judge. Call for reservation. **Rates:** $15 per night; ask for weekly rate. **Accommodations:** Best Western & Motel 6 in Kirkland, aprox. 3 miles away.

Whispering Trees Farm Bed & Breakfast **Phone: 206-788-2315**
Diane B. Simmons **or: 1-888-766-FARM**
19801 NE 155th Place [98072] **Fax: 206-844-2170**
Directions: Call for directions. **Facilities:** 3 indoor 12' x 12' matted stalls, 1 indoor 12' x 14' matted stalls, 2 outdoor stalls, 3 pastures from 1/3 to 1/2 acre each, timothy hay available, trailer parking. Located 2 blocks from unlimited riding trails. Fox hunting packages available. Reiki & equine & human massage available. **Rates:** $10 per night; $60 per week. **Accommodations:** Bed & Breakfast on site. Non-smoking inn. No pets.

YAKIMA

White Birch Stables **Phone: 509-452-3184**
Roger & Sue Hart
Ray Symmonds Road [98901] Directions: Exit 26 off of I-82. Call for easy directions. **Facilities:** 40 indoor stalls, 50 outdoor stalls, pasture & turnout, indoor arena, 2 lighted outdoor arenas, feed/hay at extra charge, trailer parking available. RV hook-ups and showers available. Natural Grow Grain Distributor. Advance notice if possible but can take on short notice. **Rates:** $10 per night; $15 per night including feed. **Accommodations:** Days Inn nearby. Call stable in advance for discount rate.

Well the turtle
won fair and square!
ol' Jake weren't exactly
the fastest gun in the
west.
SALOON
FRED LEARY

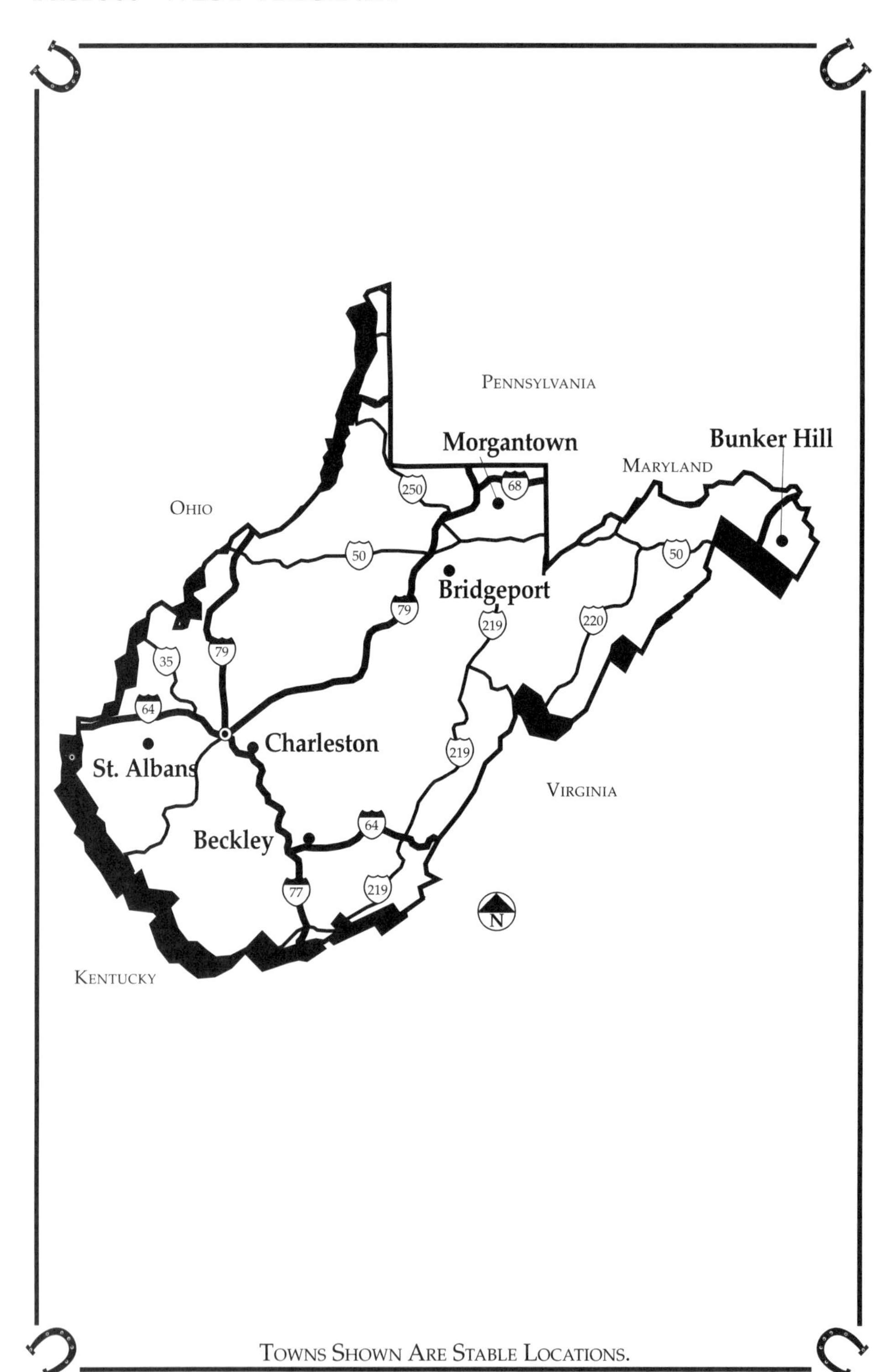

TOWNS SHOWN ARE STABLE LOCATIONS.

ALL OF OUR STABLES REQUIRE CURRENT NEG. COGGINS, CURRENT HEALTH PAPERS, & OWNERSHIP PAPERS.

BECKLEY

Graham Ranch Stables **Phone: 304-253-3570**
484 Mt. Tabor Road [25801] **mail: PO Box 7269, Sprague, WV [25926]**
Directions: From I-64/I-77: Take Harper Rd. Exit; go toward Go Mart; turn left on Mt. Tabor Rd., about 1 mile from exit. Stable is on right, 1/2 mile. **Facilities:** 26 indoor box stalls, 2-30 acres of pasture/turnout, large indoor riding arena, large outdoor arena, sweet feed & quality hay available, & trailer parking on premises. This is a boarding & breeding facility specializing in American Saddlebreds. **Rates:** $25 per night. **Accommodations:** Hampton Inn, Country Inn & Suites, Fairfield Inn, Marriott Courtyard within 1.5 miles. 18 hole golf course on site, Saddlebred Golf Club with eating facility.

BRIDGEPORT

4-T Arena **Phone: 304-592-0703**
Jeff Tucker Trina Tucker, mgr.
Rt. 3, Box 242F [26330] Directions: 4 miles off of I-79. Call for directions. **Facilities:** 30 indoor stalls, eight 20' x 40' turnout paddocks, 250' x 85' indoor arena, 300' x 200' outdoor arena, feed/hay & trailer parking on premises. Rodeos, shows, & clinics held at arena. Camper hook-up. Vet and farrier on premises. Facilities available to rent. Easy access & large parking lot. Call for reservation. **Rates:** $15 per night. **Accommodations:** EconoLodge 4 miles from stable.

BUNKER HILL

White Hall **Phone: 540-678-0948**
Edwin & Steffanie Simpson
1076 Gold Miller Road [25413] Directions: From I-81, take WV Exit 5, west on Rte 51 for 1.7 miles, left on Gold Miller Road for 2.7 miles, stable is second drive on left past stop sign. Call for further assistance. **Facilities:** 6 indoor 11' x 14' stalls with automatic water, outdoor arena, round pen, pasture/turnout, wash area, feed/hay. Easy access & parking for several semi trailers. Vet/farrier on call. Other animals accepted. American Saddlebred breeding program; stallion management and broodmare/foal programs available. Call for reservations. **Rates:** $15 per night; weekly and group rates available. **Accommodations:** Days Inn, Sheraton, Holiday Inn within 8 miles.

CHARLESTON

D & M Stable **Phone: 304-342-3751**
Roy L. & Diana Gibson
Route 2 [25314] Directions: 8 miles from I-64 & I-79. 9 miles from I-77. Call for directions. **Facilities:** 30 indoor box stalls, 150' x 90' riding ring, feed/hay, & trailer parking on premises. This is a public riding stable for trail rides that also offers monthly boarding. Reservations required. **Rates:** $20 per night. **Accommodations:** Executive Inn & Red Roof Inn both in Charleston, 8 miles from stable.

ALL OF OUR STABLES REQUIRE CURRENT NEG. COGGINS, CURRENT HEALTH PAPERS, & OWNERSHIP PAPERS.

MORGANTOWN

Appelwood Bed & Breakfast and Stables — **Phone: 304-296-2607**
Jim Humbertson — **Web: www.appelwood.com**
1749 Smithtown Road, RR 5 Box 137 [26508] — **E-mail: appelwood@aol.com**
Directions: Exit 146 off of I-79 to Goshen Road, 100 yds to stop sign at Junction 73. Turn right and go 1.5 miles. Driveway is on right across from red mailbox. **Facilities:** 9 indoor 12' x 11' stalls; 1-, 3-, and 5-acre pastures; ring; feed/hay available; trailer parking. Some trails on the 35-acre premises. **Rates:** $15 per night, $55 per week, $160 per month. **Accommodations:** Bed & Breakfast on premises, $65-$85 per night.

Meadow Green Stables — **Phone: 304-296-1979**
Ken & P.J. Neer & Tim
Rte. 10, Box 161D [26505] Directions: Exit 68E off I-79. 5-10 minutes off I-68. University Avenue Exit off of I-68. Left off interstate onto 119. Go 2 lights, right on 857. Straight at next light. Right at 4-way stop onto Kingwood Pike. 1.1 mile on left. 5 minutes from Mountaineer Mall; 10 minutes from WV University. **Facilities:** 24 - 12' x 12' stalls, 5' high board fenced paddocks, indoor/outdoor arena, individual fans, individual diet & exercise needs catered to, feed/hay, trailer parking, climate controlled lounge, boarding & training, lessons in beginner & advanced English. Warm water baths. No stallions. Coggins & shot records will be checked & are required. Please do not untrailer before papers are checked. Call for reservations. **Rates:** $25 per night. **Accommodations:** Ramada Inn, Comfort Inn within 3 miles.

ST. ALBANS — **Phone: 304-722-4630**
Sunday Stables — **304-722-4600**
Susan & Christy Sunday, owner
Stephanie Berlin, Manager — **Web: ssunday@access.k12.wv.us**
1 Twilight Lane (25177) Directions: St. Albans exit of I-64(west of Charleston). Right on ramp to rt 35. Turn left onto rt 60. Take right at Subway Sandwich shop and go 3 miles. Go through tunnel and take immediate right. Look for Sunday Stables sign. **Facilities:** 25 10x12 inside stalls, inside arena. trailer parking, feed/hay available, training, lessons, sales, breeding-ASB stallion "All Time High" (Flight Time's Son). **Rates:** $20 per night. **Accomodations:** Bed and Breakfast on site ($40-$60 per night).

STUD
FRED LEARY
1998
"So They Took Off My Hobbles
And I Never Looked Back!"

TOWNS SHOWN ARE STABLE LOCATIONS.

ALL OF OUR STABLES REQUIRE CURRENT NEG. COGGINS, CURRENT HEALTH PAPERS, & OWNERSHIP PAPERS.

BARABOO

Hilltop Riding Stables **Phone: 608-356-9667**

Shane Marsden

Hwy 159 [53913] Directions: Hwy 12 Exit off of I-94. Call for directions. **Facilities:** 22 indoor stalls, turnout pasture, feed/hay included, overnight trailer parking. Please call in advance. Located in a tourist area with many attractions. **Rates:** $10, including feed, per night. **Accommodations:** Blue & White Motel and Old Barn, plus others, 1 mile from stable.

COTTAGE GROVE

Lazy L Ranch & Horse Company **Phone: 608-873-6725**

Skip & Lila Lemanski

2189 Rinden Road [53527] Directions: From I-90, take Exit 147. Call for final directions. **Facilities:** 26 indoor stalls, round pen, paddocks, 60' x 120' indoor arena, 2 large outdoor lighted arenas. Monthly boarding & lessons. Owners on premises. Vet on call. Convenient to Madison & Wisconsin University Veterinary School. Call for reservation. **Rates:** $20 per night; call for weekly rate & availability. **Accommodations:** Motel 6 miles from stable.

EAU CLAIRE

4-J's Ranch **Phone: 715-833-1385**

Jerry Dohm

9513 Olson Drive [54703] Directions: 3 miles north of I-94 and then 7 miles east. Call for further directions. **Facilities:** 10 indoor stalls, pasture area, feed/hay & trailer parking available. As much advance notice as possible. **Rates:** Negotiable depending on number of horses. **Accommodations:** Motels within 12 miles.

Pinewood Stables **Phone: 715-834-4840**

Marit Gekin (contact Bill or Sue)

S. 5300 Hwy 37 [54701] Directions: 2 miles south of I 94 on Hwy 37. (easy access). **Facilities:** 27 indoor stalls. Outdoor boarding with shelters. Indoor riding arena. 3 Outdoor arenas; wooden round pen. No wire fences. **Rates:** $20 per night indoor stall. $15 per night outdoor shelters. **Accommodations:** Motels within 3 miles.

FOND DU LAC

Horseshoe Springs Stables **Phone: 414-921-9842**

Tom & Carol Klamrowski

N. 8054 Hwy 151 [54935] Directions: Hwy 23 Exit off of US 41. Call for directions. **Facilities:** 45 indoor stalls, pasture/turnout area, feed/hay included, trailer parking. Monthly boarding, lessons & training facility. Tack shop & farrier at stable. Located in an historic area. As much notice as possible required. **Rates:** $10, including feed, $5, w/o feed, per night. **Accommodations:** Super 8, Budgetel, & Holiday Inn about 4 miles. Camper hook-up for $15.

ALL OF OUR STABLES REQUIRE CURRENT NEG. COGGINS, CURRENT HEALTH PAPERS, & OWNERSHIP PAPERS.

KNAPP

Pinehaven **Phone: 715-643-6018**

William & Margaret Miller

E1901 890th Ave (54749) Directions: Call for directions. **Facilities:** 6 indoor box stalls. Feed available. Pasture, turnout available. Trailer parking. Trail riding available. **Rates:** $20 per night. $100 per week. **Accommodations:** American Inn, Best Western, Super 8, Motel 6 all in Menomonie 10 miles away.

MARENGO

Ashland County Fairgrounds **Phone: 715-278-3424**

Tom Richardson

Rt. 1 [54855] Directions: Take Hwy 13 to Marengo. Fairgrounds is in center of town. **Facilities:** 20 outdoor stalls, racetrack & arena. Feed/hay & trailer parking available. **Rates:** $10 per night. **Accommodations:** Super 8 & Comfort Inn in Ashland, 15 miles away.

OCONOMOWOC

Nimrod Farm **Phone: 414-567-3103**

Homer & Doty Adcock

2208 N. Summit Avenue (Hwy 67) [53066] Directions: Located 1/2 mile south of I-94 (Exit 282). 25 miles west of Milwaukee & 55 miles east of Madison on Hwy 67. Call for further directions. **Facilities:** 33 indoor box stalls, indoor arena, pasture/paddocks area, feed/hay included, horse trailer and van parking on premises. Horses boarded, trained, sold, lessons. Call for reservation. **Rates:** $25 per night includes feed. **Accommodations:** Holiday Sunspree Motel & Resort 1 mile, others slightly farther.

PLYMOUTH

Fuil Quiver Farm **Phone: 414-892-2520**

Tracy Auch

N 5073 Country Aire Road [53073] Directions: Please call for directions. **Facilities:** 4-10 indoor 10' x 11' stalls, 4 acres pasture, 300' x 150' paddock. alfalfa grass mix available, trailer parking. Family operated facility, natural horsemanship philosophy. Training and boarding available. Just 5 mi. from Kettle Moraine trails, approx. 39 miles of horse trails. **Rates:** $20 per night; $100 per week. **Accommodations:** Americ Inn Motel in Plymouth, less than 10 minutes from farm.

RACINE

Lakeview Riding Stables **Phone: 414-639-6141**

Joe Stang

4218 7 Mile Road [53402] Directions: Located conveniently off of I-94. Call for directions. **Facilities:** 40 indoor stalls, pasture/turnout area, feed/hay included, trailer parking. Open 7 days a week: 8 A.M. to 7 P.M. Horseback riding and hay rides offered at stable. Call for reservation. **Rates:** $25 per night; weekly rate negotiable. **Accommodations:** Motels within 5 miles of stable.

ALL OF OUR STABLES REQUIRE CURRENT NEG. COGGINS, CURRENT HEALTH PAPERS, & OWNERSHIP PAPERS.

REEDSBURG
Pace's Autumnwood Stables
David Pace, owner ; Ross Rote, mgr. Phone: 608-524-8110
E6993 Ski Hill Rd [53959] Directions: Call for directions. **Facilities:** Four indoor stalls, 20'x80' lean-to and 100'x150' open pasture. Fee/hay are available, trailer parking. Breeding and training of Rocky Mountain gaited horses. **Rates:** $15 per day. **Accommodations:** Super 8, Comfort Inn, The Voyaguer Inn & Conference Center 2 miles away in Reedsburg.

RHINELANDER
Triple K. Ranch, Inc. Phone: 715-282-6610
Dean Kuckkahn
7455 Firetower Road [54501] Directions: 2 miles off of Hwy 8 & 5 miles off of Hwy 51. Call for directions. **Facilities:** 30 indoor stalls in 2 barns, large pasture, paddocks, heated wash stall, large outdoor arena, 200' x 300' lunging arena, round pen, & tack shop on premises. Also, horse transportation business serving the Midwest. Call for reservation and information. **Rates:** $15 per night. **Accommodations:** Holiday Inn 9 miles from stable.

RIPON
Cedar Ridge Ranch Phone: 414-748-9394
Tom & Mary Avery, owners; Chris Schultz, mgr. or: 414-748-7597
W14471 Dartford Road [54971] Directions: In east-central Wisconsin, 80 miles from Milwaukee or Madison. Call for directions. **Facilities:** 6 indoor 12' x 12' stalls plus sheltered turnouts, 250 acres of pastures, indoor arena, feed/ hay available, trailer parking. Excellent trail riding area; very scenic and adjacent to Wisconsin's deepest lake. **Rates:** $10 per night; $60 per week. **Accommodations:** Heidel House (first-class resort) in Green Lake, 2 miles from ranch; Americ Inn in Ripon, 1.5 miles away.

SHEBOYGAN
Pine Rock Stables Phone: 414-458-1113
Pam Becker
1431 County Hwy V [53081] Directions: From I-43: Exit east on Hwy V; follow hwy. about 3/4 mile to near the end; on the right, go down driveway to barn in back. **Facilities:** 23 indoor stalls, 10 outdoor stalls with run-in shelters, 25' x 10' turnout paddocks, 3 acres of fenced pasture, 50' x 150' indoor arena, 130' x 150' outdoor arena, round pen & jump course. Located next to Kohler-Andrae State Park with campsites & riding trails. This is a boarding & training facility offering beginning lessons in English & Western and advanced lessons in Western pleasure, dressage, hunt seat, & jumping. Horses that crib must be restrained in some manner. Call for reservation. **Rates:** $15 per night; $75 per week. **Accommodations:** Camping in state park; Parkway Motel in Sheboygan, 1 mile from stable.

ALL OF OUR STABLES REQUIRE CURRENT NEG. COGGINS, CURRENT HEALTH PAPERS, & OWNERSHIP PAPERS.

WASHBURN

Misty Meadows P hone: 715-373-0331
Andrea & Joe Schomberg Fax: 715-373-0334
County Route C Box 13 [54891]

Directions: Rte 2 to Rte 13 north to County Rte. C. 2.3 miles on County C to six blue buildings on south side of road. **Facilities:** 7 indoor 12' x 12' stalls, additional temporary stalls available, 5 turnout pastures, 160' indoor arena, outdoor arena, round pen. TIZ Whiz feed & hay on site. Standing-at-stud: "Buddy He's Impressed" (AQHA stallion). Call ahead please. **Rates:** $20 per night for first horse, $15 each additional horse. . **Accommodations:** Super 8, Redwood Motel, Washburn Motel, others in Washburn, 3 miles from stable.

WHITEWATER

Echo Valley Farm Phone: 414-473-4631
Thomas P. & Rhonda L. Fuller Fax: 414-473-4635
W3218 Piper Road [53190] E-mail: fuller@idcnet.com

Directions: Off state Hwy 59 between Madison and Milwaukee. Call for specific directions. **Facilities:** 3 indoor 10' x 10' box stalls, 18' x 10' enclosed stall with attached 100' x 125' paddock, two 2-acre pastures, all access to 18' x 28' enclosed building for shelter. All fencing "Country Estate 3-Rail PVC." Feed/hay and trailer parking available. Located between Milwaukee, Madison & Janesville. Six miles from Horseman's Park, located in Southern Kettle Moraine State Forest with 50 miles of riding trails available. **Rates:** $15-$25 per night. **Accommodations:** Super 8 Motel and White Horse Inn in Whitewater, within 6 miles. Best Western in Fort Atkinson next to the nationally known Fireside Restaurant & Playhouse, within 10 miles. Also, several B&Bs in area.

NOTES AND REMINDERS

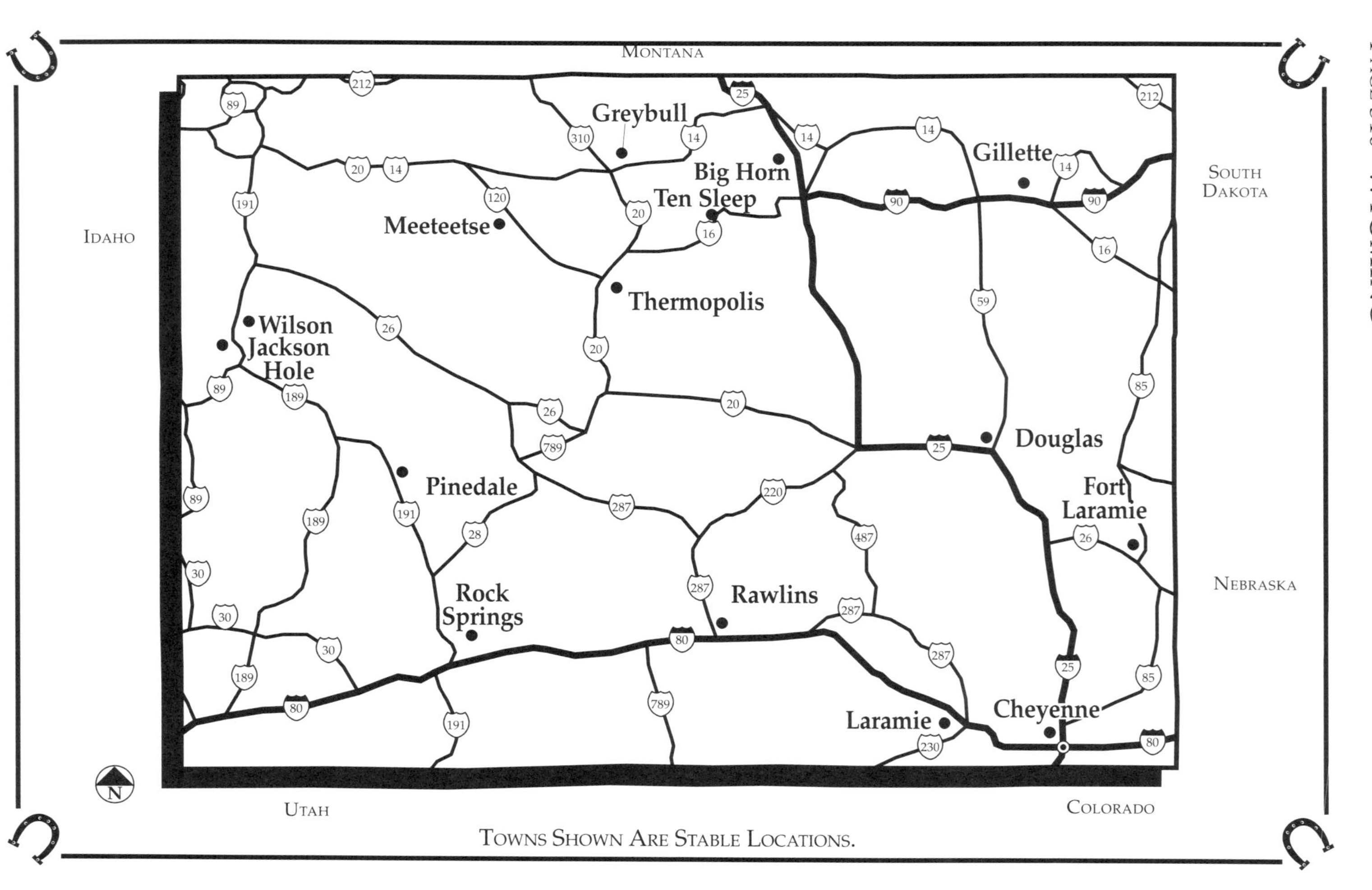
MONTANA
SOUTH DAKOTA
NEBRASKA
COLORADO
UTAH
IDAHO
Greybull
Big Horn
Ten Sleep
Gillette
Meeteetse
Thermopolis
Wilson
Jackson
Hole
Douglas
Fort
Laramie
Pinedale
Rock
Springs
Rawlins
Laramie
Cheyenne
N
TOWNS SHOWN ARE STABLE LOCATIONS.

ALL OF OUR STABLES REQUIRE CURRENT NEG. COGGINS, CURRENT HEALTH PAPERS, & OWNERSHIP PAPERS.

BIG HORN

Blue Barn Bed & Breakfast **Phone: 307-672-2381**
Carol Seidler Mavrakis
304 Hwy 335 [82833] Directions: From I-90 take Exit 25, go west to Coffeen Ave., turn south/left on Hwy 87, 4 miles to Texaco at the Y, bear right onto Hwy 335, stable 3 miles. **Facilities:** 7 indoor stalls, 4 with outdoor runs, 1 foaling stall, two 1-acre turnouts, feed/hay, trailer parking. Close to Big Horn Equestrian Center & Big Horn National Forest; King Saddlery 10 miles . Reservations required. **Rates:** $15 per night. **Accommodations:** B&B on premises. Days Inn, Holiday Inn in Sheridan, 9 miles away.

ROCK SPRINGS

Old #6 Corrals
Lance Neeff **Phone: 307-782-7912**
5000 Springs Drive [82901] Directions: Take I-80 to Elk St. exit to Stagecoach Blvd., west on Stagecoach 1 block to Spring Dr., 1/2 block north. **Facilities:** 100 year old Pole & Post Corrals; 42 corrals, each horse has plenty of room, small arena available, outdoor corrals w/ shelters. Feed/hay available, trailer parking & RV electric hookup. Cowboys w/bedrolls stay free. **Rates:** $10 per day. **Accommodations:** 3 motels within 3 blocks.

CHEYENNE

A. Drummond's Ranch B & B **Phone/Fax: 307-634-6042**
Taydie Drummond **E-mail: adrummond@juno.com**
Web: http://www.adrummond.com
399 Happy Jack Road (State Hwy 210) [82007] Directions: From I-80 Westbound: Take I-25 N; get off at Exit 10B Happy Jack Road (Hwy 210); go west to .4 mile past mile marker 22; turn left on dirt road with 18" X 3' sign which says Private Road. Private Property. No Trespassing. Call for directions from I-80 E. **Facilities:** 4 indoor 12' x 12' stalls with rubber mats, 1 run-in stall, 60' x 120' indoor arena, four 24' x 32' corrals, $5 bale for hay, trailer parking on premises. Will feed your horses in the AM. 55,000 acres for riding in Medicine Bow National Forest nearby. Reservations required. No smoking. **Rates:** $15 per night. **Accommodations:** Bed & Breakfast on premises for horse & hauler. 4 bedrooms sleep up to ten. Quiet, gracious setting of view, wildlife & garden on 120 acres. Reasonable rates vary depending on time of year.

Blue Ribbon Horse Center **Phone: 307-634-5975**
Jim Talkington
406 N. Fort Access Road [82003] Directions: 3 miles from I-80 & I-25 intersection. Call for directions & availability. **Facilities:** 6 indoor & outdoor stalls, large corral with barn, feed/hay included, pasture/turnout, trailer parking on premises. Advance notice if possible. **Rates:** $10 per night. **Accommodations:** Little American Hitching Post & Motel 6 approx. 3 miles from stable.

ALL OF OUR STABLES REQUIRE CURRENT NEG. COGGINS, CURRENT HEALTH PAPERS, & OWNERSHIP PAPERS.

CHEYENNE

Cheyenne Stockyard-Cattle & Horse Motel **Phone: 307-634-7333**
Dr. Jay Dee Fox & Clay Stecklein Fax: 307-637-4826 Toll free: 888-634-7333
Web site: www.busdir.com/cheyennest E-mail: cheysock@aol.com
350 Southwest Drive [82007] Directions: Call for directions. Within 1 mile of I-80 and I-25. **Facilities:** 21 - 20 X 25, 16 - 20 X 40 indoor stalls, 7 - 40 X 70 outdoor pens, feed/hay, trailer parking, vet/farrier/groomer on call, extended stay available. **Rates:** $15 per night. **Accommodations:** Within walking distance of restaurants, hotels, gas station, truck stop.

7XL Stables at Terry Bison Ranch **Phone: 307-634-4171**
Dan Thiel
51 I-25 Service Road East [82007] Directions: On I-25 at Wyoming-Colorado border. Take WY Exit 2 to Terry Ranch Rd. South 2 miles on Terry Ranch Road. **Facilities:** 16 indoor stalls, 12 outdoor stalls, outdoor pens available for semi loads, feed/hay available, trailer parking. Overnight and monthly stabling for horses on a 27,000-acre bison ranch. Unique guest ranch with horseback riding, horse-drawn wagon tours, fishing, RV park, guest cabins, bunkhouse, restaurant and saloon. Year-round accommodations. **Rates:** Varies. **Accommodations:** RV Park, guest cabins, & 17-room bunkhouse on site.

Singletree Stable **Phone: 800-336-0287**
Glenna Ross
4715 Thomas Road [82009] Directions: Exit 364 off of I-80. Call for directions. **Facilities:** 14 indoor 12' x 12' box stalls with auto waterers, lighted indoor & outdoor arenas, feed/hay & trailer parking available. **Rates:** $15 per night; weekly rate negotiable. **Accommodations:** Motels 3 miles from stable.

DOUGLAS

Deer Forks Ranch **Phone: 307-358-2033**
Ben and Pauline Middleton
1200 Poison Lake Road [82633] Directions: Call for directions. 25 miles south of I-25. **Facilities:** Stalls and ranch corrals available as needed, small pastures on 14,000 acres, summer pasture only, winter supplement as needed, trailer parking, RV hook-up. Ranch raises cattle, show sheep & paint horses. Big game hunting in fall. Trail rides and cattle drives available. Advance reservations recommended. **Rates:** $10 per night, $50 per week, $50 per month for pasture. **Accommodations:** B&B on premises.

ALL OF OUR STABLES REQUIRE CURRENT NEG. COGGINS, CURRENT HEALTH PAPERS, & OWNERSHIP PAPERS.

FORT LARAMIE

Carnahan Ranch **Phone/Fax: 307-837-2917**
Hal & Bess Carnahan
HC 72, Box 440 [82212] Directions: At town of Fort Laramie, turn south on Hwy 160 (follow signs to Fort Laramie National Historic Site). Go 1/2 mile past Fort entrance. Cross bridge over Laramie River, turn left into ranch drive. **Facilities:** 2400 acres. Stalls , corrals, night yard light at corrals, waterers, stock tanks & creek at corrals, alfalfa/grass hay available. Trailer parking, RV hookups (electic & water - no sewer dump or hookup),campsites, teepee rental. Rec building w/baths, full kitchen, serving counter, pool table, games, picnic tables and grills. Vet service nearby. Trail rides (on your own) on historic ranch (Oregon Trail, Mormon Trail, Deadwood/Cheyenne Stage Trail, Indian campground, etc.). Groups welcome for cattle dirves and/or historic rides, scout camps, family reunions, etc.- Spanish Colonial Mustangs, AQHA & TB horses. **Rates:** Per night- Campsites, tents & trailers $15, RV w/electric $20, horses $5. Full use of facilities included in rates. Reservations recommended. **Accommodations:** Fort Laramie Motel 3 miles in town or Bunkhouse Motel 13 in Guernsey. Restaurants available in towns.

GILLETTE

P Cross Bar Ranch **Phone: 307-682-3994**
Marion & Mary Scott **Email: pcrossbar@vcn.com**
8586 N Hwy 14-16 [82716] Directions: Call for directions. Right on US 14-16, 20 miles north of Gillette. **Facilities:** 3 indoor 15' x 20' stalls, 4 corrals with shed set-up, feed/hay, trailer parking, small fee for camper parking. Big game outfitters. Call for reservations. **Rates:** $10-$15 depending on service; weekly rates available. **Accommodations:** B&B on premises.

GREYBULL

McFadden Ranch **Phone: 307-765-9684**
Sandy McFadden **Fax: 307-769-9609**
E-mail:training@mcfaddenranch.com
Lane 30 1/2, #2480 [82426] Directions: West end of Greybull, off Hwy 20: North on Road 26, 3.3 miles to Lane 30 1/2, 4/5 mile to ranch. **Facilities:** 28 indoor 12' x 12' stalls, 5 large paddocks with shelter & water, 4 pastures with shelter and water, heated indoor riding arena (extra charge), wash racks, feed/ hay. **Rates:** $15 per night. **Accommodations:** Greybull Motel (307-765-2628) within 4 miles. Apartment available at ranch, $30 per night.

JACKSON HOLE

Moose Meadows Bed & Breakfast **Phone: 307-733-9510**
Juli James and Alan Blackburn
1225 Green Lane, Wilson [83014] Directions: Take Hwy 22 west from town of Jackson, cross Snake River. Green Lane is first road on left after the bridge. **Facilities:** 3 indoor 12' x 13' stalls, 1-acre pasture, feed/hay available, trailer parking. Juli's daughter, Jenny, was 1996 National High School Rodeo Champion in girl's cutting. **Rates:** $10 per night. **Accommodations:** B&B on premises.

ALL OF OUR STABLES REQUIRE CURRENT NEG. COGGINS, CURRENT HEALTH PAPERS, & OWNERSHIP PAPERS.

LANDER

Lander Sports Arena **Phone: 307-332-9790**
Douglas Anesi
40 Pheasant Run Drive [82520] Directions: Next to airport. Call for directions. **Facilities:** One 50' x 100' pen, seven 20' x 40' pens, no shelters, water available in 2 pens, 240' x 310' outdoor arena, 120' x 300' indoor arena with bleachers & concessions that seats 1,000. Arena used for rodeos, dressage events, & team roping events. No watchman or security. **Rates:** No charge. **Accommodations:** Holiday Lodge & Pronghorn Lodge 1 mile from arena.

Sandstone Ranch Equine Motel LLC **Phone: 307-332-2177**
Kathryn Kulcher **Fax: 307-335-9535**
2529 Sinks Canyon Road (82520) Directions: 2.5 miles from Hwy 287. Call for Directions. **Facilities:** 2 12X24, sheds with 24X24, runs, 2 12X12, with 24X12 runs, 1 12X36, shed with roughly 3600, paddock. solar water heater, salt blocks, rubber mats, 70, round pen, feed/hay available, trailer parking, may use Double A Ranch facilities (Indoor/Outdoor Arena) for $10/horse/day. **Rates:** $20 per night. **Accommodations:** Several Hotels within 3 miles.

LARAMIE

A. Drummond's Ranch B & B **Phone: 307-634-6042**
Taydie Drummond
399 Happy Jack Road (State Hwy 210) [82007]
See listing under CHEYENNE.

Delancey Training Stables **Phone: 307-742-2933**
Niki & Bernie Delancey
790 Huron [82070] Directions: 1/4 mile off I-80 & Hwy 287. Call for directions. **Facilities:** 8-12′ x12′ box stalls, 20-10′ by 20′ runs with sheds, 2 large paddocks, automatic waterers throughout. 200′ x 60′ indoor arena, 50′ x 50′ indoor round pen, 3 outdoor arenas & a round pen. Trailer parking and custom tack & repair shop on premises. No advance notice needed. **Rates:** $10 per night. Hay extra. **Accommodations:** 5 min. to Holiday Inn, Days Inn & Motel 6.

On A String Ranch **Phone: 307-742-4723**
Mernie Younger **E-mail: onastring9@aol.com**
Web: www.onastring.com
900 Howe Road [82070] Directions: 2 miles off of I-80. Call for directions. **Facilities:** 28 indoor 12' x 16' & 12' x 12' stalls, 640 acre pasture, 12 outside paddocks each holding 1-8 horses, 60' x 120' indoor arena, feed/hay & trailer parking available. Horse training available. Call for reservation. **Rates:** $15, including feed, per night. **Accommodations:** Holiday Inn & Motel, and many others nearby.

ALL OF OUR STABLES REQUIRE CURRENT NEG. COGGINS, CURRENT HEALTH PAPERS, & OWNERSHIP PAPERS.

LARAMIE

Stonehouse Stables **Phone: 307-742-7512**

Don Pratt

3070 Snowy Range Road [82070] Directions: 1-3/4 miles off of I-80. Call for directions. **Facilities:** 34 stalls, 10,000 sq. ft. indoor arena, 15,000 sq. ft. outdoor arena, hay & trailer parking available. Training & lessons in driving, reining, cutting, dressage, & Western pleasure. Call for reservation. **Rates:** $20 per night; $100 per week. **Accommodations:** Best Western & Camelot 2 miles.

MEETEETSE

Mountain Valley Horse Center **Phone: 307-868-2442**

Roxanne & Travis Richardson

P.O. Box 451 [82433] Directions: Located 30 miles south of Cody off of Hwy 120. Call for directions. **Facilities:** 6 indoor stalls, 4 outside stalls, large corrals, feed/hay & trailer parking available. Farrier on premises. Trains horses for pleasure. Riding lessons in English & Western. Horses for sale. Call for reservation. **Rates:** $15 per night. **Accommodations:** Motor home that sleeps 6 available for rent. Vision Quest Motel & Oasis Motel 2 miles from stable.

PINEDALE

Pole Creek Ranch Bed & Breakfast **Phone: 307-367-4433**

Dexter & Carole Smith

244 Pole Creek Road [82941] Directions: 1/2 mile south of Pinedale turn up Pole Creek Road and go 2.44 miles, ranch on right. **Facilities:** 5 indoor stalls, 7 acres of pastures, grass/hay available, trailer parking. Horses must be wormed. **Rates:** $5 per night, $15 per week. **Accommodations:** B&B on premises; $55 per night double occupancy. Motels in Pinedale, 3 miles from ranch.

RAWLINS

Carbon County Fairgrounds **Phone: 307-324-8101**

Spruce & Harshman [82301] Directions: 3 blocks off of I-80 & Hwy 287. **Facilities:** 48 indoor stalls, 300' x 300' outdoor arena, lunging & exercise area, no feed/hay, trailer parking available. Rodeos & demolition derby held at grandstand that seats 3,600. **Rates:** $5 per night; ask for weekly rate. **Accommodations:** Rawlins Inn, Days Inn, & Key Motel 1 mile from stable.

ROCK SPRINGS

Rock Springs KOA Campground **Phone: 307-362-3063**

Bonnie Whitley

86 Foothill Blvd. [82901] Directions: Take Exit 99 off of I-80. Follow signs east one mile. Open April-Oct. 15. **Facilities:** 3 outdoor stalls with lean-to for protection, 5-acre fenced sagebrush, feed/hay with prior notice, trailer parking. **Rates:** Horse corral free to persons renting campsite or cabin at facility.

ALL OF OUR STABLES REQUIRE CURRENT NEG. COGGINS, CURRENT HEALTH PAPERS, & OWNERSHIP PAPERS.

TEN SLEEP
Ten Broek RV Park & Cabins **Phone: 307-366-2250**
Darell D. Ten Broek
Box 10, 98 - 2nd Street [82442] Directions: 68 miles west of Buffalo on Hwy #16 off I-90. Call for directions. **Facilities:** 10 outdoor stalls, 10 indoor stalls (9' x 9') 6 large corrals, 80' x 150' pasture. 50 x 90 barn and riding area. Feed/ hay available, trailer parking with electric, water, sewer, & cable TV. Guided rides available. Fantastic scenery of Big Horn Mountains. Farrier available by advance appointment. Brand ownership required. **Rates:** $5 per night plus feed. **Accommodations:** Valley Motel 3 blocks away; Log Cabin Motel 2 blocks away.

THERMOPOLIS
Fremont Stables **Phone: 307-864-3384**
Dan Wheeler
1513 Fremont Street [82443] Directions: Take Hwy 120 West. Call for directions. **Facilities:** 4 indoor stalls, 15 outdoor pens with wind break, 80' x 200' outdoor arena. Full service feed & tack store on premises. Also buys & sells horses. Call for reservation. **Rates:** $10 per night. **Accommodations:** Holiday Inn 1 mile from stable.

WILSON
Snyder's Stable **Phone: 307-733-1790**
Ken Snyder
P.O. Box 803 [83014] Directions: Located between Teton Village & Jackson. Call for directions. **Facilities:** 24 indoor 10' x 20' stalls, 220 acres of fenced pasture, 12 outdoor 20' x 20' paddocks, riding corrals, 300' x 300' arena, feed/ hay & trailer parking available. Call for reservation. **Rates:** $10 per night. **Accommodations:** Motel 8 & Days Inn in Jackson Hole, 3 miles away.

NOTES AND REMINDERS

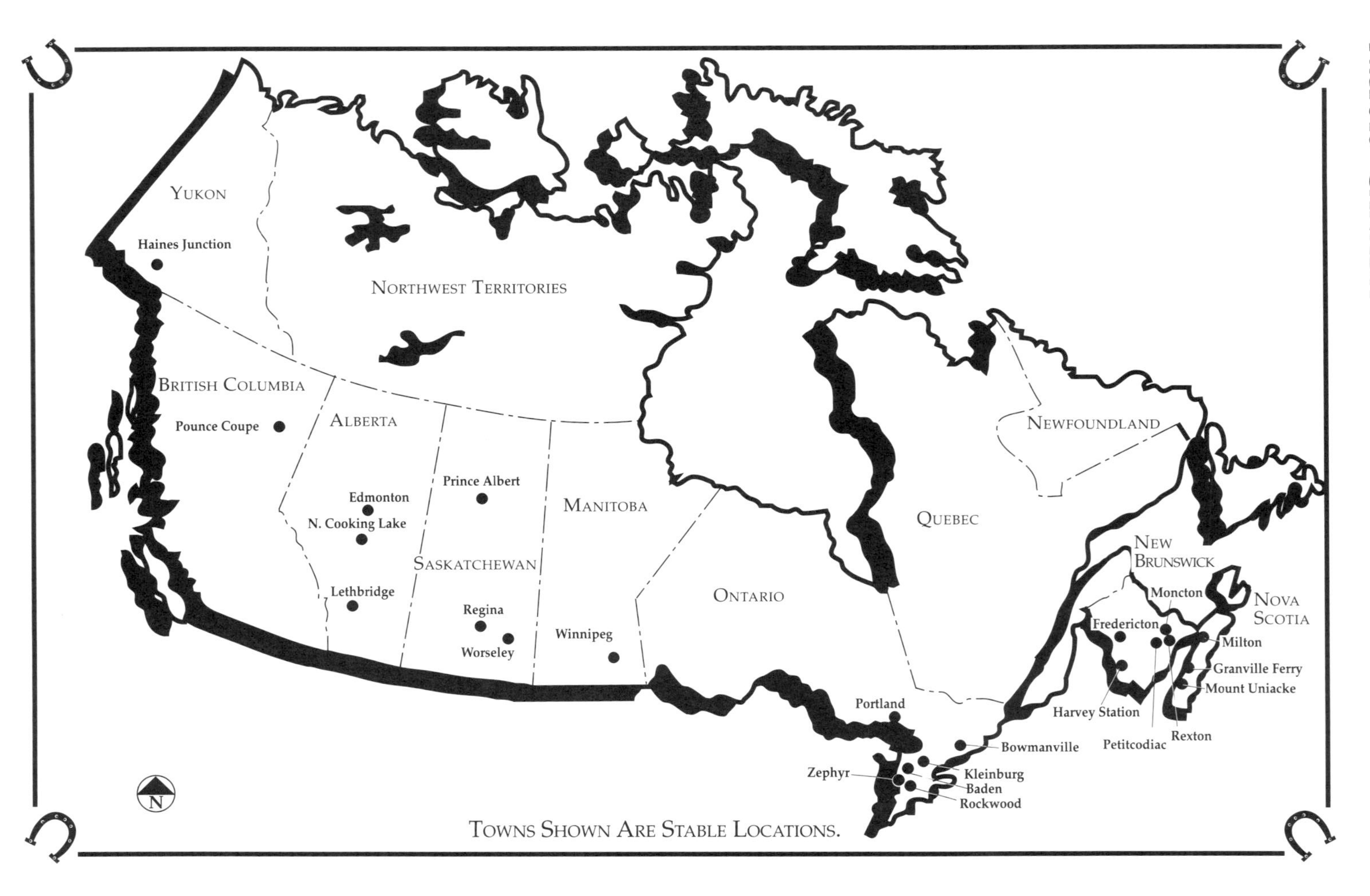

YUKON
Haines Junction
NORTHWEST TERRITORIES
BRITISH COLUMBIA
Pounce Coupe
ALBERTA
Edmonton
N. Cooking Lake
Lethbridge
Prince Albert
SASKATCHEWAN
Regina
Worseley
MANITOBA
Winnipeg
ONTARIO
Portland
Zephyr
Kleinburg
Baden
Rockwood
Bowmanville
QUEBEC
NEWFOUNDLAND
NEW BRUNSWICK
Moncton
Fredericton
Harvey Station
Petitcodiac
Rexton
NOVA SCOTIA
Milton
Granville Ferry
Mount Uniacke
N
TOWNS SHOWN ARE STABLE LOCATIONS.

ALL OF OUR STABLES REQUIRE CURRENT NEG. COGGINS, CURRENT HEALTH PAPERS, & OWNERSHIP PAPERS.

ALBERTA

EDMONTON

Double C Quarter Horses Ltd. — **Phone: 403-447-2915**
Cyril & Michelle Desjarlais — **or: 403-447-2969**
Clarence & Netta Hayes
15519 - 156 Street [T5L 4H8] Directions: 1/4 mile north of Hwy 2 (St. Albert Trail) on 156 St. 16 minutes from West Edmonton Mall. **Facilities:** 16 indoor 10' x 10' box stalls, turnout pens in varying sizes, feed/hay, trailer parking. Extensive breeding & training facility. Standing-at-stud: "Misters Razzledazzle," RSV World Champion. **Rates:** $15 per night indoors, $5 outdoors. **Accommodations:** Sleep-Inn Motel or St. Albert Inn within 1 mile of farm.

LETHBRIDGE

Rainbow Riding Centre — **Phone: 403-328-2165**
Lethbridge Handicapped Riding Association
Tracy Fyfe, Karen Shepard
RR #8 - 24 - 6 [T1J 4P4] Directions: 3 miles east of Lethbridge on 1st Avenue South (Hwy 512). **Facilities:** 27 indoor 10' x 12' stalls, heated, insulated indoor arena, outdoor arena, pasture/turnout, feed/hay, trailer parking. Riding lessons. **Rates:** $12 per night. **Accommodations:** Motels within 3 miles.

NORTH COOKING LAKE

Still Meadows Ranch — **Phone: 403-922-5566**
Rusnak Family — **or: 403-990-7229**
Range Road 210A [T0B 3N0]
Directions: Call for directions. 2-1/4 kilometers north of Hwy 14.
Facilities: 32 indoor stalls, 70' x 200' indoor arena, 2 outdoor arenas, outside pens, feed/hay, trailer parking. 160 acres with trails. Paints & quarter horses on premises. Standing-at-stud: "Chattanooga Choo Choo." **Rates:** $10 per night. **Accommodations:** Tofield Campground within 10 minutes.

ALL OF OUR STABLES REQUIRE CURRENT NEG. COGGINS, CURRENT HEALTH PAPERS, & OWNERSHIP PAPERS.

BRITISH COLUMBIA

POUCE COUPE (DAWSON CREEK area)
Red Roof Bed & Breakfast **Phone: 250-786-5581**
Laurie Embree
Box 727 [V0C 2C0] Directions: On Blockline Road, 3 miles west of Hwy 2. Call for further directions. **Facilities:** 1 log 25' round corral. 1-10 acre pasture/ turnout with barb wire fences, feed/hay, trailer parking available. **Rates:** $5 per night. **Accommodations:** B&B on premises.

MANITOBA

WINNIPEG
Poco-Razz Farm **Phone: 204-255-4717**
Jim Shapiro & Christina Eyres
130 Greenview Road [R3V 1L6] Directions: From Perimeter Hwy (#100) go 5 kilometers or 3 miles south on St. Mary's Road. Call for directions. **Facilities:** 2 indoor 9′ x 12′ stalls, 2 indoor 10′x12′ stalls, 12′ aisleway; 3 large corrals, one of which is 90′ x 110′ outdoor arena, 2 acres of pasture, oat straw bedding, water available in each stall and in corrals, hay available, no barbed wire, rubber fencing with hot wire; wash area & hot water tank, muck-out area, parking for horse trailers. Vet nearby; farrier available. Trails for riding, fishing. **Rates:** $30 per night outdoor. **Accommodations:** Comfort Inn 4.2 miles from farm in Winnipeg.

NEW BRUNSWICK

HARVEY STATION
Holiday Ranch **Phone: 506-366-3291**
Gary & Brenda Nason
Rte. 636 [E0H 1H0] Directions: Call for directions. About 4 miles off Hwy 3; 15 miles from Trans-Canada Hwy. **Facilities:** 6 box stalls, 40' x 60' indoor arena, large outdoor arena, feed/hay, trailer parking. Clinics available. Crown ground behind ranch with endless trails. Call for reservations. **Rates:** $10 per night. **Accommodations:** B&B in area. Motel in Fredricton 20 miles away.

MONCTON
Springwater Stable **Phone: 506-384-5935**
Wayne & Marsha Wilson
RR #10, K413, Ammon Road [E1C 9J9] Directions: Call for directions. 5 minutes off Trans-Canada Hwy, Gorge Road exit, turn left, 1 mile to 4 corners, turn right onto Ammon Road, 2 miles on right. **Facilities:** 15 - 8' x 10' and 10' x 12' box stalls, 3 separate pastures, round pen, 120' x 160' outside arena, feed/ hay, trailer parking. Sugarwood trails, riding lessons, coaching & training; Appaloosa stallion at stud. 5 minutes from Magic Mountain Water Park. **Rates:** $15 per night. **Accommodations:** Motels within 7 minutes.

ALL OF OUR STABLES REQUIRE CURRENT NEG. COGGINS, CURRENT HEALTH PAPERS, & OWNERSHIP PAPERS.

NEW BRUNSWICK

PETITCODIAC

Sheffield Stables **Phone: 506-756-1110**

Charles & Paula Jacob

RR #5 [E0A 2H0] Directions: Call for directions. Right off Petitcodiac exit of Trans-Canada Hwy. **Facilities:** 35 indoor 10' x 12' stalls, 250' x 150' and 150' x 150' outdoor arenas, 50' x 100' indoor arena, 9 separate paddocks, feed/hay, trailer parking available. Miles of trails, camp on trails, mechanic on property, carriage & sleigh rides. Instruction & training, 2 CEF shows a year, cross-country course. Call for reservation. **Rates:** $25 per night. **Accommodations:** Motels in Moncton, 15 miles away.

REXTON

Maple Leaf Icelandic Horse Farm. Inc. **506-523-4480**

Joan Donaher

160 Finno Lane (E4W3A1) Directions: Highway 11N to exit 134 (Ruxton). Located on rt. 116. **Facilities:** 10 indoor stables, riding area, paddock, trailer parking, feed/hay available, trail rides, negative coggins, health papers, vaccination, records. **Rates:** $15 per night **Accommodations:** Hotel on site, Cottages, B&B, Restaurant.

NOVA SCOTIA

GRANVILLE FERRY

Equus Centre **Phone: 902-532-2460**

Jennifer Gale

Box 1160, RR #1, 5613 [B0S 1K0] Directions: Call for directions. Located 5 minutes from historic Annapolis Royal. **Facilities:** 23 indoor stalls most 11' x 12', pasture/turnout, 60' x 136' indoor arena, hunt course, jumping ring, 20' x 60' dressage ring, feed/hay, trailer parking. Trails, lessons. Call for availability. **Rates:** $6 per night. **Accommodations:** Motels and B&Bs within 5-10 minutes.

MILTON

Birch Lane Farm **Phone: 902-368-1113**

John McAssey

Rte. 2 [C1E 1Z2] Directions: Call for directions. 7 kilometers west of Charlottetown. **Facilities:** 33 indoor stalls 8' x 10' to 12' x 10', turnout, 60' x 120' indoor arena, outdoor riding ring, tack room, feed/hay, trailer parking. Heated classroom, viewing lounge, lessons, training. Call for reservation. **Rates:** $10 per night. **Accommodations:** Motels in Winslow (2 kilometers) and Charlottetown (7 kilometers).

ALL OF OUR STABLES REQUIRE CURRENT NEG. COGGINS, CURRENT HEALTH PAPERS, & OWNERSHIP PAPERS.

NOVA SCOTIA

MOUNT UNIACKE

Briarwood Farm — **Phone: 902-866-1198**

Judy Covert, Vicci Fowler

Old Windsor Hwy [B0N 1Z0] Directions: Call for directions. 25 minutes out of Halifax. Off Exit 3 on Hwy 101. **Facilities:** 37 indoor stalls, paddocks, indoor & outdoor arenas, 1/2-mile track, feed/hay, trailer parking. Competition stable, riding lessons. Call for availability. **Rates:** $20 per night. **Accommodations:** Motels in Sackville, 10 miles.

ONTARIO

BADEN

North Ridge Farm — **Phone: 519-634-8595**

Sarah Banbury

RR #21 [N0B 160] Directions: Call for directions. Between Stratford, Kitchener, Waterloo. 6 kilometers from Hwy 7/8, 20 minutes north of Hwy 401. **Facilities:** 6 indoor stalls 10' x 12' or larger, limited pasture/turnout, feed/hay, trailer parking. Complete equestrian facility, outdoor arena, indoor ring, lessons. Call for reservations. **Rates:** $20 per night includes hay and bedding. **Accommodations:** Bed & breakfast on premises, $45 single, $60 double.

BOWMANVILLE

Colonial Equestrian Centre — **Phone: 905-623-7336**

Mrs. T. Ashton

3706 Rundle Road, RR #3 [L1C 3K4] Directions: Call for directions. Located close to Hwy 2 & Hwy 401. **Facilities:** 55 indoor 10' x 12' stalls, pasture/turnout, 60' x 12' indoor arena with heated viewing lounge, 3 large barns, 12 large paddocks, heated tack room, feed/hay, trailer parking. 27 acres for riding, adjoining wooded trails, hunter/jumper courses, dressage area, all wood fencing. **Rates:** $25 per night. **Accommodations:** Motels within 4 kilometers of Centre.

PORTLAND

Box Arrow Farm — **Phone: 613-272-2882**

Ruth Godwin or Shirley Prosser — **or (Summer only): 613-272-2509**

Cove Road, Big Rideau Lake [K0G 1V0] Directions: If you are coming from Kingston, turn on Hwy 15. If you are coming from Montreal, turn on Hwy 32 and it runs into Hwy 15. **Facilities:** 60 indoor 12' x 12' stalls, 6 good-sized paddocks, hay available but bring own feed, trailer parking. Lots of trails, event course. October Poker Run , Holiday Horse related place. Call for reservations. Cottages on premises, swimming, fishing. **Rates:** $25 per night; $125 per week. **Accommodations:** Cottages on the water on premises, $100 per day, $500 per week. Many lodges and small inns within 15 miles plus 4 golf courses.

ALL OF OUR STABLES REQUIRE CURRENT NEG. COGGINS, CURRENT HEALTH PAPERS, & OWNERSHIP PAPERS.

ONTARIO

ROCKWOOD

Travis Hall Equestrian Centre **Phone: 519-843-4293**
Dave and Judith Johnson **Fax: 519-843-4903**
RR #3 [N0B 2K0] **Web: www.geocities.com/travishallequestrian**
Directions: 6 kilometers north of Hwy 24 , up to Wellington 29, turn off Eramosa 30th, first farm on right. **Facilities:** 35 stalls, pasture for turnout, feed/hay, trailer parking, RV hook-ups available. Miles of gravel road for trail riding, cross-country course. Specialize in English, Western, and driving. Call for reservations. **Rates:** $20 per night. **Accommodations:** Motels in Guelph and Fergus, 15 minutes from stable.

ZEPHYR

High Fields Ranch/Country Inn & Spa **Phone: 905-473-6132**
Norma Daniel **Fax: 905-473-1044**
11570 Concession 3 [L0E 1T0] **Directions:** Don Valley Parkway north, Hwy 404 North, east on Davis Drive (last exit) for 16.5 kilometers, 7.5 kilometers north on Concession 3, west side. **Facilities:** 19 indoor 10' x 12' stalls, 2-acre pasture, hay, trailer parking. **Rates:** $20 per night. **Accommodations:** Country Inn & Spa on premises.

SASKATCHEWAN

PRINCE ALBERT

Asiil Arabians **Phone: 306-764-7900**
Jack & Harriet Lang
Box 275 SK [S6V 5R5] **Directions:** Call for directions. 8 miles southeast of town off Hwy 3. **Facilities:** 8 indoor large box stalls, pasture/turnout, outdoor riding arena, feed/hay, trailer parking. Trails close by; good riding within 10 minutes. Straight Egyptian breeding operation. Call for reservation. **Rates:** $5 per night. **Accommodations:** Motels within 5 minutes.

Red River Equestrian Center **Phone: 306-763-3434**
Gord Trueman
Box 10, Site 19, RR #5 [S6V 5R3] **Directions:** Call for directions. 3 kilometers north on Hwy 2, east at Whispering Pines Trailer Court, facility 1/2 mile on south side. **Facilities:** 47 indoor 10' x 12' box stalls and 5' x 8' tie stalls, turnout available, feed/hay, trailer parking. Nightly lessons; roping, cutting, English and Western riding, jumping. Sleigh rides and wagon rides. Call for availability. **Rates:** $10 per night. **Accommodations:** Motels within 4-5 kilometers.

ALL OF OUR STABLES REQUIRE CURRENT NEG. COGGINS, CURRENT HEALTH PAPERS, & OWNERSHIP PAPERS.

SASKATCHEWAN

REGINA

Twin Pine Stables **Phone: 306-757-4882**
Corace & Chris Pedersen
Box 405, Off Hwy 1 [S4P 3A2] Directions: Call for directions. Off Hwy 1 west of Regina. **Facilities:** 20 indoor box stalls 8' x 12' and 8' x 10', pasture/turnout, large indoor riding arena, heated barn, feed/hay, trailer parking. Riding lessons. **Rates:** $10-$20 per night depending on services. **Accommodations:** Motels within 3 kilometers in Regina.

RR3 ROCKWOOD

Travis Hall Equestrian Center **Phone: 519-843-4293**
Judith and David Johnson **Fax: 519-843-4903**
E-mail: travis_hall@sympatico.ca
Web: www.geocities.com/travishallequestrian
8159 30th Sideroad (NOB2KO) Directions: North of Hwy 24. Call for Directions. **Facilities:** 4 10'x10', turnout rings available, indoor or arena and trails on 92 acres, English, Western, Driving (training for both horse at rider), Standing Dutch Warmblood, Blacksmith on site, feed/hay available, trailer parking, negative Coggins and health papers. **Rates:** $20 per night **Accommodations:** Bed & Breakfasts, Hotels Motels within 20 minutes.

WOLSELEY

The Taylors Guest House **Phone: 306-698-2765**
Graham & Isabel Taylor **Fax: 306-698-2960**
Box 399 [S06 5H0] Directions: Call for directions. Off Hwy 1, 6 kilometers north of Wolseley. **Facilities:** 5 box stalls, 2 paddocks, hay, trailer parking. Bridle trails on property. **Rates:** $15 per night. **Accommodations:** Guest house on premises. B&B $55 per night double occupancy.

YUKON

HAINES JUNCTION

Mackintosh Lodge **Phone: 403-634-2301**
Barbara Jean Halushka **Fax: 403-634-2302**
Mile 1022 ALCAN Hwy [YOBILO] Directions: 6 miles west of Haines Junction right on ALCAN Hwy. **Facilities:** One big corral and 12-acre fenced pasture, feed/hay extra, trailer parking. Fuel service, RV parts, campground. Call for reservations. **Rates:** Free to lodge guests. **Accommodations:** Lodge on premises: $65 per night, double occupancy. Situated in majestic mountain setting, cozy lounge with fireplace, sweat lodge & cold dunk.

REESE'S
ALL-STATE HORSE EXPRESS
Service You Can Trust
Over 36 Years
Experience
We Transport Horses
Coast to Coast Every 3-4 Weeks
800-451-7696

GRAPHICSPLUS
Freelance Graphic Design
• Logo Design • Ads
• Brochures • Posters
• Newsletters • Etc.
Tel: 781-934-2818
E-mail: cherry7@gis.net

THIS REALLY HAPPENED TO ME!!! TRAILERING NIGHTMARES!!!

By: Jim Balzotti

Many years ago, in my trailering infant years, I was in the market to buy a horse. I had just sold my first horse, a lovable, old (and I do mean old) Appy gelding named Cody. Poor ol' Cody was one of these horses that you needed to put a quarter in his ear to get him to move. He was my first horse and I loved him, but I was ready to move on and get a more spiritedhorse, because, of course, months of riding Cody made me a "skilled" rider.

While reading the local paper one day, lo and behold there was an ad for a "magnificent, spirited riding horse" "Must sell" the ad enticed. Now any knowledgeable horse person might have known it really meant crazy horse, can't ride it so I must sell it!. Unsuspecting, off I went to see my new dream horse. I picked up my Mother, as she always liked adventures, and drove out to the stable. When we finally got to the stable, a very anxious owner was there to greet me. "Magnificent horse! Simply magnificent! " he repeated. "I just don't have the time to ride him anymore" Meanwhile, in the box stall behind us, I thought they were keeping a Brahma bull for all the snorting and kickin' going on. He disappeared into the stall and came out with this monstrous blue-black horse that was as tall as he was wide. Prancing and dancing, he held him tight with a curb chain. He was everything my Cody wasn't. I wanted this horse, but had to admit I was more than a little intimidated by him. I almost was afraid to even ride him, but my Mother was there. I didn't want to have my Mother think I was a wimp, did I? We led him into the arena and saddled him up. As soon as I put one foot in the stirrup, we were off to the races. We raced around the arena twice, dust flying, going so fast that my eyes were watering and I could barely see. Wow! He finally calmed down a bit and I was able to get my second foot in the stirrup. As we rode up to the seller and my mother, I just about had my money out. I bought the horse and made arrangements to pick him up that Saturday.

I had a truck and a beautiful Kingston trailer, but alas, not a lot of trailering skills. I did know a guy by the name of Ed in my town that had pigs and sheep and trailered them around. Close enough! On Saturday, Ed and I drove up to get my new horse. The stable was located on busy Route 1 in Saugus, Massachusetts, and Saturday traffic made it into a raceway. We pulled up in front of the barn and the owner went in to get the horse. We dropped the trailer ramp and soon were joined by the owner leading my new horse out on a curb chain. He handed him over me and I proceeded to lead him on as if I had done it a hundred times. All went perfectly fine. The horse walked on without a moments hesitation, but as I went to tie him up, all hell broke loose. Before Ed had a chance to put up the rear bar and ramp, the horse reared up, smashing his head into my new horse trailer. He began thrashing and kicking, and in a split second, bolted out the rear door running headlong onto Route 1 in the middle of traffic. Ed, the owner and I watched horrified as he disappeared down the road. We quickly unhitched the trailer and took off in the truck down Route 1 in search of my horse. After going not more than a mile, we saw him standing in the middle of the median. I got out of the truck and approached him cautiously. I was able to get a lead on him and calm him down, finally leading him back to the stable. We were finally able to get him on the trailer. I had that horse for years and I must say he really taught me how to ride. He threw me so many times in so many ways that I lost count. I don't have that horse anymore, but every time I see that big dent in the roof of my trailer, I think of him.

INTERVIEW WITH A FARRIER

By Ed McCollum

Joe DeLowery, Farrier
68 Bowen's Lane, Rochester, MA 02770
508-763-1741

You've heard of "Interview with a Vampire". Well, after one full year of searching through stables, farms and even circuses, I was finally able to locate one of New England's most elusive and sought after farriers, Joe DeLowery. Joe is legendary in New England for his knowledge of horses and his ability to keep them on their feet. Joe is known for his down-home sense of humor and low-key style. He is also known as a miracle worker because of his ability to get "lame" horses back on their feet and out riding again. He is the farrier for the internationally famous Big Apple Circus. We had to hobble Joe long enough to conduct this interview, but it was worth it. This is the first time Joe has agreed to a personal interview.

Ed: Joe, how did you come to be a farrier?

Joe: My grandfather was a farrier way back when not that many people made a profession out of it. In fact, I still have and use some of his tools. My folks came to the United States from Ireland and always had horses. My father shoed horses after WWII and I remember he used to charge a buck a foot. I came out of the service and used the G.I. Bill to go to the Martinsville Farrier School back in 1970. After school, I hooked up with a Scotsman and stayed with him for five years before branching off on my own. He was a great guy and a great farrier who taught me alot.

Ed: What's the secret of your legendary success working with horses? I'm sure not all of them are eager to be shod.

Joe: No, not at all. First of all, it's important to understand the animal so I try to be one of them. I start my day very early in the morning. I muck out my ten horses, turn them out and make sure they're doing fine. Then I clean my barn and do my shop work. By the time I get to my first appointment, you might say I smell like one of the herd. Because if they feel you're one of them, they'll stand around and trust you and feel they're in good hands. And that is what's most important. I may be getting paid by the client, but the horse is the one that I'm working for and they get all my attention. Their sense of smell is so strong. I want to smell like an old friend before I shoe that first horse. It may sound strange to the novice, but it's an old trick. I knew an old smithie that had one shirt that he shod in and he shoed seven or eight horses a day in that one shirt. At the end of the day, he took that shirt off and put it in the truck. The next day he'd come and put that same shirt on. You couldn't stand to be around this guy, but the horses loved him and he never had a problem shoeing a horse. Sometimes, it only takes a bad after-shave lotion or a strong deodorant or even some medicine that makes them think you're a veterinarian that causes a problem.

Ed: How has the industry changed in the past 25 years?

Joe: The amount of work has gone up considerably. Even though there are shoeing schools pumping out farriers, the number of horses out there is growing. Consider that just in the U.S. there are six million thoroughbreds registered as a fresh crop on January first. There are more quarter horses in this country than thoroughbreds not to mention all the Morgans, Arabs and other breeds. Just in the state of Tennessee there are over 40,000 registered Tennessee Walking horses. You have 70,000 registered quarter horses in the state of Texas and that's not counting all the working horses the cowboys use. I see people in their 30's and 40's who never had a horse but dreamed of it are picking them up, starting to enjoy riding and forming riding clubs.

Ed: This new group of farriers, are they sticking it out or finding it's tough work and giving up?

Joe: They're finding out it's not an easy dollar and it's a long dollar. If they're in it to make a quick buck, they'll find out what hard work it is. You have to be shoeing alot of horses to make a living at it. For me, it's peace of mind because I just enjoy doing this. It's the secret of a good life to enjoy what you do for work.

Ed: What advice to you give people who are trailering their horses. Should the horses be shod or should they pull the shoes?

Joe: If someone is traveling for a day or to a riding spot, usually they'll

have their horses shod in a nice flat steel. They'll want shoes on their horses because when they come off the trailer, most likely they be riding them right away. But you want to be careful about shoeing your horse with any kind of shoe that's apt to get hung up in the trailer. Now, if you are using a national transportation company, ask their advice, but you may want to ship them barefooted. Also, check the condition of the trailer to make sure there are no cracks or crevices that the horses can get hung up on. I like rubber all around myself.

Ed: Are steel shoes and rubber mats a bad combination? Will the horses slip on the surface?

Joe: Oh no, steel and rubber are excellent. Bare hoof on rubber is also great. There is a suction factor there. The heat of the horse combines with the rubber while they're standing and it makes for a nice suction cup on the rubber mat. But if they're going to ride the horses when they get out of the trailer in the mountains or in the desert, they'll need a set of shoes on them. I recommend going flat and never mind the traction devices when you're traveling because it can be a booby trap and you're asking for an accident to happen.

Ed: Do you recommend wrapping the legs to keep them from skinning themselves up inside the trailer?

Joe: Absolutely. A good wrapping with just regular cotton is better than nothing at all, but they're making some great shipping boots now that will do the trick. They'll cover the vital parts of the leg, the cannon bone, the ligaments and the lower vascular system. A good pair of shipping boots is a lot cheaper than replacing your horse.

Ed: Do you recommend that people use a hay bag when traveling?

Joe: Always! I always like to keep a bag tied high, not where they can get their legs caught up in it. It gives them something to do and it's a pacifier. It calms them down.

Ed: If a rider is on a trail ride and a shoe comes loose, what should they do?

Joe: Well, there are some tools that horsemen can buy and they should ask their farrier how to use them. There are simple little clinching devices that can help reclinch a shoe and tighten some nails that have been sprung. I don't recommend doing any kind of sharp paring or removing and sole, but these new clinchers can tighten up the shoe and get you home.

Ed: What if the shoe can't be clinched?

Joe: That same tool is also a pair of pulloffs that you can very gently pry the

shoe off and if you are carrying a neoprene pad you can make a temporary shoe that cowboys have been making for years that will get you home. Just fasten it with some vet wrap or good adhesive tape and off you go. A good idea is to ask your farrier to cut you a set of pads that match your horse and keep them with you when you ride. It's a quick fix for a shoe that's been sprung or been pitched or left in the mud.

Ed: What about these temporary boots I see on the market?

Joe: There are alot of brands out there like Easy Boots. They are a great invention, but make sure you get them to fit properly. Also, if the apparatus is made of hard rubber, then it should be made to fit one size larger than the horse's normal size. For instance if the horse wears a size one, fit them to a size two. It's a quick fix, and you're not going to use it for very long, but it will get you through the trail ride until you get to the next blacksmith's shop.

Ed; How do you tell your horse's shoe size?

Joe: You can ask your farrier, but it is stamped on the hoof bearing side of each shoe. You'll see a number set in the steel.

Ed: You hear alot about corrective shoeing. Is this relatively new? Did they do corrective shoeing back in the 1880's?

Joe: Absolutely. Some of the old shoes I have in my collection have a piece of steel welded on one side of it to correct a fault in a horse where he was dragging his hind leg or where he was a plow horse and you might see a cock welded on the inside heel to correct a horse that dragged his toe. In these creatures, there's no two alike and there's no horse with perfect limbs and perfect hooves so the smithies would try to help the guy who owned the horse and he could read from the old shoes where the horse was dragging. It would be worn in a certain area so he could take his new shoe and maybe double it over and make it twice as thick on one side. But corrective shoeing has been around since the first burros carried Mary and Joseph. Horse shoeing has been documented back to 900 B.C. The first shoes that have been found in Europe were of Celtic nature. They figured out that the only reason Hanibal was able to conquer as much as he did was because he had his blacksmiths and armor makers put steel on the hooves of his horses to protect their feet to cross the Alps. He conquered tribes that couldn't compete with the distances Hanibal covered in conquering Europe. There are old shoes to be seen in the Smithsonian and they are basically the same as we are using today. A shoe is just iron ore that's been hammered out, nails punched and nails made. Nailed the same way, through the hoof capsule and bent over in the form of a clinch.

Ed: How about the Native American Indians? Did they use iron shoes?

Joe: Not to our best knowledge. They covered the horses hooves and used a lot of deer hide, bear hide and buck skins. They had the right idea but they didn't have the implements. They didn't do a lot with steel. The were always careful to try to protect the hoof capsule. They knew that with the foot protected, they would get 30 years out of their horses if they were also fed right and used right.

Ed: Are horses really used for work anymore?

Joe: Sure, but not like they used to be. You'll still see the plow horse here and there, and of course the working cow horse but you have to remember one hundred years ago horses were driving this country. The Teamsters had leather in their hands, not steering wheels. Horses were pulling wagons in every little town in America.

Ed: Has the horses physical condition changed since then?

Joe: Back then the horses work load was immense. You could always see ribs. If you looked at pictures of those horses, you'd say they weren't fed, and by today's standards that's probably right. We over feed them. The vets say founder is way up. We're feeding our horses like beef cows. In the old days ,the horses were out working. The horse would be taking the father to work in the morning and bringing him home. The kids would probably use the horse to go visit their friends and at night they might harness up to go out to church or someplace to socialize. The hay and oats he was being feed was being worked off. Nowadays, it seems we're working for the horse. We've got automobiles in the driveway so we don't need them for transportation. So we over feed them out of kindness and end up killing them with kindness.

Ed: For the new horse owner, how would they go about getting comfortable with their new horse, especially if it's on the frisky side?

Joe: It's just time. If you got one that isn't finished yet and isn't quite sure of everything in its environment, you just can't quit. Give that horse two or three hours a day of your time if you can. It's a daily routine that I highly recommend to these folks. You'll gain that horses confidence and see a big improvement when he goes out on the trail or is in the trailer.

Ed: How do you shoe a horse that is big, mean and ornery?

Joe: Just had one the other day. If you lose your temper and start physically manhandling the horse, you're going to lose. In the old days, you'd apply a Yorkshire Twitch which was put on her nose and gave her something to think about while you'd go about getting her shod. Most new horse owners now don't like that method, so if a feed bag won't do it you have to

call your friend the veterinarian. There is a new drug on the market called Demozodan that I highly recommend. It's safe to use and will make the heaviest and nastiest horse comply without any problem or long term effects. As the horse gets used to being shod, you might try something milder like granules of Promosine that is sold at feed stores. Very important, take the time getting your horse used to someone handling their feet by making it a part of the grooming process.

Ed: You shoe for the famous Big Apple Circus when it comes to Boston. What's that like?

Joe: I really look forward to it every year. As you know, the Big Apple Circus is probably the best circus in the United States now. They are based in Florida and every spring come to Boston. Only the cream of the crop in talent gets hired as performers. The Big Apple was Max Shuman's circus. He's a Danish horse trainer and circus owner from Europe, and now his daughter Katja and her husband Paul Binder run it. They have the best jugglers, the best dog acts, the most amazing trapeze, and of course, their world famous horse act.

Ed: What kind of horses do they use?

Joe: Max Shuman has a set of Liberty horses that are the most skilled horses I've ever seen perform in the United States. Max Shuman is the best horse trainer that's out there. There's never a line attached and they do everything at his command. Usually they'll have eight to ten horses and to see them running in a circle the way they do shows how Max gets these horses to show their true intelligence to mankind. I was lucky to hook up with them. They're all a great bunch of people.

Ed: Joe, being a farrier for all these years, have you ever been hurt?

Joe: Oh yeah! In fact, the old expression of "having the crap kicked out of you" is about blacksmiths! You take your kicks and bites, go home chew on some aspirin and jump into a streaming hot bath. I've never had a broken bone though. Got a few cracks. Got a few kicks in the femurs and tibia that warranted taking an Advil on a cool day. It can be a dangerous profession. We had a fellow farrier that got kicked and was found dead under a horse. The horse still was on cross ties when they found this poor smithie. So you have to be careful. It's a good idea to shoe with someone around.

Ed: Last question. What's your favorite breed of horse?

Joe: Well, Ed, it's like women. I love 'em all.

Interview with a Veterinarian

By Jim Balzotti

Bruce E. Chase, DVM, PC
66 East Grove Street, Middleboro, MA 02346
508-947-9400 Fax 508-947-0030

In this edition, I was thrilled to ask one of New England's most prominent vets some of the questions most of our readers ask time and time again, namely what do I do if... Dr. Bruce, as he's known here in the northeast, is both a teaching and practicing veterinarian who is at the top of his field. Here are some of his answers to questions every rider wants to know.

Jim: If while on a trail ride, my horse suffers a cut or puncture wound, what should I do?

Dr. Chase: The first step in any wound is to control bleeding, which is generally accomplished by the application of direct pressure. When hemorrhage is controlled, rinse the wound with water to remove contaminants followed by the application of topical antiseptics. The wound should be evaluated as to whether sutures are appropriate. It is generally best to make the decision to suture or not during the first twelve to twenty four hours after the injury. If the wound is on the lower leg, bandaging should be considered.

Jim: While riding and possibly miles away from my trailer, my horse begins to limp a bit. Is it safe to ride him back to the trailer or should I get off and walk back?

Dr. Chase: At the first suspicion of a possible lameness problem, it is always wise to evaluate the foot for injury or the presence of a foreign object. A quick review of the rest of the leg for obvious injuries or swellings is also important. Whether to continue the ride back or not would depend on the cause and severity of lameness. It is always wise to err on the side of caution and not ride a lame horse due to the potential for further injury or the risk of stumbling resulting in the fall of the rider.

Jim: What are the signs that show whether or not you are overworking your horse?

Dr. Chase: The most accurate means of evaluating exhaustion has generally been to evaluate the time required for the pulse and respiration rates return to the normal resting level. This first requires an appreciation for the normal pulse and respiratory rate values. If there is a concern of exhaustion, the horse should be stopped and monitored every five to ten minutes. For most horses a proper recovery to resting values occurs within ten minutes. If recovery extends to more than twenty or thirty minutes, excess stress has occurred.

Advanced exhaustion is reflected by extreme weakness, staggering and incoordination. This represents an extreme emergency and requires immediate medical attention.

Jim: What's the best way to get your horse into shape, especially if he/she hasn't worked all winter.

Dr. Chase: A routine conditioning program by monitoring pulse and respiratory rate recoveries is the most reliable means of assessing your horse's condition. This system allows the evaluation of each horse's individual needs in a conditioning program.

Jim: Do horses get cold?

Dr. Chase: Yes. Generally hypothermia or decreased body temperature would be reflected by shivering similar to the human population. The horse is blessed with an excellent form of insulation and therefore hypothermia is a relatively rare condition. Monitoring the horse's rectal temperature is the most important step in the diagnosis and management of the problem.

The greater concern for horses is the occurrence of hyperthermia or heat exhaustion. During extended periods of work, the process of metabolism releases energy in the form of heat which must be dissipated through sweating and other cooling processes. During extensive work on a warm day a rider should be constantly aware of the horse's body temperature and condition.

Jim: What are the signs of a horse in distress and what should I do about it? What if I am miles away from home?

Dr. Chase: The most common medical emergency experienced by the horse is the syndrome of colic. A variety of activities can be associated with colic such as loss of appetite, sweating and restless behavior including pawing the ground, lying down, rolling and other signs of abdominal pain. The most important step is to attempt to stabilize the horse and send for assistance. If at all possible, getting the horse to a location for treatment is the most important consideration.

Jim: Do you recommend bringing along medical supplies on an extended trail ride? What items? What about a day's ride?

Dr. Chase: A first aid kit would take into consideration that the two most common equine emergencies involve lacerations or the occurrence of colic. Bandaging materials for leg injuries and first aid therapy for colic as suggested by your veterinarian are therefore the most important items in a first aid kit.

Jim: What is the best way to relieve your horse from pain?

Dr. Chase: There are several medications available for pain relief, however, phenylbutazone or "Bute" is still an economical and reliable anti-inflammatory. Colic pain can be relieved by several medications used on a first aid basis. One of the most popular medications at this time is flunixin meglumine or "Banamine" which comes in an oral paste as well as an injectable form and can be administered safely for a wide variety of medical conditions.

Jim: Do horses really need electrolytes?

Dr. Chase: Yes. The electrolytes required in greatest amount for the proper metabolic performance are sodium, potassium and chloride. Sodium chloride is also known as common salt. These electrolytes are often found in adequate amounts in commercially mixed rations and are rarely found to be deficient. They can be lost in significant amounts through sweating and therefore, horses that sweat extensively in their performance should be supplemented. Supplementation can occur by the addition of trace mineralized salt to the diet or the addition of a commercially prepared electrolyte mixture.

ABOUT THE CARTOON ARTIST

Fred Leary started drawing at the age of three and has never put down the pencil:

His illustrations and cartoons have been published in many Boston area publications over the last twenty years.

He has started a small comics business called Belfry Comics- thusly named because the main office is in a church belfry built in 1866 and saved from demolition by Fred and rehabbed into an office at his house.

He does comics both comical and political and illustrations with pencil and pen and ink for any occasion.

He attended the University at Bridgeport in Connecticut.

He has a background in newsprint and journalism.

He lives in Pembroke, Massachusetts with his wife, four daughters and two cats.

ANECDOTES

We asked our readers to submit humorous or curious personal stories so we could let others know that we all do pretty silly things for, around and with our horses. We frequently hear about very funny situations which we are now offering to provide to our world. SO!! Here are three we received. We hope you enjoy them, and if you have others you would like considered for our next issue, please send them to us.

Thanks,
Jim Balzotti

COWBOY FIRST AID

"One of the reasons I love owning our Barn Bed & Breakfast is the wide range of information we exchange with our guests when they stay with us. One such exchange was: A couple was telling us about going out to the barn where they used to keep their horses and finding the owners of the barn hosing a horse's leg off while blood was pouring from a wound and the blood was gathering in a pool on the ground. They couldn't stop the bleeding. An old cowboy walked up to them and very casually suggested taking some cobwebs off the wall and placing them on the wound. Having tried almost every other suggestion, they ran into the barn and scooped some cobwebs off the wall and placed them on the horse's leg. The bleeding immediately stopped. Our guests said that if they hadn't seen it for themselves, they wouldn't have believed it. So every time guests come to stay with us, I tell them this story and remind them that the cobwebs in my barn are for medicinal purposes!"

contributed by Donna Robinson
Robinson Quarter Horses
Oklahoma City, OK

ANECDOTES

TRUST YOUR HORSE TO TAKE YOU WHERE YOU NEED TO BE

"Feeling sorry for myself, alone and lonely on Christmas, I decided to go caroling on my horse, "Mrs. Robinson", and my dog, "Bandit". A neighbor invited me in for hot chocolate. Forgetting time and weather in the warm cozy kitchen, dark had descended and a snow storm had brewed. My faithful dog met me at the door and guided me to my horse for the 1/4 mile trip home. Because of the inclement weather, I could not see my way clear. It was so bad that I couldn't even see my hand in front of my face, let alone see my dog. So, trusting "Mrs. Robinson" to get us home, I laid down on her back and said it was up to her to get us back home. Well, about the time I figured we should be home, she stopped. Not having left any lights on, I was still very much disoriented as to where we were so I slid off her back to find myself knee-deep in manure! Either this was payback or the warmest place she could find near our home!"

contributed by Alice Ferguson
R & R Dude Ranch
Otto, NY

HAPPY FARM ZOO

"One cold winter night, a boarder arrived. Inside their large stock trailer were: two horses, a pen of exotic chickens, a miniature donkey, two miniature goats, a llama named Sally and two Great Danes. We stabled everything but the chickens and the Great Danes!"

contributed by Happy Farm
Trinidad, CO

ABOUT THE PUBLISHER

Photograph by J. David Congalton, M. Photog, ASP, Pembroke, Massachusetts

Jim Balzotti lives in Massachusetts with his three children, horses, and dog. He is an avid horseman and has always enjoyed introducing new people to the joys of riding. Jim has traveled thousands of miles transporting his horses and others across the country. He is pictured here with his longtime traveling companion, Wynston, who is out of Park Arabians of Bainbridge Island, Washington, and who made the long trek to Massachusetts to his new home.

NOTES AND REMINDERS